PHYSICS RESEARCH AND TECHNOLOGY

RECENT ADVANCES IN COSMOLOGY

PHYSICS RESEARCH AND TECHNOLOGY

Additional books in this series can be found on Nova's website
under the Series tab.

Additional e-books in this series can be found on Nova's website
under the e-book tab.

SPACE SCIENCE, EXPLORATION AND POLICIES

Additional books in this series can be found on Nova's website
under the Series tab.

Additional e-books in this series can be found on Nova's website
under the e-book tab.

PHYSICS RESEARCH AND TECHNOLOGY

RECENT ADVANCES IN COSMOLOGY

ANDERSON TRAVENA
AND
BRADY SOREN
EDITORS

nova publishers
New York

Library of Congress Cataloging-in-Publication Data

ISBN: 978-1-62417-943-3

Published by Nova Science Publishers, Inc. † New York

CONTENTS

PREFACE

In this book the authors gather and present current research in the study of cosmology. Topics discussed include the mysteries of the geometrization of gravitation; relativistic viscous Universe models; cosmology and science; a discussion on whether science has established if the cosmos are physically comprehensible?; Mach, Einstein, dark matter and knowability; implementing Hilltop F-term hybrid inflation in supergravity; and relative velocities, geometry and the expansion of space.

Chapter 1 – Most scientists would hold that science has not established that the cosmos is physically comprehensible – i.e. such that there is some as-yet undiscovered true physical theory of everything that is *unified*. This is an empirically untestable, or metaphysical thesis. It thus lies beyond the scope of science. Only when physics has formulated a testable unified theory of everything which has been amply corroborated empirically will science be in a position to declare that it has established that the cosmos is physically comprehensible.

But this argument presupposes a widely accepted but untenable conception of science whichthe author shall call *standard empiricism*. According to standard empiricism, in science theories are accepted solely on the basis of evidence. Choice of theory may be influenced for a time by considerations of simplicity, unity, or explanatory capacity, but not in such a way that the universe itself is permanently assumed to be simple, unified or physically comprehensible. In science, no thesis about the universe can be accepted permanently as a part of scientific knowledge independently of evidence. Granted this view, it is clear that science cannot have established that the universe is physically comprehensible.

Standard empiricism is, however, as the author has indicated, untenable. Any fundamental physical theory, in order to be accepted as a part of theoretical scientific knowledge, must satisfy two criteria. It must be (1) sufficiently empirically successful, and (2) sufficiently *unified*. Given any accepted theory of physics, endlessly many empirically more successful disunified rivals can always be concocted – disunified because they assert that *different* dynamical laws govern the diverse phenomena to which the theory applies. These disunified rivals are not considered for a moment in physics, despite their greater empirical success. This persistent rejection of empirically more successful but *disunified* rival theories means,the author argues, that a big, highly problematic, implicit assumption is made by science about the cosmos, to the effect, at least, that the cosmos is such that all seriously *disunified* theories are false.

Once this point is recognized, it becomes clear, the author argues, that we need a new conception of science which makes explicit, and so criticizable and improvable the big, influential, and problematic assumption that is at present implicit in physics in the persistent preference for unified theories. This conception of science, which the author calls *aim-oriented empiricism*, represents the assumption of physics in the form of a hierarchy of assumptions. As one goes up the hierarchy, the assumptions become less and less substantial, and more and more nearly such that their truth is required for science, or the pursuit of knowledge, to be possible at all. At each level, that assumption is accepted which (a) best accords with the next one up, and (b) has, associated with it the most empirically progressive research programme in physics, or holds out the greatest hope of leading to such an empirically progressive research programme. In this way a framework of relatively insubstantial, unproblematic, fixed assumptions and associated methods is created, high up in the hierarchy, within which much more substantial and problematic assumptions and associated methods, low down in the hierarchy, can be changed, and indeed improved, as scientific knowledge improves.

One assumption in this hierarchy of assumptions,the author argues, is that the cosmos is physically comprehensible – that is, such that some yet-to-be-discovered unified theory of everything is true.

Hence the conclusion: improve our ideas about the nature of science and it becomes apparent that science has already established that the cosmos is physically comprehensible – in so far as science can ever establish anything theoretical.

Chapter 2 – At first sight it seems that the general theory of relativity is inspired by the so called Mach´s principle. To what extent does Einstein in his theory digress from the concepts of Ernst Mach? How can we apply Mach´s principle in physics today mainly in relation to the dark matter phenomenon?

Kurt Gödel, gravedigger of Hilbert´s program, created the model of rotating, homogeneous, stationary and anisotropic universe. To what extent is it compatible with Mach´s principle?

Dark matter is one of the basic problems of modern physics. The article attempts to look at this problem from the viewpoint of the theory of knowledge and uses the same viewpoint to compare it with the problem of knowability in mathematics.

Chapter 3 – Theory of general relativity has been agreed upon by experts for almost a century and is believed to describe accurately all gravitational phenomena ranging from the solar system to the Universe. However, there is a price for this success which is often ignored. More than 95 percent of the content of the energy-stress tensor has to be dark, in the form of inflaton, dark matter and dark energy, which do not have any non-gravitational or laboratory evidence. Moreover, the dark energy poses a serious confrontation between fundamental physics and cosmology.

The present situation reminds us of Einstein's 'biggest blunder'when he forced his theory to predict a static Universe. It seems that we are making a similar blunder by forcing the energy-stress tensor into the field equations while the observations indicate that it is not needed. We seem to have a deep-rooted conviction that the spacetime will remain empty unless we 'fill' it by the energy-stress tensor. However, we have been ignoring numerous evidences telling earnestly that the so-called vacuum field equations $R^{\mu\nu} = 0$ do not necessarily represent an empty spacetime. It is shown that the energy, momenta and angular momenta of the gravitational and the material fields do exist in the spacetime of $R^{\mu\nu} = 0$, whose effects are revealed through the geometry.

This may appear orthogonal to the usual understanding. Nevertheless, it is in striking agreement with the theory and observations at all scales, without requiring the dark sectors of the standard approach. Moreover, the new finding evades the long-standing problems of the standard cosmology, besides providing natural explanations to some unexplained puzzles.

Chapter 4 – The research on relativistic universe models with viscous fluids is reviewed. Viscosity may have been of significance during the early inflationary era, and may also be of importance for the late time evolution of the Universe. Bulk viscosity and shear viscosity cause exponential decay of anisotropy, while nonlinear viscosity causes power-law decay of anisotropy. The authors consider also the influence from turbulence, in connection with future singularities of the universe (Big Rip and Little Rip). Finally, the authors review some recent developments of causal cosmology theories.

Chapter 5 – The general expectation that the acclaimed observation of a Higgs boson has finally unified the standard theoretical models of cosmology and particle physics is critically examined. On the negative side it is found that the cosmology model consists of several independent, mutually exclusive models, used indiscriminately to address different observations as needed. There is no single model to account for the conservation laws of physics, nature of the gravitational field, space-time symmetry, nucleogenesis, anomalous redshifts, cosmic radiation, microwave background, an expanding universe, dark matter, anti-matter and gamma-ray bursts. A special formulation of the theory is required for each different aspect. Apart from the Gamow-Lemaˆıtre hot big bang, there is an inflationary model with a Higgs field and a cold-dark-matter (ΛCDM) alternative, with practically no common ground between them. In particular, the different event chronologies are not synchronized. The resulting mismatch causes a discrepancy in the value of the cosmological constant (Λ) that differs by a factor of 1046 as required by the two "symbiotic" standard models.

The cause of all these problems is traced back to two seminal assumptions – the Doppler interpretation of cosmological redshifts and the point model of elementary matter. Without these assumptions there is no need of a big-bang model or a Higgs field. Without the false vacuum of particle physics – an artifact of Euclidean space – there is no basis for inflationary cosmology. An alternative cosmology based on projectively curved space-time is described. It features simple accounts of antimatter, nucleogenesis, the periodicity of atomic matter, black-hole singularities and cosmic self-similarity. The main ideas are conveyed with a minimum of mathematical formalism.

Chapter 6 – F-term hybrid inflation (FHI) of the hilltop type can generate a scalar spectral index, ns, in agreement with the fitting of the seven-year Wilkinson microwave anisotropy probe data by the standard power-law cosmological model with cold dark matter and a cosmological constant, ΛCDM. The authors investigate the realization of this type of FHI by using quasi-canonical Kähler potentials with or without the inclusion of extra hidden-sector fields. In the first case, acceptable results can be obtained by constraining the coefficients of the quadratic and/or quartic supergravity correction to the inflationary potential and therefore a mild tuning of the relevant term of the Kähler potential is unavoidable. Possible reduction of ns without generating maxima and minima of the potential on the inflationary path is also possible in a limited region of the available parameter space. The tuning of the terms of the Kähler potential can be avoided with the adoption of a simple class of string-inspired Kähler potentials for the hidden-sector fields which ensures a resolution to the η problem of FHI and allows acceptable values for the spectral index, constraining the coefficient of the quartic supergravity correction to the inflationary potential. Performing a four-point test of the analyzed models, the authors single out the most promising of these.

Chapter 7 – What does it mean to say that space expands? One approach to this question is the study of relative velocities. In this context, a non local test particle is "superluminal" if its relative velocity exceeds the local speed of light of the observer. The existence of superluminal relative velocities of receding test particles, in a particular cosmological model, suggests itself as a possible criterion for expansion of space in that model.

In this point of view, superluminal velocities of distant receding galaxy clusters result from the expansion of space between the observer and the clusters. However, there is a fundamental ambiguity that must be resolved before this approach can be meaningful.

The notion of relative velocity of a nonlocal object depends on the choice of coordinates, and this ambiguity suggests the need for coordinate independent definitions. In this chapter, the authors review four (inequivalent) geometrically defined and universal notions of relative velocity: Fermi, kinematic, astrometric, and spectroscopic relative velocities. The authors apply this formalism to test particles undergoing radial motion relative to comoving observers in expanding Robertson-Walker cosmologies, and include previously unpublished results on Fermi coordinates for a class of inflationary cosmologies.

The authors compare relative velocities to each other, and show how pairs of them determine geometric properties of the spacetime, including the scale factor with sufficient data.

Necessary and sufficient conditions are given for the existence of superluminal recessional Fermi speeds in general Robertson-Walker cosmologies. The authors conclude with a discussion of expansion of space.

In: Recent Advances in Cosmology
Editors: Anderson Travena and Brady Soren
ISBN: 978-1-62417-943-3

Chapter 1

HAS SCIENCE ESTABLISHED THAT THE COSMOS IS PHYSICALLY COMPREHENSIBLE?

Nicholas Maxwell[*]
University College London, London, UK

ABSTRACT

Most scientists would hold that science has not established that the cosmos is physically comprehensible – i.e. such that there is some as-yet undiscovered true physical theory of everything that is *unified*. This is an empirically untestable, or metaphysical thesis. It thus lies beyond the scope of science. Only when physics has formulated a testable unified theory of everything which has been amply corroborated empirically will science be in a position to declare that it has established that the cosmos is physically comprehensible.

But this argument presupposes a widely accepted but untenable conception of science which I shall call *standard empiricism*. According to standard empiricism, in science theories are accepted solely on the basis of evidence. Choice of theory may be influenced for a time by considerations of simplicity, unity, or explanatory capacity, but not in such a way that the universe itself is permanently assumed to be simple, unified or physically comprehensible. In science, no thesis about the universe can be accepted permanently as a part of scientific knowledge independently of evidence. Granted this view, it is clear that science cannot have established that the universe is physically comprehensible.

Standard empiricism is, however, as I have indicated, untenable. Any fundamental physical theory, in order to be accepted as a part of theoretical scientific knowledge, must satisfy two criteria. It must be (1) sufficiently empirically successful, and (2) sufficiently *unified*. Given any accepted theory of physics, endlessly many empirically more successful disunified rivals can always be concocted – disunified because they assert that *different* dynamical laws govern the diverse phenomena to which the theory applies. These disunified rivals are not considered for a moment in physics, despite their greater empirical success. This persistent rejection of empirically more successful but *disunified*

[*] Corresponding author: Nicholas Maxwell. E-mail: nicholas.maxwell@ucl.ac.uk.

rival theories means, I argue, that a big, highly problematic, implicit assumption is made by science about the cosmos, to the effect, at least, that the cosmos is such that all seriously *disunified* theories are false.

Once this point is recognized, it becomes clear, I argue, that we need a new conception of science which makes explicit, and so criticizable and improvable the big, influential, and problematic assumption that is at present implicit in physics in the persistent preference for unified theories. This conception of science, which I call *aim-oriented empiricism*, represents the assumption of physics in the form of a hierarchy of assumptions. As one goes up the hierarchy, the assumptions become less and less substantial, and more and more nearly such that their truth is required for science, or the pursuit of knowledge, to be possible at all. At each level, that assumption is accepted which (a) best accords with the next one up, and (b) has, associated with it the most empirically progressive research programme in physics, or holds out the greatest hope of leading to such an empirically progressive research programme. In this way a framework of relatively insubstantial, unproblematic, fixed assumptions and associated methods is created, high up in the hierarchy, within which much more substantial and problematic assumptions and associated methods, low down in the hierarchy, can be changed, and indeed improved, as scientific knowledge improves.

One assumption in this hierarchy of assumptions, I argue, is that the cosmos is physically comprehensible – that is, such that some yet-to-be-discovered unified theory of everything is true.

Hence the conclusion: improve our ideas about the nature of science and it becomes apparent that science has already established that the cosmos is physically comprehensible – in so far as science can ever establish anything theoretical.

1. Introduction

Many scientists probably hold, as a matter of personal belief, that the cosmos is physically comprehensible. Most, however, would vehemently deny that science has already established that it is comprehensible. That will only be established when physics comes up with a "theory of everything" that is (a) unified, and (b) sufficiently empirically corroborated to be regarded as a part of theoretical scientific knowledge. The best candidate we have for such a theory is string theory (or M-theory, as it is sometimes called). But string theory has not yet received a satisfactory formulation, and it certainly has not been empirically corroborated. In the absence of a physical theory satisfying (a) and (b), all we have is the bald thesis that the cosmos is physically comprehensible – that is, the thesis that the cosmos is such that there is some yet-to-be-discovered theory of everything that is unified and true. But this thesis is metaphysical – too vague to be empirically testable.[1] Being metaphysical, it cannot be a part of theoretical scientific knowledge. It is incapable of being empirically corroborated, and it is actually incompatible with existing theoretical knowledge, in that our current best fundamental theories, the standard model and general relativity, are incompatible with one another, and thus incompatible with the assertion that there is some true, unified "theory of everything".

This argument seems decisive. But it is not. It presupposes a conception of science, called by me *standard empiricism*, which is untenable. Once the unacceptability of standard

[1] Throughout this essay, "metaphysical" means "factual thesis – thesis about the world – that is not empirically testable".

empiricism is appreciated, it becomes clear that we need to adopt a better conception of science, which I call *aim-oriented empiricism*. This holds that science has established that the cosmos is physically comprehensible – even though this thesis is metaphysical, and incompatible with current theoretical knowledge in physics.

The argument against standard empiricism (SE), and for aim-oriented empiricism (AOE), has been developed by me in a series of publications during the last forty years.[2] The argument began as an attempt to improve Karl Popper's falsificationist conception of science in order to overcome a difficulty fatal to that view. Subsequently it broadened into an argument which claimed to show that AOE is a dramatic improvement over all contending views about science. In one way, as we shall see, AOE is highly Popperian in character. But in other respects, it differs starkly from Popper's view, most notably in holding that metaphysical theses form an integral part of theoretical scientific knowledge.

Despite the many publications expounding the argument for AOE, the view itself, and the argument for it, have been almost entirely ignored so far by philosophers of science. Three criticisms of AOE have been published, two very bad,[3] one good,[4] but none valid.[5] Apart from that, books that I have published that expound and argue for AOE have received some glowing reviews.[6] Otherwise, silence.

Here, I will give as succinct a resumé as I can of the argument against SE and for AOE. I will then consider what the implications are of rejecting SE and accepting AOE in its stead for cosmology, for physics, for science more generally, for philosophy of science, and for the nature of rationality.

2. Standard Empiricism: Exposition

By "standard empiricism" I mean the doctrine that it is exclusively, or almost exclusively, evidence that decides what theories are accepted and rejected in science. Non-empirical considerations such as the simplicity, unity or explanatory character of a theory may influence choice of theory as well, for a time at least, but not in such a way that nature herself, or the phenomena, are presupposed to be simple, unified or comprehensible. *In science, no factual thesis about the world, or about the phenomena, can be accepted as a part scientific knowledge independently of empirical considerations.*

This rather thin doctrine is a central component of almost all views about science that philosophers of science have come up with. It is common ground for logical positivism, inductivism, logical empiricism, hypothetico-deductivism, conventionalism, constructive empiricism, pragmatism, realism, induction-to-the-best-explanationism, and the views of Karl Popper, Thomas Kuhn and Imre Lakatos.[7] There is a sense in which even Paul Feyerabend,

---

[2] See Maxwell (1972a; 1974; 1984, chs. 5 and 9; 1993; 1998; 2000a; 2002; 2004a; 2005; 2006a; 2007, chs. 5, 9 and especially 14; 2008; 2011a; 2012a; 2012b; 2013).

[3] See Miller (2006, p. 92-94) and Muller (2008).

[4] See Vicente (2010).

[5] For my replies to Miller, Muller and Vicente see Maxwell (2006b; 2009a; 2010b, pp. 667-677).

[6] See for example Kneller (1978, pp. 80-87 and 90-91), Harris (1980, pp. 25-26), Chakravartty (1999), Smart (2000), Juhl (2000), McHenry (2000). Weinert (2000), Roush (2001), Muller (2004), McNiven (2005), Iredale (2005), Grebowicz (2006).

[7] For discussion of the claim that Kuhn and Lakatos defend versions of SE see Maxwell (1998, p. 40), and Maxwell (2005). Bayesianism might seem to reject SE, in acknowledging both prior and posteriori probabilities. But

and even social constructivist and relativist sociologists and historians of science uphold SE as the best available ideal of scientific rationality.[8] *If* science can be exhibited as rational, they hold (in effect), then this must be done in a way that is compatible with SE. The failure of science to live up to the rational ideal of SE is taken by them to demonstrate that science is not rational. That it is so taken demonstrates convincingly that they hold SE to be the only possible rational ideal for science (an ideal which cannot, it so happens, in their view, be met).

SE is more or less unthinkingly taken for granted by the vast majority of working scientists – so much so that it is rather rare to find the doctrine being explicitly formulated, let alone defended. Scattered throughout the writings of scientists one can, nevertheless, find affirmations of the view. Thus Planck once remarked "Experiments are the *only* means of knowledge at our disposal. The rest is poetry, imagination" (Atkins, 1983, p. xiv). Or, as Poincaré (1952, p. 140) put it "Experiment is the sole source of truth. It alone can teach us something new; it alone can give us certainty."[9]

Furthermore, scientists do what they can to ensure that science conforms to the doctrine. As a result, it exercises a widespread influence over science itself. It influences such things as the way aims and priorities of research are discussed and chosen, criteria for publication of scientific results, criteria for acceptance of results, the intellectual content of science, science education, the relationship between science and the public, science and other disciplines, even scientific careers, awards and prizes.[10]

3. STANDARD EMPIRICISM: REFUTATION

If SE is valid, then the argument with which we began, which presupposes SE, is sound: it cannot be held that science has, today, established that the cosmos is physically comprehensible. But SE is not valid, as the following simple argument demonstrates.

Physics only ever accepts theories that are (more or less) *unified*, even though endlessly many empirically more successful disunified rivals can always be concocted. This persistent failure even to consider, let alone accept, empirically more successful disunified rivals means that physics makes a big, permanent assumption about the nature of the cosmos independent of all empirical considerations: the cosmos is such that all seriously disunified theories are false, whatever their empirical success might be. There is some kind of underlying unity in nature – to some extent at least. This in turn implies that SE is false, since SE asserts that science makes no persistent assumption about the cosmos independent of empirical considerations.

But is it the case that, given any accepted, unified theory, endlessly many empirically more successful, disunified rivals can always be concocted? Here is my demonstration that this is indeed the case.

Bayesianism tries to conform to the spirit of SE as much as possible, by regarding prior probabilities as personal, subjective and non-rational, their role in theory choice being reduced as rapidly as possible by empirical testing: see Maxwell (1998), p. 44.

[8] See Feyerabend (1975); Bloor (1991); Barnes, Bloor and Henry (1996).

[9] For more detailed discussion of the point that SE is widely taken for granted see Maxwell (1984), chs. 2 and 6; Maxwell (1998), ch. 2; Maxwell (2004a), pp. 13-14, note 14.

[10] For a more detailed exposition of SE, see Maxwell (2007), pp. 32-51. For grounds for holding scientists do, by and large, accept SE, see Maxwell (1998, pp. 38-45; 2007, pp. 145-156; 2004a, pp. 5-6, note 5).

Let T be any accepted fundamental physical theory – Newtonian theory, classical electrodynamics, quantum theory, general relativity, quantum electrodynamics, quantum electroweak theory, quantum chromodynamics, or the standard model. There are, to begin with, infinitely many disunified rivals to T – T_1, T_2, ... T_∞ – that are *just as empirically successful* as T. In order to concoct such a rival, T_1 say, all we need to do is modify T in an entirely *ad hoc* way for phenomena that occur after some future date. Thus, if T is Newtonian theory (NT), NT_1 might assert: everything occurs as NT predicts until the first moment of 2050 (GMT) when an inverse cube law of gravitation comes into operation: $F = Gm_1m_2/d^3$. Infinitely may such disunified rivals can be concocted by choosing infinitely many different future times for an abrupt, arbitrary change of law. These theories will no doubt be refuted as each date falls due, but infinitely many will remain unrefuted.[11] We can also concoct endlessly many disunified rivals to T by modifying the predictions of T for just one kind of system that we have never observed. Thus, if T is, as before, NT, then NT_2 might assert: everything occurs as NT predicts except for any system of pure gold spheres, each of mass greater than 1,000 tons, moving in a vacuum, centres no more than 1,000 miles apart, when Newton's law becomes $F = Gm_1m_2/d^4$. Yet again, we may concoct further endlessly many equally empirically successful disunified rivals to T by taking any standard experiment that corroborates T and modifying it in some trivial, irrelevant fashion – painting the apparatus purple, for example, or sprinkling diamond dust in a circle around the apparatus. We then modify T in an *ad hoc* way so that the modified theory, T_3 say, agrees with T for all phenomena except for the trivially modified experiment. For this experiment, not yet performed, T_3 predicts – whatever we choose. We may choose endlessly many different outcomes, thus creating endlessly many different modifications of T associated with this one trivially modified experiment. On top of that, we can, of course, trivially modify endlessly many further experiments, each of which generates endlessly many further disunified rivals to T.

Each of these equally empirically successful, disunified rivals to T – T_1, T_2, ... T_∞ – can now be modified further, so that each becomes *empirically more successful* than T. Any accepted fundamental physical theory is almost bound to face some empirical difficulties, and is thus, on the face of it, refuted – by phenomena A. There will be phenomena, B, which come within the scope of the theory but which cannot be predicted because the equations of the theory cannot (as yet) be solved. And there will be other phenomena (C) that fall outside the scope of the theory altogether. We can now take any one of the disunified rivals to T, T_1 say, and modify it further so that the new theory, T_1*, differs further from T in predicting, in an entirely *ad hoc* way, that phenomena A, B and C occur in accordance with empirically established laws L_A, L_B and L_C. T_1* successfully predicts all that T has successfully predicted; T_1* successfully predicts phenomena A that ostensibly refute T; and T_1*

[11] The famous arguments of David Hume (1959) concerning causation have long rendered familiar the point that, for all we can ever know for certain, physical laws may, at some future time, abruptly change – or, to put the same point in an equivalent way, given any physical theory, however empirically successful, there will always be infinitely many rival theories that fit all available evidence just as well but which postulate an abrupt change in physical laws at some future time. What I go on to argue in the text is that these "Humean" theories, as we may call them, that postulate an abrupt change in physical law at some future time are just special cases of a much wider class of disunified theories that postulate an abrupt change in physical law as conditions change in some way other than time, such as kind of system, or experimental conditions. And furthermore, I argue, all these disunified theories can be further doctored to be empirically more successful than the theories we accept.

successfully predicts phenomena B and C that T fails to predict. On empirical grounds alone, T_1* is clearly more successful and better corroborated, than T. And all this can be repeated as far as all the other disunified rivals of T are concerned, to generate infinitely many empirically more successful disunified rivals to T: T_1*, T_2*, ... T_∞*.[12]

But even though all of T_1*, T_2*, ... T_∞* are more successful empirically than T, they are all, quite correctly, ignored by physics because they are all horribly disunified. They postulate different laws for different phenomena in a wholly *ad hoc* fashion, and are just assumed to be false. But this means physics makes a big, implicit assumption about the cosmos: it is such that all such grossly disunified, "patchwork quilt" theories are false.

If physicists only ever accepted theories that postulate atoms even though empirically more successful rival theories are available that postulate other entities such as fields, it would surely be quite clear: physicists implicitly assume that the cosmos is such that all theories that postulate entities other than atoms are false. Just the same holds in connection with unified theories. That physicists only ever accept unified theories even though endlessly many empirically more successful, disunified rival theories are available means that physics implicitly assumes that the cosmos is such that all such disunified theories are false.

But SE holds that no permanent thesis about the world can be accepted as a part of scientific knowledge independent of evidence (let alone against the evidence). That physics does accept permanently (if implicitly) that there is some kind of underlying unity in nature thus suffices to refute SE. SE is, in short, untenable.[13] Physics makes a big implicit assumption about the nature of the cosmos, upheld independently of empirical considerations – even, in a certain sense, in violation of such considerations: the cosmos possesses some kind of underlying dynamic unity, to the extent at least that it is such that all sufficiently disunified physical theories are false. This is such a secure tenet of scientific knowledge that empirically successful theories that clash with it are not even considered for acceptance.

This argument establishes that SE is untenable. It demonstrates that it is a part of current theoretical knowledge in physics that the cosmos is physically comprehensible to the extent, at least, that all sufficiently disunified, "patchwork quilt" physical theories are false. This cosmological thesis, though metaphysical and thus not directly corroborated empirically, is nevertheless a central tenet of current scientific knowledge, so much so, indeed, that all physical theories (of the kind considered above) that clash with it are rejected whatever their empirical success might be.

We have taken a giant step towards our goal, but we are not there yet. I have shown:

A. Science accepts, as a part of theoretical knowledge, that the cosmos is such that all sufficiently disunified physical theories are false.

But what I need to show is the rather more substantial thesis:

B. Science accepts, as a part of theoretical knowledge, that the cosmos is such that there is a yet-to-be-discovered physical "theory of everything" that is (a) unified, and (b) true.

---

[12] Even if no phenomena ostensibly refute T, each of T1*, T2*, ... T∞* is still more successful empirically than T.

[13] For earlier refutations of SE along these lines, see Maxwell (1998, ch. 2; 2004a, ch.1; 2005; 2007, ch. 9). For my rebuttal of sixteen objections to the validity of this refutation of SE see Maxwell (2012c).

In order to get from (A) to (B), two crucial questions need to be answered.

1. What exactly does it mean to say of a physical theory that it is "unified"? How are degrees of unity, or disunity, to be assessed?
2. Granted that science accepts, at least, that the cosmos is such that all sufficiently disunified theories are false, what metaphysical thesis about the cosmos ought science to adopt?

In what follows I take these two questions in turn.

4. DEGREES OF DISUNITY

What does it mean to say of a physical theory that it is *unified*? How can one capture this notion of the unity of a theory when an apparently beautifully unified theory can always be reformulated so that it becomes horribly disunified, and *vice versa*, a horribly disunified theory can be reformulated to become unified?[14] How are *degrees* of unity to be specified? These problems of "unification", or of "simplicity" as they are sometimes called, are widely understood to be fundamental problems of the philosophy of science.[15]

It is worth noting in passing that Einstein fully recognized the importance of the problem of what the unity of a theory is, and yet did not know how to solve it. This is apparent from a passage of Einstein's "Autobiographical Notes" during which he discusses ways in which physical theories, quite generally, can be critically assessed (Einstein, 1949, pp. 20-25).

Einstein emphasizes that theories need to be critically assessed from *two* distinct points of view: from the standpoint of their empirical success, and from the standpoint of their "inner perfection", the "naturalness" or "logical simplicity" of the postulates. Einstein stresses that the theories that need to be considered are those "whose object is the *totality* of all physical appearances", that is, "theories of everything" in modern parlance. And Einstein acknowledges that "an exact formulation" of the second point of view (having to do with the "inner perfection" of theories) "meets with great difficulties" even though it "has played an important role in the selection of theories since time immemorial". Einstein continues:

> The problem here is not simply one of a kind of enumeration of the logically independent premises (if anything like this were at all unequivocally possible), but that of a kind of reciprocal weighing of incommensurable qualities. ... The following I reckon as also belonging to the "inner perfection" of a theory: We prize a theory more highly if, from the logical standpoint, it is not the result of an arbitrary choice among theories which, among themselves, are of equal value and analogously constructed (Einstein, 1949, p. 23).

---

[14] Richard Feynman has provided the following amusing illustration of this point (Feynman et al., 1965, 25-10 – 25-11). Consider an appallingly disunified, complex theory, made up of 10^{10} quite different, distinct laws, stuck arbitrarily together. Such a theory can easily be reformulated so that it reduces to the dazzlingly unified, simple form: $A = 0$. Suppose the 10^{10} distinct laws of the universe are: (1) $F = ma$; (2) $F = Gm_1m_2/d^2$; and so on, for all 10^{10} laws. Let $A_1 = (F - ma)^2$, $A_2 = (F - Gm_1m_2/d^2)^2$, and so on. Let $A = A_1 + A_2 + \dots + A_{10^{10}}$. The theory can now be formulated in the unified, simple form $A = 0$. (This is true if and only if each $A_r = 0$, for $r = 1, 2, \dots 10^{10}$).

[15] See for example Salmon (1989). Elsewhere, I have formulated seven problems of unity or simplicity: see Maxwell (1998, pp. 104-105).

And Einstein comments:

> I shall not attempt to excuse the meagre precision of these assertions ... by lack of sufficient printing space at my disposal, but confess herewith that I am not, without more ado, and perhaps not at all, capable of replacing these hints with more precise definitions. I believe, however, that a sharper formulation would be possible (Einstein, 1949, p. 23).

Since Einstein's day, many attempts have been made to solve the problem, none successful.[16] How is the problem to be solved – a problem which baffled even Einstein?

The key to the solution is to appreciate that standard attempts to solve the problem have been looking at entirely the wrong thing. They have been looking at the *theory itself*, its axiomatic structure, its number of postulates, its formulation, its characteristic derivations, the language in which it is formulated. (Even Einstein made this mistake.) But all this is wrong. What one needs to look at is not the theory itself, but rather what the theory says about the world, the *content* of the theory in other words. One needs to look, not at the theory, but at the world *as depicted by the theory*. At a stroke the worst aspect of the problem of what unity *is* vanishes. No longer does one face what may be called the *terminological* problem of unity – the problem, namely, that the extent to which a theory is unified appears to be highly dependent on the way it is *formulated*.

Suppose we have a given theory T, which is formulated in N different ways, some formulations exhibiting T as beautifully unified, others as horribly complex and disunified, but all formulations being interpreted in precisely the same way, so as to make precisely the same assertion about the world. If unity has to do exclusively with *content*, then *all these diverse formulations of T, having the same content, have precisely the same degree of unity*. The variability of apparent unity with varying formulations of one and the same theory, T, (given some specific interpretation) – which poses such an insurmountable problem for traditional approaches to the problem (see Salmon, 1989; Maxwell, 1998, pp. 56-68) – poses no problem whatsoever for the thesis that unity has to do with *content*. Variability of formulation of a theory which leaves its content unaffected is wholly irrelevant: the unity of the theory is unaffected.

But now we have a new problem: How is the unity of the *content* of a theory to be assessed? What exactly does it mean to assert that a dynamical physical theory has a unified content?

What it means is that the theory has *the same* content throughout the range of possible phenomena to which the theory applies. Unity, in other words, means that there is just *one* content throughout the range of possible phenomena to which the theory applies. If the theory postulates *different* contents, *different* laws, for different ranges of possible phenomena, then the theory is *disunified*, and the more such *different* contents there are so the more *disunified* the theory is.

Thus "unity" means "one", and "disunity" means "more than one", the disunity becoming worse and worse as the number of different contents goes up, from two to three to four, and so on. Not only does this enable us to distinguish between "unified" and "disunified" theories;

---

[16] For my criticisms of proposals put forward by Jeffreys and Wrinch (1921), Popper (1959, pp. 62-70 and 126-145), Friedman (1974), Kitcher (1981) and Watkins (1984, pp. 203-213), see my (1998, pp. 56-68). For criticisms of more recent proposals see my (2004b).

it enables us to assign "degrees of unity" to theories, or to partially order theories with respect to their degree of unity.[17]

All this can be illustrated by considering the disunified "patchwork quilt" theories indicated in Section 3 above. Whereas an accepted, unified theory, T, asserts that the same dynamical laws (as far as their content is concerned) apply in ranges of phenomena A, B and C, a "patchwork quilt" rival, T*, asserts that distinct laws, L_A, L_B, and L_C, apply in A, B and C, T* being disunified as a result.

To take a specific example, Newton's theory of gravitation, $F = GM_1M_2/d^2$ is unified in that what it asserts is *the same* throughout all possible phenomena to which it applies. Bodies attract each other in proportion to their masses, and inversely proportional to the square of the distance between them, and this remains the same, according to the theory, throughout all the phenomena to which the theory applies, whatever the masses of the bodies may be, whatever their constitution, shape, relative velocity, distance apart, place and time. One kind of disunified version of Newton's theory would assert that Newton's law becomes an inverse cube law after some definite time t_o, the theory asserting that $F = GM_1M_2/d^2$ for times $t \leq t_o$ and $F = GM_1M_2/d^3$ for times $t > t_o$. This theory is disunified because what it asserts is blatantly *not* the same throughout the range of possible phenomena to which it applies.

Note that special terminology could be introduced to make Newtonian theory look *disunified*, and the disunified version of Newtonian theory look *unified*. All we need do is interpret "d^N" to mean "d^N if $t \leq t_o$ and d^{N+1} if $t > t_o$". In terms of this (admittedly somewhat bizarre) terminology, the disunified theory has the form "$F = GM_1M_2/d^2$", and Newtonian theory has the disunified *form* "$F = GM_1M_2/d^2$ for times $t \leq t_o$ and $F = GM_1M_2/d$ for times $t > t_o$". But this mere *terminological* reversal of disunity does not affect the *content* of the two theories: the content of Newtonian theory remains unified, and the content of the disunified version (which looks unified) remains disunified. For unity, in other words, we require that the theory is *terminologically* invariant throughout the range of possible phenomena to which it applies when *terminology*, used to formulate the theory, is itself invariant throughout the range of possible phenomena (so that *terminological* invariance implies *content* invariance).

This account of unity needs to be refined further. In assessing the extent to which a theory is disunified we may need to consider *how* different, or *in what way* different, one from another, the different contents of a theory are. A theory that postulates different laws at different times and places is disunified in a much more serious way than a theory which postulates the same laws at all times and places, but also postulates that distinct kinds of physical particle exist, with different dynamical properties, such as charge or mass. This second theory still postulates *different* laws for different ranges of phenomena: laws of one kind for possible physical systems consisting of one kind of particle, and slightly different laws for possible physical systems consisting of another kind of particle. But this second kind of difference in content is much less serious than the first kind (which involves different laws at different times and places). In other words, the more *different*, one form another, the different contents of a theory are, so the more seriously *disunified* the theory is.

What this means is that there are different *kinds* of disunity, different *dimensions* of disunity, as one might say, some more serious than others, but all facets of the same basic idea. Eight different facets of disunity can be distinguished as follows.

---

[17] For earlier accounts of my proposed solution to the problem of unity of physical theory see Maxwell (1998, chs. 3 and 4; 2004a, chs. 1-2 and appendix; 2004b; 2007, pp. 373-386; and 2011a).

Any dynamical physical theory can be regarded as specifying a space, S, of possible physical states to which the theory applies, a distinct physical state corresponding to each distinct point in S. (S might be a set of such spaces.) For unity, we require that the theory, T, asserts that *the same* dynamical laws apply throughout S, governing the evolution of the physical state immediately before and after the instant in question. If T postulates N distinct dynamical laws in N distinct regions of S, then T has disunity of degree N. For unity in each case we require that N = 1.

1. T divides space-time up into N distinct regions, $R_1...R_N$, and asserts that the laws governing the evolution of phenomena are the same for all space-time regions within each R-region, but are different within different R-regions. Example: the *ad hoc* version of Newtonian theory (NT) indicated above: for that theory, N = 2, in a type (1) way.[18]
2. T postulates that, for distinct ranges of physical variables (other than position and time), such as mass or relative velocity, in distinct regions, $R_1,...R_N$ of the space of all possible phenomena, distinct dynamical laws obtain. Example: T asserts that everything occurs as NT asserts, except for the case of any two solid gold spheres, each having a mass of between one and two thousand tons, moving in otherwise empty space up to a mile apart, in which case the spheres attract each other by means of an inverse cube law of gravitation. Here, N = 2 in a type (2) way.
3. In addition to postulating non-unique physical entities (such as particles), or entities unique but not spatially restricted (such as fields), T postulates, in an arbitrary fashion, M distinct, unique, spatially localized objects, each with its own distinct, unique dynamic properties. In this case, T is disunified to degree N = M + 1, in a type (3) way. Example: T asserts that everything occurs as NT asserts, except there is one object in the universe, of mass 8 tons, such that, for any matter up to 8 miles from the centre of mass of this object, gravitation is a repulsive rather than attractive force. The object only interacts by means of gravitation. Here, N = 2, in a type (3) way.
4. T postulates physical entities interacting by means of N distinct forces, different forces affecting different entities, and being specified by different force laws. (In this case one might require one force to be universal so that the universe does not fall into distinct parts that do not interact with one another.) Example: T postulates particles that interact by means of Newtonian gravitation; some of these also interact by means of an electrostatic force $F = Kq_1q_2/d^2$, this force being attractive if q_1 and q_2 are oppositely charged, otherwise being repulsive, the force being much stronger than gravitation. Here, N = 2 in a type (4) way.

[18] As I have formulated it here, (1) is open to two somewhat different interpretations. First, for N = 1 we require only that the same law operates throughout space in the sense that this would be true even if the law in question asserted that all objects experience a force directed at a unique point in space, and inversely proportional to their distance from that point. Second, for N = 1, we require that the same law operates throughout space in the sense that a mere change of position in space of an isolated physical system has no effect on the way the system evolves. An analogous distinction arises in connection with time. In what follows I adopt the second interpretation of (1), which means that a theory which is unified with respect to (1) exhibits symmetry with respect to spatial location, and time of occurrence. As far as the *ad hoc* version of NT is concerned, N = 2 for both versions of (1).

5. T postulates N different kinds of physical entity,[19] differing with respect to some dynamic property, such as value of mass or charge, but otherwise interacting by means of the same force. Example: T postulates particles that interact by means of Newtonian gravitation, there being three kinds of particles, of mass m, 2m and 3m. Here, N = 3 in a type (5) way.
6. Consider a theory, T, that postulates N distinct kinds of entity (e.g. particles or fields), but these N entities can be regarded as arising because T exhibits some symmetry (in the way that the electric and magnetic fields of classical electromagnetism can be regarded as arising because of the symmetry of Lorentz invariance, or the eight gluons of chromodynamics can be regarded as arising as a result of the local gauge symmetry of SU(3)). If the symmetry group, G, is not a direct product of subgroups, we can declare that T is fully unified; if G is a direct product of subgroups, T lacks full unity; and if the N entities are such that they cannot be regarded as arising as a result of some symmetry of T, with some group structure G, then T is disunified.[20] Example: T postulates the classical electromagnetic field, composed of the electric and magnetic fields, obeying Maxwell's equations for the field in the vacuum. The symmetry of Lorentz invariance unifies these two fields (see below). Here, N = 1 (in a type (6) way).
7. If (apparent) disunity of there being N distinct kinds of particle or distinct fields has emerged as a result of cosmic spontaneous symmetry-breaking events, there being manifest unity before these occurred, then the relevant theory, T, is unified. If current (apparent) disunity has not emerged from unity in this way, as a result of spontaneous symmetry-breaking, then the relevant theory, T, is disunified. Example: Weinberg's and Salam's electroweak theory, according to which at very high energies, such as those that existed soon after the big bang, the electroweak force has the form of two forces, one with three associated massless particles, two charged, W^- and W+, and one neutral, W^o, and the other with one neutral massless particle, V^o. According to the theory, the two neutral particles, W^o and V^o, are intermingled in two different ways, to form two new, neutral particles, the photon, γ, and another neutral massless particle, Z^o. As energy decreases, the W+, W- and Z^o particles acquire mass, due to the mechanism known as spontaneous symmetry-breaking (involving the hypothetical Higgs particle), while the photon, γ, retains its zero mass. This theory unifies the weak and electromagnetic forces as a result of exhibiting the symmetry of local gauge invariance; this unification is only partial, however, because the symmetry group is a direct product of two groups, U(1) associated with V^o, and SU(2) associated with W^-, W+ and W^o.[21]

---

[19] Counting entities is rendered a little less ambiguous if a system of M particles is counted as a (somewhat peculiar) field. This means that M particles all of the same kind (i.e. with the same dynamic properties) is counted as one entity. In the text I continue to adopt the convention that M particles all the same dynamically represent one kind of entity, rather than one entity.

[20] An informal sketch of these matters is given in (Maxwell, 1998, ch. 4, sections 11 to 13, and the appendix). For rather more detailed accounts of the locally gauge invariant structure of quantum field theories see: (Moriyasu, 1983; Aitchison and Hey, 1982, part III; and Griffiths, 1987, ch. 11). For a non-technical discussion of the role of symmetry and group theory in physics, see (Maxwell, 1998, pp. 257-265); for somewhat more technical introductory accounts of group theory as it arises in physics see (Isham, 1989; or Jones, 1990).

[21] For accounts of spontaneous symmetry breaking see (Moriyasu, 1983; Mandl and Shaw, 1984; Griffiths, 1987, ch. 11).

8. According to general relativity, Newton's force of gravitation is merely an aspect of the curvature of space-time. As a result of a change in our ideas about the nature of space-time, so that its geometric properties become dynamic, a physical force disappears, or becomes unified with space-time. This suggests the following requirement for unity: space-time on the one hand, and physical particles-and-forces on the other, must be unified into a single self-interacting entity, U. If T postulates space-time and physical "particles-and-forces" as two fundamentally distinct kinds of entity, then T is not unified in this respect. Example: One might imagine a version of string theory without strings, different vibrational modes (perhaps) of empty, compactified six-dimensional space giving rise to the appearance of particles and forces, even though in reality there is only 11 dimensional space-time. Or one might imagine that the quantization of space-time leads to the appearance of particles and forces as only apparently distinct from empty space-time. In either case, N = 1: there is just the one self-interacting entity, empty space-time.

For unity, in each case, we require N = 1. As we go from (1) to (5), the requirements for unity are intended to be cumulative: each presupposes that N = 1 for previous requirements. As far as (6) and (7) are concerned, if there are N distinct kinds of entity which are not unified by a symmetry, whether broken or not, then the degree of disunity is the same as that for (4) and (5), depending on whether there are N distinct forces, or one force but N distinct kinds of entity between which the force acts.[22]

(8) does not introduce a new kind of unity, but introduces, rather, a new, more severe way of counting different kinds of entity. (1) to (7) require, for unity, that there is one kind of self-interacting physical entity evolving in a distinct space-time, the way this entity evolves being specified, of course, by a consistent physical theory. According to (1) to (7), even though there are, in a sense, two kinds of entity, matter (or particles-and-forces) on the one hand, and space-time on the other, nevertheless N = 1. According to (8), this would yield N = 2. For N = 1 in an (8) type way we require space-time and matter to be just one basic entity (unified by means of a spontaneously broken symmetry, perhaps).

As we go from (1) to (8), then, requirements for unity become increasingly demanding, with (6) and (7) being at least as demanding as (4) and (5), as explained above.[23]

(1) to (8) may seem very different requirements for unity. In fact they all exemplify the same basic idea: disunity arises when *different* dynamical laws govern the evolution of physical states in different regions of the space, S, of all possible physical states. For example, if a theory postulates more than one force, or kind of particle, not unified by symmetry, then in different regions of S different force laws will operate. If different fields, or different kinds of particle, are unified by a symmetry, then a symmetry transformation may

---

[22] Some kinds of disunity can be decreased in a trivial way by splitting a theory into two or more distinct theories. Thus it may be possible to split a theory postulating two distinct forces into two distinct theories, each postulating one force. This trivial manoeuvre can be outlawed by insisting that unity applies to all fundamental physical theories applying to the full range of physical phenomena – and to empirical laws themselves if some phenomena are without a theory that applies to them. A new theory is then acceptable, on unity grounds, if its acceptance improves the unity of the totality of fundamental theory.

[23] The account of theoretical unity given here simplifies the account given in (Maxwell, 1998, chs. 3 and 4), where unity is explicated as "exemplifying physicalism", where physicalism is the metaphysical thesis asserting that the universe has some kind of unified dynamic structure. Explicating unity in that way invites the charge of circularity, a charge that is not actually valid (see Maxwell, 1998, pp. 118-23 and 168-72). The account given here forestalls this charge from the outset.

seem to change the relative strengths of the fields, or change one kind of particle into another kind, but will actually leave the dynamics unaffected. In this case, then, the same fields, or the same kind of particle are, in effect, present everywhere throughout S. If (8) is not satisfied, there is a region of S where only empty space exists, the laws being merely those which specify the nature of empty space or space-time. The eight distinct facets of unity, (1) to (8) arise, as I have said, because of the eight *different* ways in which content can vary from one region of S to another, some differences being *more* different than others. Some of the above requirements for unity are suggested by developments in 20th century physics. This is true in particular of (6) to (8). The important point, however, is that all these requirements for unity, (1) to (8), exemplify the same basic idea: a theory, in order to be unified, must assert that the same laws apply throughout the possible phenomena to which it applies. (1) to (8) in effect represent different, increasingly subtle ways in which a theory can fail to be unified in this sense, granted that $N > 1$ in each case.

One point demonstrated by (1) to (8) is that there is no sharp distinction between the very crudest kind of unity or non-*ad hocness*, specified by (1) or (2), and the most demanding kind of theoretical unity, indicated by (8). One might, perhaps, draw a line between (1) to (3) on the one hand, and (4) to (8) on the other hand, and declare (1) to (3) to be essential to physics, and (4) to (8) as being desirable but optional: but any such distinction is more or less arbitrary.[24]

Symmetry plays a role in almost all of the above eight facets of unity.

What this role is, already indicated above, can be illustrated further by means of the example of classical electrodynamics and the symmetry of special relativity: Lorentz invariance. One might be inclined to hold that the electromagnetic field is made up of two distinct entities: the electric and the magnetic fields. But, as a result of being Lorentz invariant, the way the electromagnetic field divides up into the electric and magnetic fields depends on choice of reference frame. And, according to special relativity, nothing of fundamental dynamical significance can depend on choice of (inertial) reference frame. Any specific way we might choose to divide the field into the electric and magnetic fields would be arbitrary; it would be equivalent to choosing, arbitrarily, one frame, one state of motion to constitute absolute rest, from the infinitely many equally available, and all equivalent, according to special relativity. This would violate special relativity. Thus special relativity (Lorentz invariance) requires that we regard the electromagnetic field as *one* entity with two aspects, and not as *two* distinct entities.

I now consider briefly four questions that may be asked in connection with this proposed solution to the problem of unity of physical theory.

First, what of "simplicity"? Is this the same as "unity", or something distinct? The "simplicity" of a theory can be interpreted as having to do, not with whether the *same* laws

[24] This point is of fundamental importance for the problem of induction. Traditionally, the problem is interpreted as the problem of justifying exclusion of empirically successful theories that are *ad hoc* in sense (1): How can evidence from the past provide grounds for any belief about the future? This makes the problem seem highly "philosophical", remote from any problem realistically encountered in scientific practice. But the moment it is appreciated that the problem of justifying exclusion of empirically successful theories that are *ad hoc* in sense (1) is just an extreme, special case of the more general problem of excluding empirically successful theories that are *ad hoc* in senses (1) to (8), it becomes clear that this latter problem is a scientific problem, a problem of theoretical physics itself. For the implications of this crucial insight, and for a proposal as to how the problem of induction is to be solved exploiting it, see Maxwell (1998, especially chs. 4 and 5; 2004a, pp. 205-220; 2007, pp. 400-430). See also Maxwell (2006a).

apply throughout the space of possible phenomena predicted by the theory in question, but rather with the *nature* of the laws, granted that they are the same. Some laws are simpler than others. In order to overcome the objection that simplicity is formulation dependent it is essential, as in the case of unity, to interpret "simplicity" as applying to the *content* of theories, and not to their *formulation*, their *axiomatic structure*, etc. For details, see Maxwell (1998, pp. 157-9). It is a great success of the account of theoretical unity given here that it succeeds in distinguishing sharply between these two aspects of the problem of what the explanatory character of a physical theory *is*, namely the *unity* aspect, and the *simplicity* aspect, and succeeds in solving both.

On the face of it, mere terminological unity can play no important heuristic or methodological role in physics at all because, given any *unified* theory, it can be made as simple or complex as we like by appropriate choice of terminology. But what is paradoxical about the role of unity in physics is that terminological unity does, in practice, seem to be highly significant heuristically and methodologically. How is this paradox to be resolved? The answer is that, if terminological unity and *content* unity – or, as we may call it, *physical* unity – are in one-to-one correspondence with one another, so that the degree and kind of unity of one echoes precisely that of the other then, in these circumstances, terminological and physical unity are of equal significance (the former being an accurate guide to the latter). One way of arranging matters so that terminological and physical unity mirror each other is to ensure that symmetries of a theory can be interpreted in both "passive" and "active" senses – so that corresponding to a passive symmetry operation that is no more than a change in the way a physical system is described, there is a corresponding active symmetry operation that changes the physical system in question itself (changes its spatial position, orientation, or uniform velocity, for example, as opposed to the spatial position, orientation or velocity of the coordinate system used to describe the system).[25]

Second, does the question of whether laws governing a range of phenomena *remain the same* throughout their range of application have an unambiguous answer, in view of Goodman's "grue" and "bleen" paradox (Goodman, 1954)? Adapting Goodman's notions slightly, an object is grue if it is green up to the last moment of the 21st century, and blue thereafter; an object is bleen if it is blue up to the last moment of the 21st century, and green thereafter. Are not grue and bleen just as good predicates as blue and green? If the colours of objects change dramatically at the end of the 21st century, so that blue objects become green, and green objects blue, can we not, with equal legitimacy, say that there is no change, objects continue to be grue and bleen? This much discussed paradox is, in my view, very largely a red herring. On the face of it, the distinction made above, between formulation and content, suffices to dismiss the paradox. The sentence "This object is grue" (S) may, as far as its written form is concerned, be invariant through the end of the 21st century, but what this sentence asserts, its content, is not invariant. To this, the reply may be made that the *content* of S may be regarded as being invariant. But this is not what is ordinarily meant by "invariant" or "remain the same": the above account of unity of theory appeals to the ordinary meaning of "invariant" or "remains the same", and not the perverse grue and bleen meaning. Two additional points. It should be noted that the Goodman paradox implicitly accepts the ordinary meaning of "remains the same" in employing the terminology of "grue" and "bleen", terminology which remains the same, in the ordinary sense, throughout the end of the 21st

[25] For further details see Maxwell, 1998, pp. 110-113.

century. Second, that the content of grue and bleen is not invariant with respect to the passage of time – unlike the content of blue and green which is invariant – is demonstrated by the point that if objects really are grue and bleen, and a person is convinced of this, then he can tell, by looking at grass and sky, whether or not the 21st century has ended, whereas the same is not true with respect to green and blue. Grue and bleen implicitly refer to a specific time in a way in which green and blue do not.

Third, Goodman's point concerning the ambiguity of "remains the same" may seem to gain support from the mathematical notion of a function as a rule which takes one from one set of numbers to another. According to this notion, the two functions (1) $y = 3x$ for all x, and (2) $y = 3x$ for $x \leq 2$ and $y = 4x$ for $x > 2$, are equally good functions. Both functions "remain the same" as x increases and passes through the value $x = 2$. Clearly, we need a narrower notion of function than this if we are to be able to distinguish between functional relationships which do, and which do not, "remain the same" as values of variables change. We need to appeal to what may be called "invariant functions", functions which specify some fixed set of mathematical operations to be performed on "x" (or its equivalent) to obtain "y" (or its equivalent). In the example just given, (1) is invariant, but (2) is not. (2) is made up of two truncated invariant functions, stuck together at $x = 2$. Functions that appear in theoretical physics are *analytic*; that is, they can be represented as a power series (Penrose, 2004, pp. 112-4). Analytic functions are repeatedly differentiable. Such functions have the remarkable property that from any small bit of the function, the whole function can be reconstructed uniquely, by a process called "analytic continuation". All analytic functions are thus invariant. The latter notion is however a wider one, and theoretical physics might, one day, need to employ this wider notion explicitly, if space and time turn out to be discontinuous, and analytic functions have to be abandoned at a fundamental level.[26]

Fourth, can it really be meaningful to say that what a theory T asserts *remains the same* as one moves through a range of possible phenomena to which the theory applies? In order for this to make sense, does one not need to specify some aspect, some physical property or feature *with respect to which* what T asserts remains the same? The answer here is straightforward. It can be put like this. T is unified if what it asserts remains the same (as we move through possible phenomena to which T applies) *with respect to what determines (perhaps probabilistically) how phenomena evolve in space and time*. The aspect or feature with respect to which what T asserts must not change (for unity) is *that aspect of things which determines how events evolve*. Thus classical electrodynamics is unified because what it asserts about the electromagnetic field which determines the way the field evolves in space and time remains the same as we move through the space, S, of all possible states of the field.

This concludes my discussion of what it means to say of a theory that it is simple and unified.

This account of theoretical unity not only explicates and makes precise what it means to assert of a dynamical physical theory, T, that it is unified. It solves problems associated with the notion of theoretical unity, reveals that there are at least eight different kinds of unity, all exemplifying the same underlying notion, and provides the means for assessing theories with respect to their degree of unity. Furthermore, it explicates what we ought to mean when we

---

[26] For a fascinating discussion of the problems that arise in connection with the wider notion of what I have called 'invariant function', see Roger Penrose's discussion of what he calls the 'Eulerian' notion of function: Penrose (2004, 6.4).

declare of a physical theory that it is *explanatory*, for we can hold that the more (a) unified a theory is, and the greater its (b) empirical content (or predictive power), so the more *explanatory* it is.

And the above account of theoretical unity does more. It explicates what we ought to mean when we declare of the cosmos that it is *physically comprehensible*. And it brings order to the otherwise chaotic infinite realm of possible physically comprehensible universes – whether partially or wholly comprehensible. Let T stand for a "theory of everything" – a physical theory, that is, which, together with relevant initial and boundary conditions, in principle (not in practice) predicts accurately all physical phenomena (indeed, all physically possible phenomena). The thesis that the cosmos is at least partially physically comprehensible – a doctrine we may call *physicalism* – can be interpreted to mean that there is some yet-to-be-discovered theory of everything, T, which is (a) true, and (b) unified to some degree at least. But, as we saw above, there are eight different kinds of unity, n = 1, 2,... 8, each kind having some degree of unity, N = 1, 2 ... ∞. For perfect unity, perfect physical comprehensibility, we require that n = 8 and N = 1. What the above account of theoretical unity enables us to do, then, is to throw a two-dimensional grid over all possible *partially physically comprehensible* universes – the grid partially ordering universes with respect to kind and degree of unity or physical comprehensibility. We can interpret physicalism(n,N) to assert: the cosmos is such that the true theory of everything, T, is disunified in an (n) type way to extent N, with n = 1 ... 8, and N = 1 ... ∞, the cosmos being perfectly physically comprehensible when n = 8 and N = 1.

5. Metaphysical Assumptions of Physics

In section 3 we saw that standard empiricism (SE) is untenable. In persistently accepting only *unified* theories even though empirically more successful disunified rivals can always be concocted, physics in effect makes a substantial metaphysical assumption about the cosmos: it is such that all seriously disunified theories are false. At once the all-important question arises: Given that physics accepts, as a part of theoretical knowledge, *some* assumption concerning the unity, the physical comprehensibility, of the cosmos – *some* version of physicalism – what precisely ought this assumption to be? In the last section we have seen that, given that the cosmos is physically comprehensible to some extent at least, there is, before us, a vast range of possibilities, as n varies from 1 to 8 (or 4 to 8), and N varies from 1 to some number less than 100 (we may perhaps assume). And there are, no doubt, infinitely many possible different partially or wholly physically comprehensible universes for any fixed values of n and N. What considerations ought to guide our choice, and what ought our choice to be?

This is a question that it is both profoundly important for physics to answer correctly, and at the same time almost impossible to answer correctly. The answer that physics gives – whether implicitly or explicitly – will have a massive affect over both the theories that theoretical physics seeks to develop, and the theories that physics accepts. Progress in theoretical physics is almost bound to be affected dramatically by how good the choice of assumption is that physics adopts concerning the physical comprehensibility of the cosmos. At the same time, it is almost impossible for us, in our present state of knowledge and ignorance, to give the correct answer, for what is at issue is that of which we are most ignorant, the ultimate nature of the cosmos.

Most physicists and cosmologists, all too aware of the impossibility of specifying correctly today the ultimate nature of the cosmos, even in general terms, come to the conclusion that no attempt should be made to do so (unless it can be formulated as a testable theory). But this is not an option, as we saw in section 3. If we really did allow evidence alone to determine what theories we accept, we would persistently accept empirically successful but horribly disunified theories, which would sabotage all hope of progress in theoretical physics and cosmology. In quite correctly excluding empirically successful but grossly disunified theories from consideration, and thus making progress in physics possible, inevitably a big implicit assumption about the nature of the cosmos is made.

The choice before us is thus not whether we should adopt some assumption about the ultimate nature of the cosmos, or refrain from making any such assumption. Rather it is whether we make the assumption we adopt *explicit*, so that it can be critically assessed, so that alternatives can be developed and criticized in the hope of *improving* the assumption we adopt or, on the other hand, whether we leave the assumption we adopt *implicit*, and thus uncriticizable, an unacknowledged dogma. Implicit, unacknowledged assumptions are much more likely to retard scientific progress than assumptions explicitly acknowledged for the obvious reason that the former cannot be subjected to critical scrutiny, to sustained attempts to improve them.

A glance at the history of physics bears this out. In practice, at any stage in its development, theoretical physics makes far more substantial and restrictive metaphysical assumptions about the nature of the cosmos than that indicated in section 3 above. These assumptions are made, either explicitly, or implicitly as a consequence of theoretical and mathematical concepts employed or symmetry principles adopted. The failure of physics to subject these implicit metaphysical assumptions to sustained criticism and attempted improvement has, on occasions, delayed discovery and acceptance of good new theories.

An example is Newton's theory of gravitation. When Newton put forward the theory in his *Principia* in 1686, the reigning metaphysical idea among the small band of natural philosophers at the time was the corpuscular hypothesis. This holds that nature is made up of tiny corpuscles which interact only by contact. This doctrine functioned as a standard of acceptability, even intelligibility: no fundamental physical theory could claim to be acceptable, even intelligible, if it could not be interpreted in terms of the corpuscular hypothesis. The impossibility of interpreting Newton's theory of gravitation in terms of the corpuscular hypothesis initially led some of Newton's most eminent contemporaries to reject Newton's theory. Thus Huygens, in a letter to Leibniz, writes: "Concerning the Cause of the flux given by M. Newton, I am by no means satisfied [by it], nor by all the other Theories that he builds upon his Principle of Attraction, which seems to me absurd. . . I have often wondered how he could have given himself all the trouble of making such a number of investigations and difficult calculations that have no other foundation that this very principle" (Koyré, 1965, pp. 117-8). Newton in a sense agreed, as is indicated by his remark: "That gravity should be innate, inherent and essential to matter, so that one body may act upon another, at a distance through a vacuum, without the mediation of anything else. . . is to me so great an absurdity, that I believe no man who has in philosophical matters a competent faculty of thinking can ever fall into it" (Burtt, 1932, pp. 265-6). The impossibility of interpreting the law of gravitation in terms of the corpuscular hypothesis, in terms of action-by-contact, led

Newton to interpret the law instrumentalistically, as specifying the way bodies move without providing any kind of explanation for the motion, in terms of unobservable forces.

A somewhat more explicitly critical attitude towards the corpuscular hypothesis might have led Huygens and his contemporaries to appreciate that this view, far from embodying an ideal of comprehensibility, is actually a rather arbitrary special case of something much more general and comprehensible. For the corpuscular hypothesis, viewed exclusively from the standpoint of physics, without any implicit appeal being made to intuitive notions (arising, for example, from familiarity with billiard balls), can be seen to be a very arbitrary special case of something much more general. The corpuscle is a rigid sphere, on the surface of which there is an infinitely repulsive force that is zero elsewhere, so that if corpuscles collide they instantly rebound. This can be regarded as an arbitrary special case of the much more general, and more mathematically and physically tractable, idea of a point-particle surrounded by a rigid, spherically symmetrical field of force which varies smoothly with distance from the central point-particle, alternatively attractive and repulsive perhaps at different distances, tending to infinitely repulsive as the centre is approached. Just such an idea was eventually proposed by Boscovich in 1758.[27] Reject the corpuscular hypothesis and adopt, instead, Boscovich's much more general, and more acceptable metaphysical view, and Newton's theory ceases to be incomprehensible and unacceptable, and becomes the very model of comprehensibility.[28]

Another example is provided by James Clerk Maxwell's theory of electrodynamics. Maxwell himself, and most of his contemporaries and immediate successors, sought to interpret the electromagnetic field in terms of a material substratum, the hypothetical aether, itself to be understood in Newtonian terms. A tremendous amount of effort was put into trying to understand Maxwell's field equations in terms of the aether. Faraday, who appreciated that one should take the electromagnetic field as a new kind of physical entity, and explain matter in terms of the field rather than try to explain the field in terms of a kind of hypothetical matter (the aether), was ignored. The unrealistic character, and ultimate failure, of mechanical models of the electromagnetic field led many to hold that the real nature of the field must remain a mystery. The most that one could hope for from Maxwell's equations, it seemed, was the successful prediction of observable phenomena associated with electromagnetism. This instrumentalistic attitude remained even after the advent of Einstein's special theory of relativity in 1905, which might be interpreted as giving credence to the idea that it is the field that is fundamental. Gradually, however, the view came to prevail that one should see the field as a new kind of physical entity, quite distinct from the corpuscle and point-particle.

Once again, misguided effort to interpret the new theory – in this case Maxwellian electrodynamics – in terms of the current metaphysics – Newtonian or Boschovichean point-atomism – might have been avoided if the then current metaphysics had been subjected to critical scrutiny. For then it might have been realized – as Faraday evidently did realize – that Boscovich's point-atom view is an arbitrary special case of something more general. Point-atomism assumes that the force field associated with each point-particle is rigid. The point-

[27] See Boscovich (1966).

[28] Strictly speaking, Newton's theory clashes with Boscovich's metaphysics, because Newton's theory implies that the force between point-particles becomes infinitely attractive as the distance between them becomes zero. Boscovich's view can easily be reformulated to remove this clash. Point-particles can be assigned different charges, electrical, and perhaps other charges in addition to gravitational charge.

atom remains at the centre of its associated force field however rapidly it accelerates in any direction. A more general view would postulate that changes in the force field travel at some finite velocity – for example, the velocity of light – rather than at infinite velocity. But this more general view becomes horrendously complicated if each point-particle continues to have, associated with it, its own distinct field. Granted Boscovichean point-atomism, in order to specify the instantaneous state of any physical system, one needs only to specify the instantaneous positions and velocities of the point-atoms of which the system consists. The infinite rigidity of the force-field of each point-atom means that the instantaneous state of the field, associated with any point-atom, is determined by the general law which species how the field varies with distance, plus the instantaneous state of the relevant point-particle. Once infinite rigidity is relaxed, however, this can no longer be done. The instantaneous state of each field, associated with each point-particle, must be specified independently for each spatial point. All this becomes much simpler if all the distinct fields of the distinct point-particles are amalgamated to form just one field – an entity that acquires an existence independent of the point-particles that are its source. Thus is the Faraday/Einstein notion of the field born. Subject Boscovichean point-atomism to critical scrutiny, to generalization, in other words, and one arrives at the metaphysical idea of the field which enables one to see Maxwellian electrodynamics as an entirely comprehensible, acceptable theory, exemplifying as it does the notion of the field. The need to interpret the theory in terms of the aether disappears.[29]

Much the same happened, in my view, in connection with quantum theory, as I will recount below. It may be thought that the lesson to be learned from these episodes of bad, old, implicit metaphysics obstructing the acceptance and correct interpretation of a new theory is that physics should abjure *all* metaphysics, and insist that empirical considerations *alone* can decide whether a new theory is to be accepted or rejected. We have seen, however, in section 3, that it is just this option that is not possible. Non-empirical considerations must, inevitably, decide what theories are accepted and rejected, in addition, of course, to empirical considerations. Whatever these non-empirical considerations are, inevitably there will be metaphysical principles inherent in them, whether implicit or explicit. A vital key to rigour, success and progress in theoretical physics is to make implicit metaphysical assumptions explicit, so that they can be critically assessed and, we may hope, improved.

But how is this to be done? As I have already in effect remarked, the question of what metaphysical assumption about the nature of the cosmos to make is both *profoundly important* and *almost impossible for physics to get right.* On the one hand it is in our interests to choose an assumption that is as insubstantial as possible, to increase our chances of our chosen assumption being *true.* (The less you say, the greater the chances, other things being equal, of what you say being true.) But the more insubstantial the assumption that we choose, so the less well it helps constrain what theories we accept, and the avenues we pursue in seeking to develop new and better theories. On the other hand, if we choose an assumption that is very substantial, theories we accept and seek to develop will indeed be much more tightly constrained. This is excellent if the assumption is true, but a disaster if it is false, since it means we will constantly be accepting, and trying to develop, the wrong kind of theory.

[29] It must be admitted, however, that the field introduces new problems, especially if charged point-particles are embedded in the field. Problems of infinite self-interaction arise, problems that haunt, not just classical electrodynamics, but quantum electrodynamics, and other quantum field theories as well.

Progress in physics will be stultified. We must, and must not, make our choice of metaphysical assumption as substantial, and so as restrictive, as possible. How can we do justice to these conflicting desiderata?

6. Aim-Oriented Empiricism

The solution is to adopt, and put into scientific practice, a doctrine that I have called *aim-oriented empiricism*. The key idea of aim-oriented empiricism (AOE) is that we need to represent the metaphysical assumptions of physics in the form of a hierarchy of assumptions, as one goes up the hierarchy the assumptions becoming less and less substantial, and more nearly such that their truth is required for science, or the pursuit of knowledge, to be possible at all. In this way we create a relatively unproblematic framework of assumptions, and associated methodological rules, high up in the hierarchy, within which much more substantial, problematic assumptions, and associated methodological rules, low down in the hierarchy, can be critically assessed and, we may hope, improved, in the light of the empirical success they lead to, and other considerations: see figure 1.

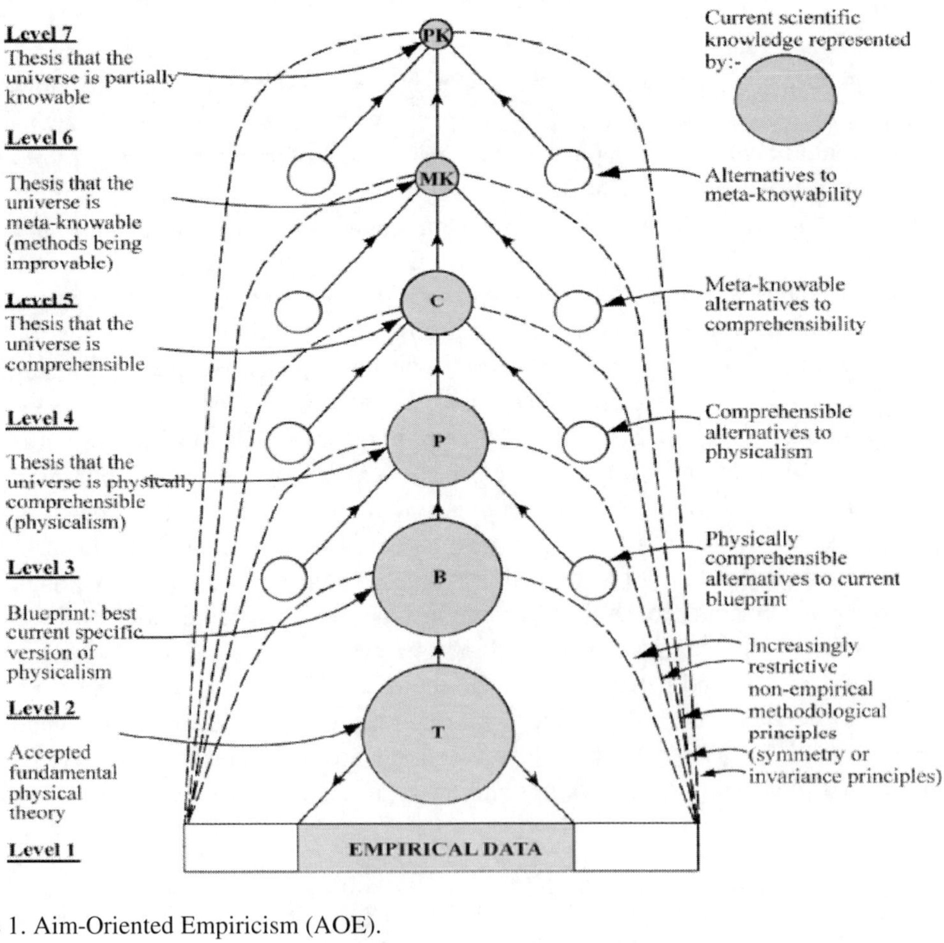

Figure 1. Aim-Oriented Empiricism (AOE).

At the top there is the relatively insubstantial assumption that the cosmos is such that we can acquire some knowledge of our local circumstances. If this assumption is false, we will not be able to acquire knowledge whatever we assume. We are justified in accepting this assumption permanently as a part of our knowledge, even though we have no grounds for holding it to be true. As we descend the hierarchy, the assumptions become increasingly substantial and thus increasingly likely to be false.

At level 6 there is the more substantial thesis that there is some rationally discoverable thesis about the nature of the cosmos which, if true and if accepted, makes it possible progressively to improve methods for the improvement of knowledge. "Rationally discoverable", here, means at least that the thesis is not an arbitrary choice from infinitely many analogous theses. At level 5 we have the even more substantial thesis that the cosmos is *comprehensible* in some way or other, whether physically or in some other way. This thesis asserts that the cosmos is such that there is *something* (God, tribe of gods, cosmic goal, physical entity, cosmic programme or whatever), which exists everywhere in an unchanging form and which, in some sense, determines or is responsible for everything that changes (all change and diversity in the world in principle being explicable and understandable in terms of the underlying unchanging *something*). A cosmos of this type deserves to be called "comprehensible" because it is such that everything that occurs, all change and diversity, can in principle be explained and understood as being the outcome of the operations of the one underlying *something*, present throughout all phenomena. At level 4 we have the still more substantial thesis that the cosmos is *physically* comprehensible in some way or other. This asserts that the cosmos is made up one unified self-interacting physical entity (or one kind of entity), all change and diversity being in principle explicable in terms of this entity. What this amounts to is that the cosmos is such that some yet-to-be-discovered unified physical theory of everything is true. In terms of the terminology of section 4 above, this thesis is physicalism(8,1). At level 3, we have an even more substantial thesis, the best, currently available specific idea as to how the cosmos is physically comprehensible. This asserts that everything is made of some specific kind of physical entity: corpuscle, point-particle, classical field, quantum field, convoluted space-time, string, or whatever. Given the historical record of dramatically changing ideas at this level, and given the relatively highly specific and substantial character of successive assumptions made at this level, we can be reasonably confident that the best assumption available at any stage in the development of physics at this level will be false, and will need future revision. Here, ideas evolve with evolving knowledge. At level 2 there are the accepted fundamental theories of physics, currently general relativity and the standard model. Here, if anything, we can be even more confident that current theories are false, despite their immense empirical success. This confidence comes partly from the vast empirical content of these theories, and partly from the historical record. The greater the content of a proposition the more likely it is to be false; the fundamental theories of physics, general relativity and the standard model have such vast empirical content that this in itself almost guarantees falsity. And the historical record backs this up; Kepler's laws of planetary motion, and Galileo's laws of terrestrial motion are corrected by Newtonian theory, which is in turn corrected by special and general relativity; classical physics is corrected by quantum theory, in turn corrected by relativistic quantum theory, quantum field theory and the standard model. Each new theory in physics reveals that predecessors are false. Indeed, if the level 4 assumption is correct, then all current physical theories are false, since this assumption asserts that the true physical theory of everything is unified, and the totality of current fundamental physical theory, general relativity plus the standard model, is notoriously disunified. Finally, at level 1 there are accepted empirical data, low level, corroborated, empirical laws.[30]

The idea is to separate out what is most likely to be true, and not in need of revision, at and near the top of the hierarchy, from what is most likely to be false, and most in need of criticism and revision, near the bottom of the hierarchy. Evidence, at level 1, and assumptions high up in the hierarchy, are rather firmly accepted, as being most likely to be true (although still open to revision): this is then used to criticize, and to try to improve, theses at levels 2 and 3 (and perhaps 4), where falsity is most likely to be located.

In order to be acceptable, an assumption at any level from 6 to 3 must (as far as possible) be compatible with, and a special case of, the assumption above in the hierarchy; at the same time it must be (or promise to be) empirically fruitful in the sense that successive accepted physical theories increasingly successfully accord with (or exemplify) the assumption. At level 2, those physical theories are accepted which are sufficiently (a) empirically successful and (b) in accord with the best available assumption at level 3 (or level 4). Corresponding to each assumption, at any level from 7 to 3, there is a methodological principle, represented by sloping dotted lines in the diagram, requiring that theses lower down in the hierarchy are compatible with the given assumption.

When theoretical physics has completed its central task, and the true theory of everything, T, has been discovered, then T will (in principle) successfully predict all empirical phenomena at level 1, and will entail the assumption at level 3, which will in turn entail the assumption at level 4, and so on up the hierarchy. As it is, physics has not completed its task, T has not (yet) been discovered, and we are ignorant of the nature of the cosmos. This ignorance is reflected in clashes between theses at different levels of AOE. There are clashes between levels 1 and 2, 2 and 3, and 3 and 4. And the two fundamental theories at level 2, the standard model and general relativity, clash as well. The attempt to resolve these clashes drives physics forward.

In seeking to resolve clashes between levels, influences can go in both directions. Thus, given a clash between levels 1 and 2, this may lead to the modification, or replacement of the relevant theory at level 2; but, on the other hand, it may lead to the discovery that the relevant experimental result is not correct for any of a number of possible reasons, and needs to be modified. In general, however, such a clash leads to the rejection of the level 2 theory rather than the level 1 experimental result; the latter are held onto more firmly than the former, in part because experimental results have vastly less empirical content than theories, in part because of our confidence in the results of observation and direct experimental manipulation (especially after repetition and expert critical examination). Again, given a clash between levels 2 and 3, this may lead to the rejection of the relevant level 2 theory (because it is disunified, *ad hoc*, at odds with the current metaphysics of physics); but, on the other hand, it may lead to the rejection of the level 3 assumption and the adoption, instead, of a new

[30] I first expounded and defended a version of this hierarchical view in Maxwell (1974). It was further elaborated in Maxwell (1984 and 1993). A more elaborate version still is expounded and defended in great detail in Maxwell (1998). For a more detailed defence of the version indicated here, see Maxwell (2004a, chs. 1 and 2, and appendix). In Maxwell (2005) I argue that this view is a sort of synthesis of the views of Popper, Kuhn and Lakatos, but an improvement over the views of all three. In Maxwell (2007, ch. 14) I give a detailed exposition of AOE, and argue, in some detail, that AOE succeeds in solving major problems in the philosophy of science, including the problems of induction, simplicity and verisimilitude. For an account of how I developed AOE, partly as a result of criticizing Popper's falsificationism, see Maxwell (2012a).

assumption (as has happened a number of times in the history of physics, as we saw in section 5 above). The rejection of the current level 3 assumption is likely to take place if the level 2 theory, which clashes with it, is highly successful empirically, and furthermore has the effect of increasing unity in the totality of fundamental physical theory overall, so that clashes between levels 2 and 4 are decreased. In general, however, clashes between levels 2 and 3 are resolved by the rejection or modification of theories at level 2 rather than the assumption at level 3, in part because of the vastly greater empirical content of level 2 theories, in part because of the empirical fruitfulness of the level 3 assumption (in the sense indicated above).

It is conceivable that the clash between level 2 theories and the level 4 assumption might lead to the revision of the latter rather than the former. This happened when Galileo rejected the then current level 4 assumption of Aristotelianism, and replaced it with the idea that "the book of nature is written in the language of mathematics" (an early precursor of our current level 4 assumption). The whole idea, however, is that as we go up the hierarchy of assumptions we are increasingly unlikely to encounter error, and the need for revision. The higher up we go, the more firmly assumptions are upheld, the more resistance there is to modification.

The idea of representing the metaphysical presuppositions of physics (concerning the nature of the cosmos) as a hierarchy of theses, increasingly insubstantial as one goes up the hierarchy, gains some support from the fact that something somewhat similar exists informally at the empirical level – level 1 of the figure – and for much the same reason. There are, at the lowest level, the results of experiments performed at specific times and places. Then, above these, there are low-level experimental laws, asserting that each experimental result is a repeatable effect. Next up, there are empirical laws such as Hooke's law, Ohm's law or the gas laws. Above these there are such physical laws as those of electrostatics or of thermodynamics. And above these there are theories which have been refuted, but which can be "derived", when appropriate limits are taken, from accepted fundamental theory – as Newtonian theory can be "derived" from general relativity. This informal hierarchy at the empirical level exists for the same reason we need the hierarchy at the metaphysical level: so that relatively contentless and secure theses (at the bottom of the empirical hierarchy) may be distinguished from more contentful and insecure theses (further up the hierarchy) to facilitate pinpointing what needs to be revised, and how, should the need for revision arise. That such a hierarchy exists at the empirical level provides some support for my claim that we need to adopt such a hierarchy at the metaphysical level.

AOE, as depicted in figure 1, provides physics with a meta-methodology which facilitates improvement of the metaphysical assumptions, and associated methods, as physics advances, in the light of which seem to be the most fruitful empirically, and other considerations. As knowledge in physics improves, so metaphysical assumptions and methods improve as well or, in other words, knowledge about how to improve knowledge. There is something like positive feedback between improving knowledge of the cosmos, and improving aims and methods, improving knowledge about how to improve knowledge. Everyone would acknowledge that this kind of positive feedback goes on at the empirical level. New empirical knowledge can lead to new methods, via the development of new instruments, new experimental techniques, which in turn lead to further acquisition of new knowledge. AOE provides methods which facilitate such positive feedback at the metaphysical and theoretical level as well. As we increase our scientific knowledge and understanding of the cosmos, we increase our knowledge of how to increase knowledge – the very nub of scientific rationality which helps explain the explosive, and apparently ever accelerating, growth of scientific knowledge. We adapt the nature of science to what we find out about the nature of the cosmos. All this has gone on in science to some extent *implicitly*: what the transition from SE to AOE does is to make the implicit *explicit* – science becoming even more successful as a result.

A potential objection to what has been said so far is that, since Galileo and Kepler, physics has been extraordinarily successful in developing a succession of theories that satisfy the almost contradictory requirements of (a) enormously increasing the scope of empirical predictions, and simultaneously (b) bringing ever greater unity to fundamental theory in physics. This immense success has been achieved, for most of the time, with the physics community taking SE for granted. It is just not conceivable that this success could have been achieved with physicists accepting and implementing a conception of science that is irrational, unworkable and untenable, as section 3 above claims. The above refutation of SE must be invalid.

My reply, as I have just indicated, is that physics has met with such extraordinary success because AOE has been implicitly put into practice despite official allegiance being paid to SE. Success has been achieved *despite*, not *because of*, allegiance to SE. Fortunately, acceptance of SE has been somewhat hypocritical; it is that which has allowed something close to AOE to be put into scientific practice, which in turn has made it possible for physics to achieve such astonishing success. But, as I have indicated above, and as I shall point out again below, physics would be even more successful if it repudiated SE and explicitly adopted and implemented AOE. The attempt to keep alive the myth of SE obstructs, to some extent, the full implementation of AOE. It is of interest to note that Einstein, whose mature view of science included elements of AOE,[31] was aware of a mismatch between what theoretical physicists *say* they do, and what they *actually* do. He wrote "If you want to find out anything from the theoretical physicists about the methods they use … don't listen to their words, fix your attention on their deeds" (Einstein, 1973, p. 270).

As I have argued in some detail elsewhere (see note 31), AOE almost became explicit with Einstein's discovery of special and general relativity. Both theories arose out of Einstein's search for theoretical unity, the unification of Newton's and Maxwell's theories. In both cases Einstein was led to the new theory via the formulation of general principles: the restricted principle of relativity and the light postulate, in the case of special relativity, and Lorentz invariance and the principle of equivalence in the case of general relativity. Remarkably, Einstein first discovered general relativity in the form of a metaphysical idea: gravitation is the variable curvature of space-time induced by matter and energy. He then had to work hard to turn this into a precise, testable physical theory. This, according to AOE, is the manner in which new theories in physics should be created, as we shall see below. It is strikingly different from the way Newton and Maxwell created their theories: in these cases, not only did the then current metaphysical ideas not lead to the new theory – they actually obstructed the correct interpretation of the new theory once it had been formulated, as we have seen. Below I argue that this happened too with the discovery and interpretation of quantum theory. It is significant that Einstein believed passionately in the physical comprehensibility of the universe, and this belief played a significant role in his discovery of special and general relativity. He devoted his life to trying to discover how the cosmos is

[31] See Maxwell (1993, pp. 275-305).

comprehensible, in the form of a testable, unified theory. And in his later years Einstein almost came to advocate a key component of AOE: physics cannot be understood without the assumption that the universe is physically comprehensible.[32] Eugene Wigner has pointed out, correctly, that the methods Einstein employed to discover his theories of relativity have had a profound impact on subsequent physics (Wigner, 1970, p. 15). But Einstein's work has not yet led to the repudiation of SE and the acceptance of AOE, as it perhaps should have done. It deserves to be noted, nevertheless, that the roots of AOE lie in physics, not in philosophy.[33]

7. Post-Popperian Kantianism

It has been established, in section 3 above, that physics must accept some metaphysical thesis concerning the (at least partial) physical comprehensibility of the cosmos. What arguments are there for holding that physics ought to accept, as a part of theoretical knowledge, that the cosmos is perfectly physically comprehensible as an integral part of AOE, as depicted in figure 1?

The first vital point to appreciate is that there is no argument to the effect that any of the metaphysical theses at levels 3 to 7 in figure 1 are *true*, or true with some degree of probability. They are all wholly conjectural in character. All that I attempt to establish is that physics is rationally entitled to accept these theses as a part of theoretical knowledge as the best available granted (a) physics seeks to improve knowledge of truth about the cosmos, and (b) some metaphysical thesis concerning the (at least partial) physical comprehensibility of the cosmos must be held to be a part of theoretical knowledge given that physics persistently only accepts unified theories, even though empirically more successful disunified rivals can always be concocted (and physics must accept some such metaphysical thesis in order to proceed at all).

I accept Karl Popper's point that, in the end, all our knowledge is conjectural in character – the vital task being to try to improve our conjectural knowledge by means of criticism.[34] There is, however, a crucial general consideration that Popper overlooked. We need to be *critically* critical and not, as Popper tended to be, merely *dogmatically* critical. The whole

---

[32] See Einstein (1973, p. 357). See, also, Maxwell (1993, pp. 275-305).

[33] AOE can be generalized in a number of ways. First, it can be reformulated so that it depicts a hierarchy of aims and methods, rather than assumptions and methods. We can interpret the aim of physics, at each level, from 3 to 7, to be to turn the assumption, at that level, into the precise, true physical theory of everything. Second, we need to recognise that there are, not just metaphysical assumptions inherent in the aims of science, but also value assumptions, and humanitarian or political assumptions. The level 4 aim of seeking explanatory truth – truth presupposed to be explanatory – is a special case of the more general aim of seeking truth that is of value, intellectually or practically. And this in turn is sought so that it may be used by people to enrich life, in either cultural or practical ways. These value and political assumptions inherent in the aims of science are almost more problematic than metaphysical assumptions. Here, too, aims and assumptions need to be made explicit so that they may be critically assessed and, we may hope, improved. Third, AOE can be generalized so that it applies to all the diverse branches of natural science, and not just to theoretical physics. Each science has its own specific, problematic aims which need to be made explicit and subjected to sustained critical scrutiny. Finally, it is not just in science that aims are problematic; this is the case in life too, at all levels, from the individual and personal, to the institutional, social and global. AOE needs to be generalized to become a conception of rationality – aim-oriented rationality – which stresses that problematic aims need to be represented in the form of a hierarchy, so that aims can be improved as we act, as we live. For these generalizations of AOE, and their implications for academic thought and life, see: Maxwell (1976a; 1984 or 2007; 2004a; 2010a; 2012a).

[34] Popper (1959; 1963).

raison d'etre for criticism is to *improve* our knowledge, improve the extent to which it is truthful and explanatory. When criticism can be shown to be incapable of improving knowledge, it becomes irrational, and deserves to be ignored.

This has the important consequence that if we can show that there are items of conjectural knowledge which are such that, subjecting them to criticism cannot help the improvement of knowledge *in any circumstances whatsoever*, *even if they are false*, then there are rational grounds for holding that these items constitute permanent constituents of our knowledge *even though there is no argument or evidence to support their truth*. We have here the possibility of identifying what may be called *a priori conjectures* – items of conjectural knowledge which we are entitled to accept as a part of knowledge because we can show that accepting them as such can only aid, and can never obstruct or undermine, the search for knowledge *in any circumstances whatsoever*, *whether they are true or false*. It may be, for example, that a factual proposition is such that, if it is false, all factual knowledge becomes impossible. If so, then we have nothing to lose in accepting it as a part of knowledge even though we have no reason whatsoever to suppose it is true, because if it is false we cannot have any knowledge whatever we assume.

Other possibilities emerge from appreciating that the whole *raison d'etre* of criticism is to promote the growth, the improvement, of knowledge, criticism becoming irrational when it cannot do this. It may be that we can show that, *in present circumstances*, accepting some factual proposition as a part of our knowledge can only aid, and cannot hinder, the improvement of knowledge even if the proposition is false. We are rationally entitled to accept such a proposition as a part of our knowledge *as long as the relevant circumstances persist*. Again, it may be that we can show that subjecting items of knowledge P to criticism is more likely to contribute to the improvement of knowledge than is the criticism of items Q. In this case, we have rational grounds for directing our critical fire at P in preference to Q. Some items of conjectural knowledge become more conjectural than others. It may be that some thesis about the nature of the cosmos, though metaphysical and untestable, is nevertheless empirically fruitful in that it is associated with an empirically progressive research programme. Attempts to capture the thesis in the form of a testable theory have led to the development of a succession of theories that have met with great empirical success – and the thesis deserves to be accepted as a part of scientific knowledge on these grounds.

These are the kind of considerations I appeal to in arguing for AOE, and for the claim that science has already established that the cosmos is physically comprehensible.

The *critically* critical position I have just indicated might be called *post-Popperian Kantianism*. It is Popperian because it appreciates Popper's fundamental points that (a) all our knowledge is ultimately conjectural, and (b) we can nevertheless improve our conjectural knowledge by subjecting it to criticism. The position modifies Popperianism, however, in recognizing that some items of our conjectural knowledge play such a crucial role in the pursuit of knowledge that we are rationally entitled to accept them, and render them either entirely or partially immune to criticism, even though we have no grounds whatsoever for holding them to be true. These items of knowledge are quasi-Kantian in character, in that they amount to *a priori* items of knowledge, established by means of what might be called, after Kant, a "transcendental" argument, entirely, or almost entirely, independent of empirical considerations. On the other hand, these *a priori* items of knowledge are profoundly non-Kantian in character in that they are purely *conjectural* in character, and are not to any extent whatsoever known to be true (about our experience) with apodictic certainty – as Kant required of any "synthetic *a priori*" statement (a factual statement established by reason alone, by means of a transcendental argument). There is another crucial difference between Kant's and my *a priori* statements. Kant's were all about, not the real world (the *noumenal* world, as he called it) but the world of experience. The *a priori* statements with which we are concerned here are all conjectures about the *real* world – Kant's *noumenal* world.[35] And it is not just that the epistemological status and interpretation of the two sets of statements are quite different: the statements themselves are quite different – and the arguments for their acceptance are different.

There are, in short, elements of Kant and of Popper in *post-Popperian Kantianism*, but at the same time the doctrine differs dramatically from, and improves on, both Kant and Popper.

8. Why Accept Aim-Oriented Empiricism?

I now spell out arguments for accepting the metaphysical theses of AOE, from level 7 to 4, beginning at the top. These arguments are all of the kind indicated in the previous section.

They all stem from being *critically* critical. They seek to establish that we are rationally entitled to accept the thesis in question, as a part of theoretical scientific knowledge (even though we have no grounds for holding it to be true, or probably true), granted *some* metaphysical thesis is implicit in the methods of physics, and must be accepted. We are rationally entitled to accept the thesis in question because, for example, if it is false we can have no knowledge at all (and so acceptance can only help and cannot retard the pursuit of knowledge), or because it has led to a more empirically successful research programme than any rival thesis – or holds out a greater promise of doing so than any rival thesis.

Level 7: Partial Knowability

The cosmos is such that we can acquire some knowledge of our local circumstances. As I have already indicated, we are rationally entitled to accept this thesis because, if this thesis is false, we can acquire no knowledge whatsoever.

Assuming the thesis to be true cannot in any way whatsoever, in any circumstances, sabotage or hinder the quest for knowledge. We are rationally entitled to accept the thesis as a

[35] Kant argued that we can know nothing about the noumenal world except that it exists. But his argument is not just invalid; his position is inconsistent. To assert that the noumenal world is such that we can know nothing about it is to assert that we do know something about it, namely that it is such that it is unknowable. Kant was presupposing a version of Cartesian dualism, much discredited since Kant's time, according to which appearances form an impenetrable barrier between us and reality, preventing us from acquiring any knowledge of reality. We cannot possibly know that we cannot know anything about the nature of reality (unless we interpret "know" in an illegitimately epistemologically strong way to mean something like "know with absolute certainty"). How, we may wonder, could Kant possibly be so certain that reality is such that it is unknowable? We need to be more epistemologically modest, and admit that reality may, for all we know, be such that it is knowable. Here is an item of knowledge about ultimate reality, Kant's noumenal world, which we can be reasonably confident is true: the cosmos is not a chicken. Just possibly this is false, and the milky way and the other galaxies we observe are giant molecules in the wattle of the cosmic chicken. Almost certainly, this is not the case. We can be reasonably certain of the truth of "the cosmos is not a chicken" because the statement says so little. There are infinitely many ways in which the cosmos can contrive not to be a chicken.

permanent part of scientific knowledge even though we have no grounds for holding it to be true.

It deserves to be noted that even our most trivial items of local, common sense knowledge make cosmological presuppositions. I would ordinarily be said to know that the desk now before me on which rests the computer I am using to write this sentence will continue to exist for the next two minutes. This very modest item of common sense knowledge about a particular object in my immediate surroundings nevertheless makes a substantial presupposition about the entire cosmos.

It presupposes that, nowhere in the cosmos is an explosive event occurring which will spread with near infinite speed to engulf the earth, me, and my desk in 30 seconds time.

In so far as we have any factual knowledge of anything, we have some knowledge of the entire cosmos. Nothing could demonstrate more dramatically Popper's point that *all* our knowledge is conjectural in character, however humble and restricted in scope it may be.

Science undoubtedly corrects and refines common sense knowledge, but it also depends on it. Bereft of common sense knowledge of objects around us, we would not be able to set up and perform the sophisticated experiments and observations upon which modern physics and cosmology depend. Thus the cosmological presuppositions of ordinary common sense knowledge are also presuppositions of modern science.

But might not science reveal that our local circumstances will soon become such that all knowledge, and life itself, will come to an end, perhaps because of the advent of a supernova in our immediate neighbourhood? Would not scientific knowledge then *negate* just that which it is supposed to presuppose? True.

But our ability to acquire our knowledge of our imminent doom still depends on our local circumstances being such that we can continue to acquire knowledge of it. When our ability to acquire knowledge of our local circumstance disappears, we disappear too – and with us, science.

Level 6: Meta-Knowability

The cosmos is such that, not only can we improve our knowledge of it: we can improve our knowledge about how to improve knowledge. Or rather, slightly more modestly, the cosmos is such that, if we are deceived in our attempts to improve our knowledge about how to improve knowledge, we can always discover that we have been deceived. This does not guarantee that attempts to improve knowledge about how to improve knowledge will meet with success. But what it does guarantee is that if such attempts meet with empirical success, and the success is illusory, then we can discover, by empirical investigation, that it is illusory. As Einstein once put it "*Raffiniert ist der Herrgott, aber boshaft ist er nicht*", sometimes freely translated as "The universe is sublime, but not malicious".[36] The cosmos is not such that we are lured into thinking we are making great scientific progress when actually it is all illusory, and it would be quite impossible for us to discover that it is illusory (except, perhaps at some later date when the discovery may be catastrophic).

The idea, here, is that it may be within our power, given our existing knowledge, methods and capabilities, to formulate and adopt some "rationally discoverable" metaphysical thesis

[36] Remark Einstein made during his first visit to Princeton University in 1921.

about the nature of the cosmos which, if adopted as indicating the kind of theory physics should seek to develop and accept, provides the means to achieve genuine progress in theoretical knowledge in physics – the means to develop a genuinely empirically progressive research programme in physics. The idea, in other words, is that the cosmos may be knowable in the sense that knowledge about how to improve knowledge can be progressively improved, there being positive feedback between improving knowledge and improving knowledge about how to improve knowledge, along the lines promised by AOE – the final outcome being the true, physical theory of everything.

The notion of a cosmological thesis being "rationally discoverable" is problematic. It means, at least, that a "rationally discoverable" metaphysical thesis about the cosmos is one which can be formulated given our current knowledge and capabilities, and is such that it is not an arbitrary choice from infinitely many comparable, equally viable theses.

It is to be expected, of course that, as the hunt for the true theory of everything proceeds, AOE will be modified. Empirical data at level 1 will be amplified, new theories will emerge at level 2 which will expand on, correct and unify our current theories, the thesis at level 3 will, almost certainly, be modified, perhaps quite radically, and it is possible that this might happen to the thesis at level 4, and even, perhaps, at level 5. In order to take such future, and past, versions of AOE into account, we need to appeal to a view which I shall call *generalized aim-oriented empiricism* (GAOE). This asserts that we need to represent our knowledge and methods in physics in the form of a hierarchy, like the hierarchy of AOE, with the same theses as AOE at levels 7 and 6, and with the best available theses at levels 5, 4 and 3, the best available physical theories at level 2, and the best available empirical data at level 1.

We can now interpret meta-knowability to assert: the cosmos is such that there is some version of GAOE which we can formulate and implement which is such that, if it meets with empirical success and the success is illusory then, at any time, it is possible for us to discover that it is illusory.

I have two arguments in support of accepting meta-knowability as a part of scientific knowledge. First, however, I must consider an apparently lethal argument against AOE and GAOE, and thus against the basic thesis of this essay. Both AOE and GAOE hold that, as physics proceeds, we can progressively improve accepted metaphysical theses about the nature of the cosmos in the light of which seem best able to lead to the development of empirically successful theories, these theories being accepted in part because they accord with accepted metaphysical theses, there being positive feedback between improving metaphysical theses and empirically successful theories.

But this would seem to be a viciously circular procedure. We justify accepting a metaphysical thesis because it leads to an empirically successful theory, and then justify accepting the theory because it accords with our accepted metaphysical thesis. As Bas van Fraassen has put it, "From Gravesande's axiom of the uniformity of nature in 1717 to Russell's postulates of human knowledge in 1948, this has been a mug's game" (van Fraassen, 1985, pp. 259-60). How does AOE escape this charge?

One point can be made straight away. AOE makes no attempt to justify the truth (or probable truth) of any of its metaphysical theses. They are all irredeemably *conjectural* in character. A basic idea behind AOE is that metaphysical theses that are implicit in physics need to be made explicit so that they can be *criticized* – the very opposite of *justified*. A key feature of AOE is that theories, at level 2, are incompatible with metaphysical theses, at levels

3, 4 and 5. It hardly makes sense to claim the empirical success of theories justifies metaphysical theses when they are incompatible with them!

But even when all this has been conceded, there is still a problem. Even if AOE does not seek to justify the (probable) truth of any metaphysical thesis, it does seek to justify *acceptance* of these theses in the interests of the pursuit of truth.[37] And here we do seem to run into a version of the viciously circular argument which van Fraassen correctly condemns. Metaphysical thesis M is accepted because it supports acceptance of the empirically successful theory, T; and T is accepted, in part, because it is in accord with M. How can AOE be rescued from this apparently lethal charge of vicious circularity?

My proposed solution involves, crucially, an appeal to the level 6 thesis of meta-knowability.

Permitting metaphysical assumptions to influence what theories are accepted, and at the same time permitting the empirical success of theories to influence what metaphysical assumptions are accepted, may (if carried out properly), *in certain sorts of cosmos*, lead to genuine progress in knowledge. Meta-knowability is to be interpreted as asserting that *this is just such a cosmos*. And furthermore, quite essentially, reasons for accepting meta-knowability make no appeal to the success of science. In this way, meta-knowability legitimises the potentially invalid circularity of generalized AOE (GAOE), and of AOE.

Relative to an existing body of knowledge and methods for the acquisition of new knowledge, possible universes can be divided up, roughly, into three categories: (i) those which are such that the meta-methodology of GAOE or AOE can meet with no success, not even apparent success, in the sense that new metaphysical ideas and associated methods for the improvement of knowledge cannot be put into practice so that success (or at least apparent success) is achieved; (ii) those which are such that AOE can appear to be successful for a time, but this success is illusory, this being impossible to discover during the period of illusory success; and (iii) those which are such that GAOE, and even AOE, can meet with genuine success (it being possible to discover, at any time, that illusory success is indeed illusory, by ordinary methods of scientific investigation). Meta-knowability asserts that our universe is a type (i) or (iii) universe; it rules out universes of type (ii).

Meta-knowability asserts, in short, that the universe is such that AOE can meet with success and will not lead us astray in a way in which we cannot hope to discover by normal methods of scientific inquiry (as would be the case in a type (ii) universe). If we have good grounds for accepting meta-knowability as a part of scientific knowledge – grounds which do not appeal to the success of science – then we have good grounds for adopting and implementing generalized AOE (GAOE). Meta-knowability, if true, does not guarantee that GAOE will be successful. Instead it guarantees that GAOE will not meet with illusory success, the illusory character of this apparent success being such that it could not have been discovered by any means whatsoever before some date is reached.

---

[37] Some Popperians argue that any attempt to argue for the acceptance of a thesis or theory goes against Popper's falsificationist, critical rationalist philosophy. But this is wrong. An argument intended to justify acceptance of a thesis sets out to do something quite different from an argument intended to justify the truth (or probable truth) of a thesis. Popper himself argues for the acceptance of theory that has greater empirical content than its predecessor, and has survived severe testing – a central claim of Popper (1959). Here, the more acceptable a theory is, according to Popper, so the less likely it is to be true – just because its empirical content is greater. One could hardly highlight more dramatically the distinction between arguing for acceptance of a thesis, and arguing for its truth.

If AOE lacks meta-knowability, its circular procedure, interpreted as one designed to procure knowledge to the extent that this is possible, becomes dramatically invalid, as the following consideration reveals. Corresponding to the succession of accepted fundamental physical theories developed from Newton down to today, there is a succession of severely disunified rivals which postulate that gravitation becomes a repulsive force from the beginning of 2150, let us say. Corresponding to these disunified theories there is a hierarchy of disunified metaphysical theses, all of which assert that there is an abrupt change in the laws of nature at 2150. The disunified theories, just as empirically successful as the theories we accept, render the disunified versions of physicalism just as scientifically fruitful as unified versions of physicalism are rendered by the unified theories we actually accept. The circularity inherent in AOE is invalid because it can be employed so as to lead to the adoption of *disunified* theories and metaphysical theses just as legitimately as it can be employed to lead to the adoption of *unified* theories and metaphysical theses. This is the case, at least, if AOE is bereft of meta-knowability. But if we have good reasons to accept meta-knowability as a part of scientific knowledge, then we have good reasons to reject *disunified* versions of physicalism: these lack the crucial requirement of rational discoverability. If we have good reasons to accept meta-knowability as an item of scientific knowledge, and these reasons make no appeal to the success of science, then the circularity inherent in AOE ceases to be invalid: meta-knowability asserts that the universe is such that empirical success achieved by implementing AOE will not be illusory *in a way which could not discovered by any means before a certain date.*

But what reasons have we for accepting meta-knowability that make no appeal to the success of science? One argument is simply this. If we accept meta-knowability, we have little to lose,[38] and may have much to gain. In accepting meta-knowability we decide, in effect, that it is worthwhile to try to improve knowledge about how to improve knowledge. We take seriously the possibility that the cosmos is such that we can discover something rather general about its nature which will enable us to *improve* our methods for improving knowledge. Not only do we hope to learn about the world; we hope to learn about how to learn about the world, and we are prepared to implement a meta-methodology (GAOE) which capitalizes on this possibility should it turn out to be actual. To fail to try to *improve* methods for improving knowledge on the grounds that apparent success might prove to be illusory is surely to proceed in a cripplingly over-cautious fashion. Any attempt at improving knowledge may unexpectedly fail, including the attempt to improve methods for improving knowledge. But eschewing the attempt to learn because it may fail cannot be sound: such an excuse for not making the attempt always exists. In accepting meta-knowability we do not assume, note, that the universe *is* such that GAOE will meet with success. We assume, merely, that it is such that *if* GAOE or AOE appears to meet with empirical success, this success will not be illusory in a way which could not have been discovered prior to the illusory character of the success becoming apparent. But this is an entirely sensible assumption to make. Nothing is to be gained from foregoing the attempt to acquire knowledge because of the fear that future, inherently unpredictable changes in the laws of nature may occur which render knowledge acquired obsolete.

---

[38] What might we have to lose? If meta-knowability is false, then efforts to improve knowledge about how to improve knowledge cannot succeed, and will have turned out to have been for nothing. On the other hand, the only way we could discover that meta-knowability is false is to try to improve knowledge about how to improve knowledge, and fail in the attempt.

Can anything more be said? I think it can. We can argue as follows. As the pursuit of knowledge, and science, have developed over the millennia, GAOE has in fact been put into practice. Metaphysical presuppositions have been revised in the light of what seems to meet with the greatest empirical success – from myths, religious views, the ideas of the Presocratics, the ideas of Plato, Aristotle, Galileo, Boyle, Newton, and Boscovich, to metaphysical ideas concerning the underlying unity of the cosmos implicit in currently accepted symmetry principles, such as Lorentz invariance (associated with special relativity), gauge invariance (associated with quantum field theory and the standard model) and even, perhaps, supersymmetry (associated with string theory). Even empirical methods have been revised in the light of metaphysical revisions.

For example, given Aristotelian metaphysics, with its denial that precise mathematical laws govern natural phenomena, there is little point in performing precise experiments to decide between rival theories. This changes dramatically once Galileo's metaphysics is accepted, according to which "the book of nature is written in the language of mathematics" (an early statement of physicalism). Suddenly, it becomes highly pertinent to perform precise experiments, of the kind performed by Galileo involving, for example, rolling balls down inclined planes, to try to determine what precise mathematical law governs the fall of bodies near the surface of the earth. Granted, then, that GAOE has been put into practice over the millennia and right up to the present, science is more rigorous if the metaphysical assumption, implicit in this practice, is made explicit. This is the case even if this explicit thesis remains a conjecture *with no other reasons being given for its acceptance* over and above that it is implicit in scientific practice. Meta-knowability is, in short, implicit in the scientific endeavour.

Science becomes more rigorous if this implicit assumption is made explicit, and is accepted as a part of scientific knowledge even though no other argument is forthcoming for its acceptance.[39] Indeed, if this is not done, and meta-knowability is repudiated, then science becomes viciously circular, and thus inherently irrational – in so far as GAOE is implemented. Instead of being directed against Gravesande and Russell, van Fraassen's "mug's game" argument becomes redirected, in an entirely valid way, against science itself! In order to avoid this charge, science must accept meta-knowability as a (conjectural) item of knowledge, even if there is no other argument for its acceptance.[40]

These two arguments, taken together, provide strong grounds for accepting meta-knowability as a permanent item of scientific knowledge.

Level 5: Comprehensibility

This is the thesis that the universe is comprehensible in some way or other, there being *something*, or an aspect of something (kind of physical entity, God, society of gods, cosmic purpose, cosmic programme or whatever) that runs through all phenomena, and in terms of

[39] No circularity is involved here, because no attempt is made to justify acceptance of meta-knowability by an appeal to the success of science. The argument is that we should make meta-knowability explicit even if GAOE science is entirely unsuccessful. We should make explicit metaphysical assumptions implicit in our methodological practice whether this practice has met with real success or only apparent success.

[40] But there is another argument, as we have seen. In accepting meta-knowability, we have little to lose and may have much to gain.

which all phenomena can, in principle, be explained and understood. Almost all (perhaps all) cultures possess a myth, cosmology or religious view taken to explain natural phenomena, presupposed by attempts to improve knowledge. Almost all of these are personalistic, animistic or purposive in character: natural phenomena are explained in terms of the actions of gods, spirits, God, or purposes. Acceptance of some version of comprehensibility is often combined, however, with a clause that places strict limits on knowability (this clause being required, perhaps, to protect the thesis against criticism, and to explain away the lack of success of the view in promoting acquisition of knowledge). Thus God is said to be mysterious and unknowable. That the universe is held to be (more or less) comprehensible in almost all cultures is not, however, a good reason to hold it to be worthy of acceptance.

The comprehensibility thesis does, however, perfectly exemplify the basic idea of meta-knowability. Here, indeed, is a thesis about the nature of the cosmos which, if true, holds out the promise[41] that we can improve knowledge about how to improve knowledge by proposing and testing explanatory conjectures about phenomena, progressively modifying our idea as to how the cosmos is comprehensible in the direction of those explanatory conjectures which meet with the greatest empirical success.

Comprehensibility is, however, by no means the only thesis to exemplify the basic idea of meta-knowability. The cosmos might be meta-knowable even though only partially comprehensible, or even incomprehensible. It might be such that a series of theoretical revolutions are required in order to make progress, the number of distinct unified theories increasing by one each time there is a revolution, there being no end to this process. We might discover (or guess) that the cosmos has this structure after a few revolutions, and this guess might well help the endeavour to improve knowledge, even though the cosmos is ultimately incomprehensible.

What grounds are there for preferring the thesis of perfect comprehensibility to rival cosmological theses that also exemplify meta-knowability? Before the advent of modern science, they are very weak. We might argue that perfect comprehensibility is to be preferred because, if true, it promises to offer greater help with improving knowledge about how to improve knowledge than any rival comparable thesis. We might also argue that perfect comprehensibility seems intuitively more plausible than any partial comprehensibility thesis, or incomprehensibility thesis one can dream up.

These other theses seem to come close to violating the requirement of "rational discoverability" of meta-knowability. To say that there are six, or twelve, distinct, basic entities that govern all phenomena – or six or twelve theories that can be unified no further – does seem to make an arbitrary choice between there being N such entities (or theories), N being any integer greater than one.

Before the advent of modern science, grounds for holding that the universe is perfectly comprehensible seem very weak. This is because no version of the conjecture led to spectacular empirically successful theorizing. It is only when modern science does exhibit spectacular empirically successful theorizing that strong grounds for holding that the cosmos is perfectly comprehensible do arise. It is only when we go down a level to *physical* comprehensibility that such strong grounds arise, as I now try to demonstrate.

---

[41] Comprehensibility may promise meta-knowability, but it does not imply it. The cosmos might be comprehensible but unknowable to us; or it might be comprehensible in such a way that it is not possible for us to put GAOE into practice in an empirically successful way.

Level 4: Physicalism (8,1)

This is the thesis that the universe is physically comprehensible, everything being made up of just one kind of physical entity (or perhaps just one entity), all change and diversity being in principle explicable in terms of this one kind of entity. This thesis asserts that the universe is such that some yet-to-be-discovered physical "theory of everything" is unified in a type 8 way, and true.

Granted meta-knowability, we are justified in accepting that thesis, other things being equal, which holds out the greatest promise, if true, for progress in empirical knowledge. Physicalism(8,1) satisfies this requirement better than any rival thesis at this level, in that it places more demanding restrictions on any testable theory that is to be ultimately acceptable. (Such a theory must, in principle, predict and explain all physical phenomena, and must be unified in a type 8 way – the most demanding requirement for unity.) Physicalism(8,1) also indicates a path along which physics may proceed in order to improve empirical knowledge: testable theories need to be put forward and tested that, as far as possible (a) predict ever wider ranges of phenomena, and (b) are ever more unified. In order to develop good new theories, the attempt needs to be made to resolve clashes between existing empirically successful, unified theories. In short, physicalism(8,1), if true, indicates that AOE needs to be put into scientific practice.

But it is not just that physicalism(8,1) holds out the promise of progress; it has been associated, implicitly, with all the great advances in theoretical knowledge and understanding in physics at least since Galileo's time.

All advances in theory in physics since the scientific revolution have been advances in unification, in the sense of (1) to (8) above. Thus Newtonian theory (NT) unifies Galileo's laws of terrestrial motion and Kepler's laws of planetary motion (and much else besides): this is unification in senses (1) to (3). Maxwellian classical electrodynamics, (CEM), unifies electricity, magnetism and light (plus radio, infra red, ultra violet, X and gamma rays): this is unification in sense (4). Special relativity (SR) brings greater unity to CEM, in revealing that the way one divides up the electromagnetic field into the electric and magnetic fields depends on one's reference frame: this is unification in sense (6). SR is also a step towards unifying NT and CEM in that it transforms space and time so as to make CEM satisfy a basic principle fundamental to NT, namely the (restricted) principle of relativity. SR also brings about a unification of matter and energy, via the most famous equation of modern physics, $E = mc^2$, and partially unifies space and time into Minkowskian space-time. General relativity (GR) unifies space-time and gravitation, in that, according to GR, gravitation is no more than an effect of the curvature of space-time – a step towards unification in sense (8). Quantum theory (QM) and atomic theory unify a mass of phenomena having to do with the structure and properties of matter, and the way matter interacts with light: this is unification in senses (4) and (5). Quantum electrodynamics unifies QM, CEM and SR. Quantum electroweak theory unifies (partially) electromagnetism and the weak force: this is (partial) unification in sense (7). Quantum chromodynamics brings unity to hadron physics (via quarks) and brings unity to the eight kinds of gluons of the strong force: this is unification in sense (6). The standard model (SM) unifies to a considerable extent all known phenomena associated with fundamental particles and the forces between them (apart from gravitation): partial unification in senses (4) to (7). The theory unifies to some extent its two component quantum field theories in that both are locally gauge invariant (the symmetry group being

U(1)xSU(2)xSU(3)). All the current programmes to unify SM and GR known to me, including string theory or M-theory, seek to unify in senses (4) to (8).[42]

In short, all advances in fundamental theory since Galileo have invariably brought greater unity to theoretical physics in one or other, or all, of senses (1) to (8): all successive theories have increasingly successfully exemplified and given precision to physicalism(8,1) to an extent which cannot be said of any rival metaphysical thesis, at that level of generality. The whole way theoretical physics has developed points towards physicalism(8,1), in other words, as the goal towards which physics has developed. Furthermore, what it means to say this is given precision by the account of theoretical unity given in section 4 above.

In response to this claim it may be objected that theoretical physics could equally well be regarded as pointing towards a less restrictive version of physicalism – one which does not require matter and space-time to be unified, or one which demands only that the true theory everything is no more disunified than in a type 4 way to an extent N = 3, let us say (so that the true theory postulates three kinds of forces). What grounds are there for preferring physicalism(8,1) to physicalism(5,6), let us say? There are at least four, none of course decisive.

Fundamental to the whole argument for AOE is that physics needs to put into practice what may be called the *principle of intellectual integrity*: make explicit assumptions that are substantial, influential and implicit (AOE can be construed as the outcome of successive applications of this principle). In considering what thesis ought to be accepted at level 4, then, we need to consider what is implicit in those current methods of physics that influence what theories are to be accepted on non-empirical grounds – having to do with simplicity, unity, explanatoriness. There can be no doubt that, as far as non-empirical considerations are concerned, the more nearly a new fundamental physical theory satisfies all eight of the above requirements for unity, with N = 1, the more acceptable it will be deemed to be. Furthermore, failure of a theory to satisfy elements of these criteria is taken to be grounds for holding the theory to be false even in the absence of empirical difficulties. For example, high energy physics in the 1960s kept discovering more and more different hadrons, and was judged to be in a state of crisis as the number rose to over one hundred. Again, even though the standard model (the current quantum field theory of fundamental particles and forces) does not face serious empirical problems, it is nevertheless regarded by most physicists as unlikely to be correct just because of its serious lack of unity. In adopting such non-empirical criteria for acceptability, physicists thereby implicitly assume that the best conjecture as to where the truth lies is in the direction of physicalism (8,1). The principle of intellectual integrity requires that this implicit assumption – or conjecture – be made explicit so that it can be critically assessed and, we may hope, improved. Physics with physicalism (8,1) explicitly acknowledged as a part of conjectural knowledge is more rigorous than physics without this being acknowledged because physics pursued in the former way is able to subject non-empirical methods to critical appraisal as physicalism (8,1) is critically appraised, whereas physics pursued in the latter way cannot do this (or not to the same extent).

A second point to note is that it may well be that, even if some other version of physicalism(n,N) is true, with $n < 8$ and $N > 1$, nevertheless our best hope of discovering the truth may still lie in attempting to discover a theory that exemplifies physicalism (8,1), and failing in the attempt. As N becomes bigger, so the number of possible theories of everything

[42] For further discussion see Maxwell (1998, pp. 80-89, 131-140, 257-265 and additional works referred to therein).

compatible with that version of physicalism rapidly increases. (If N = 2, and the universe is made up of two distinct unified, dynamical patterns, there are, nevertheless, in general, infinitely many ways in which these two distinct patterns can be fitted together to make infinitely many different possible universes exemplifying just these two dynamic patterns. The step from one specified unified pattern to two is the step from one possible universe to infinitely many!) It makes sense to seek the simplest, most discoverable possibility, and design our methodology accordingly. As I have indicated, one can imagine a cosmos in which we might have reasons for adopting a methodological rule different from: (A) in order to be ultimately acceptable, a theory must be comprehensive and unified in a type (8) way. An example is: (B) in order to discover the true theory of everything, there need to be infinitely many theoretical revolutions, the number of forces increasing by one at each revolution. We cannot, therefore, just argue that, even if some version of physicalism other than physicalism(8,1) is true, nevertheless our best hope of discovering the truth is to adopt (A), try to discover a theory that exemplifies physicalism(8,1), and fail in the attempt. But we can argue that, in our current state of ignorance, our best bet is to adopt (A), and revise our acceptance of physicalism(8,1) if some other version of physicalism should emerge as appearing to fit the progress of physics better (such as a number of revolutions have taken place, and each time, the number of forces has gone up by one.)

There is another reason for preferring physicalism(8,1) to any other version, namely: only this version can do justice to the way general relativity unifies gravitation and space-time. This is a step towards type (8) unification in that, according to the theory, gravitation as a force disappears, and we are left with a dynamic theory of space-time. (Matter, or energy-density more generally, tells space-time how to curve: bodies then move along geodesics – the nearest things to straight lines in curved space-time.) It is above all general relativity which holds out the possibility that, not just gravitation, but all the forces and particles may be unified with space-time.

Finally, it needs to be remembered that what we are discussing is reasons for accepting physicalism(8,1) at level 4 within the context of AOE. If physicalism(8,1) was a candidate for the only metaphysical thesis to be accepted by science, it might well be thought to be much too specific and risky to be regarded as a part of scientific knowledge. But the whole point of AOE is that, as we descend the hierarchy, theses become increasingly specific, risky, tentative, and likely to require rejection, or at least revision. Physicalism(8,1) is bound to have a much more dubious epistemological status than partial knowability, let us say.

In section 3 we saw that physics must make some metaphysical assumption about the cosmos (whether this is explicitly acknowledged or not). In considering the acceptability of physicalism (8,1), then, what needs to be argued is not "why do we accept this thesis rather than none?", but rather "why is physicalism(8,1) more acceptable than its rivals?". One argument is that it accords better with meta-knowability than its rivals. A second argument is that it is implicit in current non-empirical requirements in physics for a theory to be unproblematically acceptable and – in accordance with the principle of intellectual integrity – this implicit assumption needs to be made explicit (so that it can be critically assessed and, if necessary, improved). A third argument is that the whole way theoretical physics has developed since Galileo down to today points towards physicalism(8,1). And a fourth argument is that, even if physicalism(8,1) is false, and n < 8 and N > 1, nevertheless it may well be that our best, even our only, hope of discovering the truth lies in assuming physicalism(8,1) and getting as close as we can to capturing this in a unified, empirically successful theory. If theoretical physics as a whole is only partially unified in that $n < 8$ and/or $N > 1$, the best – perhaps the only – grounds we can have for holding that nothing more unified exists is to strive, perhaps for centuries, to discover a more unified theory, and fail in the attempt. Only then would we have grounds for declaring, conjecturally, physicalism(8,1) to be false.

To sum up: some metaphysical thesis must be accepted as a part of theoretical scientific knowledge (as we saw in section 3). I have now shown that physicalism(8,1) is more acceptable than its rival metaphysical theses. I have shown that science has already established that the cosmos is physically comprehensible (to the extent that this is an item of scientific knowledge, and in so far as science can ever establish anything theoretical in character).[43]

I conclude this section with a few remarks about the above argument in support of this contention. To begin with, I must re-emphasize two crucial features of what I claim to have established. First, there is no argument for the truth, or probable truth, of any of the theses at the various levels of AOE. All I have argued is that we are rationally entitled to accept these theses granted our aim in pursuing science is to improve knowledge about aspects of the cosmos, in so far as this is possible. In my view, we must accept Popper's point that all our knowledge is conjectural in character – even if some parts of our knowledge are more conjectural than others. Second, in section 3 we saw that physics, and therefore the whole of natural science, must accept some metaphysical thesis about the nature of the cosmos, whether this is acknowledged or not. What is at issue in this section, then, is not "what grounds are there for accepting any metaphysical thesis?", but rather "what grounds are there for preferring physicalism(8,1) to its rivals?". It is the second issue I have addressed in this section, not the first. Third, a remark about AOE. There are two inter-related reasons why we should adopt the hierarchical view of AOE, or GAOE (as opposed to accepting a metaphysical thesis or theses at one level as, for example, Bertrand Russell advocated). The epistemological standing of the various theses at the different levels of AOE are very different. Theses at levels 7 and 6 are such that it can never be in the interests of science, or the pursuit of knowledge, to repudiate them. The arguments in support of accepting these theses make no appeal to the empirical success of science. But theses at levels 5 and 4 might, conceivably be rejected in the future by science – especially the thesis at level 4. The arguments in support of accepting these theses do appeal to the empirical success of science, and to the whole character of this success. Only if the levels are distinguished can these very different reasons for accepting these different theses be distinguished and acknowledged. Furthermore, that the epistemological status of the theses at the different levels is different has everything to do with the heuristic and methodological fruitfulness of AOE (or GAOE). Not only does AOE concentrate criticism and the development of alternatives where it is likely to be most fruitful, at levels 3 and, to a lesser extent, 4. Furthermore, the epistemological security of theses at levels 7 and 6 – their acceptance being independent of

[43] The claim that science has "established" that the universe is physically comprehensible is perhaps a bit misleading, in that it strongly suggests that science has established the (probable) truth of the thesis. What I have done, put more accurately, is to show that the thesis is a central part of current theoretical scientific knowledge. The thesis is so firmly accepted that accepted physical theories that clash with it, such as the standard model and general relativity, are regarded in science as problematic, and likely to turn out to be false. Theories that clash violently with the thesis, in that they are seriously disunified, are not even considered in physics even though they are ostensibly empirically more successful than accepted unified rival theories.

the empirical success or failure of science (at levels 1 and 2) means that these theses provide permanent constraints on what can be accepted at lower levels. This is an aid to the improvement of theses at levels 3, and perhaps 4.

9. IMPLICATIONS

I now consider some implications of rejecting (all versions of) standard empiricism (SE), and accepting aim-oriented empiricism (AOE) instead.

I. Scope of Science

There is a substantial increase in the scope of what we can legitimately take to be scientific knowledge. From the standpoint of standard empiricism, we do not (yet) possess scientific knowledge about the ultimate nature of the cosmos. From the standpoint of aim-oriented empiricism, we do. As a result of moving from SE to AOE, the metaphysical thesis that the cosmos is physically comprehensible ceases to be a somewhat obscure, untestable, and therefore unscientific proposition, and becomes instead a lucid and secure item of theoretical scientific knowledge – more secure, indeed than any physical theory, such as quantum theory, however well corroborated empirically.

II. Intellectual Significance

We gain an item of scientific knowledge that is of profound significance, namely that the cosmos is physically comprehensible. From a general intellectual and cultural standpoint, many detailed, technical theoretical and observational discoveries of science are *not* of general significance – though of course some are (such as Darwin's theory of evolution). The scientific discovery (as we may choose to call it) that the cosmos is physically comprehensible is, like the theory of evolution, one of those discoveries that has implications for the whole of thought, and for all of our culture. It means, for example, that our whole human world, imbued with such things as colour, sounds, feelings, thoughts, consciousness, free will, meaning and value, must somehow be accommodated within the general framework of a physically comprehensible universe. Very general constraints are imposed on the way we can think about human life, its meaning, nature and value.[44]

III. Quantum Theory

As a result of rejecting SE and accepting AOE, the acceptability of orthodox quantum theory (OQT) is transformed. OQT is perhaps the most empirically successful theory ever

[44] For my views as to how our human world, imbued with colours, sounds, meaning and value, consciousness and free will can be accommodated within the universe, presumed to be physically comprehensible, see Maxwell (1966; 1968a; 1968b; 1976a; 1984 or 2007, ch. 10; 2000b; 2001; 2009b; 2011b; and especially 2010a).

formulated, given the great scope and variety of its predictions, and given that the theory remains unrefuted.[45] Granted SE, OQT must be judged to be acceptable. It satisfies in abundance the only scientifically relevant requirement – empirical success. But granted AOE, OQT must be judged to be unacceptable, or at least highly problematic, because of its disunity. As a result of the failure to solve the quantum wave/particle dilemma, OQT was developed as a theory which restricts itself to making predictions about the results of performing measurements on micro systems such as electrons and atoms. But this in turn means the theory that makes predictions is very seriously disunified in that it is made up of two mutually contradictory parts: a purely quantum mechanical part, and a part which describes the measuring instrument and consists, as Niels Bohr always emphasized,[46] of some part of classical physics.[47] OQT is, in other words, disunified in a type (2) way, as formulated in section 4 above. This is a very severe kind of disunity. It means, granted AOE, that OQT is at best highly problematic, even though so highly successful empirically.

This very serious defect in OQT stems, as Einstein always emphasized,[48] from the abandonment of micro-realism, the failure to solve the wave/particle problem, required if quantum theory is to be interpreted to be about micro systems per se, and how they interact with one another, whether they are being measured or not. This failure not only results in OQT being severely disunified; it also means, as I have argued in detail elsewhere, that OQT (a) fails to specify the physical nature of micro systems, (b) is ambiguous about whether nature is fundamentally probabilistic or deterministic, (c) lacks precision, (d) lacks explanatory power, (e) obstructs unification with general relativity, and (f) is incapable of being applied to the cosmos as a whole.[49]

I suggest the failure to develop an unproblematic version of QT, even though over 80 years elapsed since it was first formulated in 1925 and 1926, is due to the failure to put AOE into practice, and subject implicit metaphysical assumptions to explicit criticism. The key question "is the electron a wave or particle?" implicitly presupposes determinism, since wave and particle are classical notions, and classical physics, at the fundamental level, is deterministic. But perhaps what QT is trying to tell us is that nature is fundamentally probabilistic. QT, after all, in general makes *probabilistic* predictions. If nature is fundamentally probabilistic, then the question "is the electron a wave or particle?" is entirely the wrong question to ask. The correct question is twofold. What kinds of unproblematic, fundamentally probabilistic entities are there, as possibilities? Can quantum entities – electrons and atoms – be construed to be some variety of unproblematic probabilistic entity – an entity which interacts with others in a probabilistic way? Elsewhere, I have demonstrated that this can be done.[50] All such fundamentally probabilistic versions of QT, however, face the problem of specifying, in precise, quantum mechanical terms, the conditions for probabilistic transitions to occur, no appeal being made to measurement, macroscopic phenomena or irreversibility. My specific proposal, here, is that probabilistic transitions occur when new particles or bound states are created as a result of inelastic interactions.

---

[45] This isn't quite correct. Relativistic phenomena refute non-relativistic OQT. These phenomena can, however, at least in principle, be predicted by relativistic quantum theory.

[46] See, for example, Bohr (1949).

[47] See Maxwell (1972b; 1976b, pp. 276-279; 1988, pp. 1-8; 1998, pp. 228-235).

[48] See Maxwell (1993, pp. 290-296).

[49] See Maxwell (1988, pp. 1-8; 1998, pp. 230-235).

[50] See Maxwell (1976b; 1982; 1988; 1994; 1998, ch. 7; 2004c; 2011c).

Superpositions of different particle states exist, but do not persist. The outcome is a fully micro realistic, fundamentally probabilistic version of QT, free of the disunity defects of OQT, able to recover all the empirical success of OQT, and able to make as yet untested empirical predictions that differ from those of OQT.[51]

My specific version of probabilistic QT may well not be correct. That the option of developing testable fundamentally probabilistic versions of QT has, by and large, been neglected is, however, in my view, somewhat disgraceful.[52] It is due, I suggest, once again to the failure to put AOE into explicit scientific practice, resulting in the failure to examine critically implicit metaphysical assumptions influencing interpretations of QT. Adopting a fundamentally probabilistic metaphysics transforms the interpretative problems of QT, but this requires that metaphysical ideas are subjected to imaginative and critical scrutiny within physics, something that SE discourages as "unscientific". In addition, of course, SE encourages the attitude that OQT, given its immense empirical success, cannot conceivably be regarded as inherently problematic when judged from a scientific standpoint. General acceptance of SE, and the failure to adopt and implement AOE, may well, in other words, have delayed the development of an unproblematic version of QT.

The issue just discussed concerning QT has major implications, it should be noted, for subsequent developments in theoretical physics and cosmology – for quantum electrodynamics, for the standard model, for string theory, and for quantum cosmology.

IV. Implications for Cosmology

Does AOE have more specific implications for cosmology? Might AOE help decide, for example, whether or not the cosmological theory that there are many, perhaps infinitely many, universes alongside ours is to be accepted as the true "theory of everything"? Might AOE have implications for the acceptability of string theory, or M-theory as it is sometimes called?

The answer is, I think, yes. We might develop a potential "theory of everything", T, that meets with empirical success and is perfectly unified except that T predicts that the cosmos evolves into infinitely many universes, each having slightly different values of 30 or so parameters from any other, there being theoretical reasons why this cannot be interpreted as a cosmological probabilistic jump into just one of these infinitely many universes (so that none of the others exist). In this case we would be confronted by two theories, T and T*, both empirically adequate in our universe. The two theories differ in two respects. First, T asserts that the cosmos consists of infinitely many universes whereas T* asserts that there is just our universe. Second, whereas T is perfectly unified, T* is, in comparison, drastically disunified in that it has 30 constants (having to do with the strength of different forces and the properties of particles) which are entirely arbitrary, and have to be determined empirically. In these circumstances, we might take AOE to imply that T is to be accepted on the grounds that T* is drastically less unified than T. By contrast, SE implies that there is no scientific basis for preferring T to T* since both fit all empirical data equally well.

---

[51] Maxwell (1994; 1998, ch. 7; 2004c; 2011c).

[52] I know of only three theoretical physicists who have sought to develop fully micro-realistic, fundamentally probabilistic versions of quantum theory. The first is Roger Penrose: see his (1986; 2004, ch. 30). The other two are Gordon Fleming and Rudolf Haag (personal communication).

As far as string theory is concerned, we may take AOE to imply that, for perfect unity, we require there to be a unification of space-time and matter so that both are aspects of one underlying entity (possibly a kind of space-time such that matter is merely an aspect of it). Current formulations of string theory do not satisfy this requirement for perfect unity (so that physicalism(8,1) is exemplified). Strings are entities *in* space-time, they are not *aspects* of space-time. String theory does not, for example, postulate that space-time itself has a "stringy" aspect which gives rise to the strings that constitute particles of matter. Thus, according to AOE, quite apart from empirical considerations, string theory, as at present formulated, is inadequate because it fails to be fully unified – a version of physicalism(8,1).

V. Best Level 3 Blueprint

AOE highlights the fundamental importance, for theoretical physics and cosmology, of formulating and adopting the best possible thesis – or "blueprint" – at level 3 in the hierarchy of theses of AOE, taking into account developments at levels 1, 2 and 4. One possibility is to accept the best available formulation of string theory. But this, as we have seen, contradicts the best available thesis at level 4, namely physicalism(8,1). Perhaps, in any case, a thesis somewhat less specific than string theory is required. One possibility is a doctrine that may be called *Lagrangianism*. All fundamental, dynamical theories accepted so far in physics – Newtonian theory (NT), classical electrodynamics, general relativity, quantum theory, quantum electrodynamics, quantum electroweak theory, quantum chromodynamics and the standard model – can be formulated in terms of a Lagrangian and Hamilton's principle of least action. In the case of NT, this takes the following form. Given any system, we can specify its kinetic energy, KE (energy of motion), and its potential energy, PE (energy of position due to forces), at each instant. This enables us to define the Lagrangian, L, equal at each instant to KE - PE. Hamilton's principle states that, given two instants, t_1 and t_2, the system evolves in such a way that the sum of instantaneous values of KE - PE, for times between t_1 and t_2, is a minimum value (or, more accurately, a stationary value, so that it is unaffected to first order by infinitesimal variations in the way the system evolves). From the Lagrangian for NT (a function of the positions and momenta of particles) and Hamilton's principle of least action, we can derive NT in the form familiar from elementary textbooks.

It is this way of formulating NT, in terms of a Lagrangian, L, and Hamilton's principle, that can be generalized to apply to all accepted fundamental theories in physics. Lagrangianism, then, asserts that the universe is such that all phenomena evolve in accordance with Hamilton's principle of least action, formulated in terms of some unified Lagrangian (or Lagrangian density), L. We require, here, that L is not the sum of two or more distinct Lagrangians, with distinct physical interpretations and symmetries, for example one for the electroweak force, one for the strong force, and one for gravitation, as at present; L must have a single physical interpretation, and its symmetries must have an appropriate group structure. We require, in addition, that current quantum field theories and general relativity emerge when appropriate limits are taken.[53]

---

[53] For accounts of Lagrangian formulations of classical and quantum mechanical theories see: Feynman et al. (1965: vol. II, ch. 19), Mandl and Shaw (1984: ch. 2), Goldstein (1980).

There are, however, developments in quantum gravity, having to do with "duality", which suggest that Lagrangianism may well be false: see Isham (1997, pp. 194-195). If physicalism(8,1) is true, and space-time has some kind of discrete, quantum mechanical character, then Lagrangianism is false, granted that it requires space-time to be continuous.

An important task for theoretical physics and cosmology would seem to be to formulate the best possible level 3 thesis that is compatible with physicalism(8,1) – compatible, that is with the idea that matter and space-time are unified, space-time, perhaps, having some kind of discontinuous, quantum mechanical character.

VI. Cosmological Physicalism

A basic motivation for making explicit metaphysical assumptions that are implicit in the methods of physics is that it provokes us into inventing new metaphysical possibilities, which we might not otherwise have considered. We are much more victims of *implicit* assumptions – of assumptions we deny making – than of assumptions we make explicit. This consideration prompts the question: Are alternatives to physicalism(8,1) conceivable? There are, of course, endlessly many possible universes less comprehensible physically than those depicted by physicalism(8,1).[54] Are there any different from physicalism(8,1), but just as comprehensible physically?

Physicalism(8,1) holds that the cosmos, at any given instant,[55] is made up of two distinct aspects, which we may call U and V. U is what is depicted by the true physical theory of everything, T. It is inherent in all phenomena, everywhere, at all times. It does not itself change, but determines (perhaps probabilistically) the way that which changes does change. V, by contrast, is what does change and vary, from moment to moment, and from one place to the next. U and V together, at one instant, determine (perhaps probabilistically) V at the next instant.

This distinction between U and V can be traced back to atomism, the very first version of physicalism put forward by Democritus some two and a half thousand years ago. Given atomism, U consists of the unchanging properties of atoms and space, while V consists of the changing (relative) positions and motions of the atoms. As modern physics developed, ideas about the nature of U and V have changed, but the distinction itself has persisted up to the present. After Newton, rigid atoms interacting only by contact were transformed into point-atoms surrounded by rigid, centrally-directed fields of force. Here, U consists of the unchanging properties of the point-atoms and their surrounding fields of force (including the way the force falls off with distance and the affect it has on other point-atoms), while V consists of the changing (relative) positions and motions of the point-particles. Then it emerged, as a result of Faraday's speculations, Maxwell's theory of the electromagnetic field and Einstein's theory of special relativity, that force fields are not rigid. Changes in the field take time to travel. This led to a new *unified field* version of physicalism, according to which everything is made up of an extended, self-interacting, unified field (matter being simply especially intense regions of the field). On the one hand there are changing, variable features

[54] Universes that just meet the requirements of physicalism(n,1) become less and less comprehensible physically as n = 7, 6 ... 2,1. And, for any given n, physicalism (n,N) with N = 1, 2,... represent universes increasingly disunified, and hence increasingly incomprehensible physically.

[55] Or on any given spacelike hypersurface, looking at things from the standpoint of general relativity.

of the field, V; and on the other, there are the unchanging features of the field, U, which determine how V changes, and which correspond to the laws of the true theory of the field. Subsequent developments have led to further changes in ideas as to what U and V are, but have not undermined the distinction itself.

It is no accident that the atomism of Democritus sharply distinguishes U and V. Atomism arose as an attempt to solve the problem of change, in particular the problem Parmenides posed with his argument that change involves a contradiction, and his view that the cosmos is a homogeneous, unchanging sphere.[56] Parmenides argued that change is impossible because the non-existent cannot exist, hence the world must be full, and hence there can be no room for movement or change. Democritus accepted the argument but rejected the conclusion. There is change, hence the non-existent must exist. The non-existent or, as we might say today, the void surrounds Parmenides's homogeneous, unchanging universe. Other Parmendian universes exist in the void. These can be shrunk down to a minute size, put in relative motion – and we have atomism. Each Democritean atom is a miniature Parmenidean universe. Atomism solves the problem posed by Parmenides by retaining as much as possible of the Parmenidean homogeneous, unchanging universe, but at the same time modifying this view just sufficiently to allow for change and diversity. Atomism solves the general problem of change – the problem of understanding how something can both remain the same *and* change – by segregating very precisely those aspects of atoms which do not change, and those which do, the key to the distinction between U and V.

But there is another possible response to Parmenides. The cosmos as depicted by Parmenides – a homogeneous unchanging sphere – is a very special, uniquely unified state of the cosmos, the big bang state. This unified, initial state of the cosmos is unstable: spontaneous symmetry breaking occurs, and the cosmos evolves into a state made up of a great number of *virtual* big bang states. Today, every space-time point is made up of just one thing: a fleetingly existent, virtual big bang state.

Quantum theory can be interpreted as asserting that for very short intervals of time there is uncertainty of energy, and this permits so-called *virtual* particles to come into existence in the vacuum and almost immediately cease to exist. According to *cosmic physicalism* – the alternative to atomism as a response to Parmenides – every miniscule space-time region is composed, not of virtual particles, but of the virtual universe in its initial, unified, Parmenidean state. Before the big bang, unity is real and all disunity is virtual. After the big bang, disunity is real and unity is virtual. In a sense, there is only the big bang state. Variety and change come from the different ways in which instantaneously existent, virtual big bang states of the universe are inter-related.

There are, then, two distinct versions of physicalism which we may call *atomistic* and *cosmic* physicalism. They can be regarded as arising as a result of giving different responses to the challenge posed by Parmenides's impossible physically comprehensible universe.

Atomistic physicalism takes the Parmenidean cosmos to depict U – that aspect of the cosmos which does not change and which determines the way that which changes, V, does change. Initially, U represented the properties of the atom. Subsequent developments in theoretical physics have transformed U, so that it may be taken to represent the invariant properties of a unified field, a quantum field, space-time of variable curvature, and so on.

[56] This story has been told brilliantly by Karl Popper: see Popper (1998).

Despite these developments, the distinction between U and V persists, and it is this which is the hallmark of atomistic physicalism.

Cosmic physicalism, by contrast, takes the Parmenidean cosmos to be a special, uniquely unified state of the cosmos – the big bang state. According to cosmic physicalism, the true theory of everything, T, specifies the properties of the cosmos in this state. At a fundamental level, the distinction between U and V does not arise. The distinction only arises when spontaneous symmetry breaking has occurred, and the cosmos consists of momentarily existing virtual big bang states. V consists of the different, changing ways in which these momentarily existing big bang states are inter-related.

There are other striking differences between these two versions of physicalism. Cosmic physicalism is inherently *cosmological* in character, whereas atomistic physicalism is not. According to cosmic physicalism, T of itself specifies the initial state of the cosmos, whereas according to atomistic physicalism, initial conditions are required in addition to T to specify the initial state of the cosmos. Cosmic physicalism is inherently *probabilistic*, since spontaneous symmetry breaking is an inherently probabilistic process, whereas atomistic physicalism may be deterministic or probabilistic. Cosmic physicalism must be quantum mechanical to the extent, at least, of incorporating the quantum mechanical distinction between *actual* and *virtual*. Atomistic physicalism makes no such demand.

The two versions of physicalism specify very different conditions for underlying unity to become apparent in as simple a way as possible. According to atomistic physicalism, this happens when the physical system being considered is as simple as possible – the vacuum, or a one particle system or, somewhat more complex, a two particle system. According to cosmic physicalism, it is exactly the opposite: underlying unity is made manifest in a system consisting of *everything* – the entire cosmos in a very special state, the initial big bang state.

Theoretical physics so far has presupposed atomistic physicalism. But it is possible that cosmic, and not atomistic, physicalism is true. Here are a few developments in theoretical physics during the 20th century which may be taken as pointing towards cosmic physicalism.

a. A basic idea of atomistic physicalism is that the physically simplest, most elemental state of affairs that can exist is the vacuum: physical states of affairs become progressively more complex as 1, or 2, or ... n atoms, or fundamental particles, are added to the vacuum. With the advent of field theory, however, this straightforward ordering of physical complexity begins to break down: here, presumably, we would have to say that the simplest state obtains when the value of the field is everywhere zero, a less simple state arising when the value of the field is everywhere a constant value, more complicated states arising with increasingly complicated variable values of the field.
b. The idea that the vacuum is the simplest state, in that it is always present in an unchanging form, breaks down further with the advent of Einstein's general theory of relativity (GR). According to GR, the curvature of space-time varies with varying amounts of matter, mass or energy-density; and curved space-time itself possesses energy. Space is no longer a bland, unchanging arena within which more or less complex physical events unfold: the variable curvature of space, or of space-time, itself takes part in dynamical evolution. (Empty, flat, unchanging space-time still arises, however, as a possible solution to the equations of GR.)

c. With the advent of quantum field theory (QFT), empty space becomes even more complex in that it is full of so-called vacuum fluctuations. These may be pictured as follows. According to one (perhaps somewhat dubious) way of interpreting the time-energy uncertainty relations, $\Delta t \Delta E \geq h/2\pi$ (here, h is Planck's constant), given any state of affairs, for very short time-intervals, Δt, there will be an uncertainty of energy, ΔE, with $\Delta E \approx h/(2\pi\Delta t)$. Thus, even in empty space, there is sufficient energy, ΔE, available to create an electron/positron pair anywhere, at any time, with the proviso only that such a pair must mutually annihilate after a time $\Delta t \approx h/(2\pi \Delta E)$ and $\Delta E \approx 2m_ec^2$, where m_e is the rest mass of the electron. Such fleetingly existing particles are called "virtual" particles. According to QFT, each minute space-time region is full of virtual processes, involving the creation and annihilation of particle/anti-particle pairs. Indeed, all possible virtual processes occur that violate no other conservation law except that of energy (understood classically). Within very tiny space-time regions so much (virtual) energy can exist that there is sufficient to create a short-lived virtual black hole.

According to QFT, in other words, the vacuum is a mass of seething activity, which averages out to nothing over sufficiently large space-time regions. One may, indeed, interpret QFT as a theory of the vacuum, all possible physical processes going on, within minute space-time regions, as *virtual* processes. In supplying discrete units of energy, ΔE_1, ΔE_2, ... ΔE_n, we merely change some of the *virtual* processes into *actual* processes. There is a sense in which the most complex physical state imaginable is no more complex than the vacuum: it is just an energetic state of the vacuum such that some virtual processes are actual processes.[57] In brief, QFT transforms the simple, elemental, unchanging vacuum of 19th century physics into a seething mass of complex, virtual processes – mirroring, in a ghost-like way, all the complexity of the most complex actual physical processes that exist when there is matter. It might be supposed that quantum vacuum fluctuations are not real physical phenomena, but only artefacts of the formalism of QFT given a certain (questionable) interpretation. But this does not take into account that vacuum fluctuations have been detected! This was done decades ago by means of the Casimir effect. According to QFT, if two metal plates are held a small distance apart, virtual processes that involve the creation and annihilation of electron/positron pairs will tend to be *suppressed* in the space between the plates. This results in there being a small pressure tending to push the plates together, due to the *unsuppressed* virtual processes taking place in the space surrounding the two plates. This minute force, due to vacuum fluctuations – the Casimir effect – has been detected and measured.

d. Cosmic physicalism requires that there is a special state of the entire cosmos which is such that all diversity and change disappears: the very distinction between space and matter, we may presume, disappears, there being just *one* homogeneous, instantaneously unchanging *something*, which is also *everything*. This is made possible by big bang cosmology. It is conceivable that the big bang state of the universe, when all of space and matter was packed into a tiny region, was the Parmenidean state of instantaneous unity and homogeneity.

[57] For an excellent account of the quantum theory of the vacuum, see Aitchison (1985).

e. According to GR, the force of gravity is not a force at all; it is rather the tendency of the curvature of space-time to be affected by the presence of matter, or energy-density. With GR, one apparent force, gravity, becomes a feature of physical geometry. This suggests that it may be possible to carry this process of "geometricizing" physics further, the eventual outcome being the unification of space-time, on the one hand, and matter or energy on the other hand. Just this is required by cosmic physicalism.
f. According to the Salam-Weinberg theory of the electroweak force, at high energies, the distinction between the electromagnetic force on the one hand, and the weak force on the other, disappears. As we go backwards in time towards the big bang, towards a time when the energy-density of the universe was sufficiently high, there existed just one unified force, the electroweak force. As the universe expanded, and the energy-density went down, the unity of the electroweak force was broken: the currently observed disunity of two forces with very different properties emerged. This is strikingly in accord with the basic idea of cosmic physicalism: as we move backwards to the original big bang state so we move towards a state of affairs of greater unity, simplicity, symmetry or homogeneity. The potentially immensely important idea of cosmic spontaneous symmetry breaking, which the Salam-Weinberg theory introduces at the level of fundamental theoretical physics, is precisely what is required to make cosmic physicalism a possibility. It is this development in theoretical physics, above all, which makes cosmic physicalism a viable possibility.
g. The idea of an initial state of high symmetry or unity evolving into something asymmetrical and disunified, is further supported by superstring theory – or M theory as it is often called. According to this theory, space has 10 dimensions. The dimensions that we do not observe are curled up into such a minute multi-dimensional "ball" that we do not ordinarily notice their existence. The idea here is that at the big bang state all 10 dimensions of space were curled up in this fashion; the subsequent evolution of the universe consists of just three spatial dimensions growing in size to become, eventually, the space in which we find ourselves. Here, an original Parmenidean unity becomes disunity as three dimensions of space become dramatically different from the rest.

VII. Criteria for Unity

One of the great merits of aim-oriented empiricism (AOE) is that it provides precise criteria for assessing degrees of unity of dynamical theories of physics. These criteria are incompatible with standard empiricism (SE). Thus, in demanding, for example that, in order to be acceptable, a theory must at least be unified in a type (3) way, with N = 1, we thereby commit physics to accepting, as a part of scientific knowledge, the metaphysical thesis that no dynamically unique, spatially restricted entities exist.

And similar points arise in connection with the other seven kinds of unity. AOE is required in order to render acceptable the eight criteria for unity spelled out in section 4 above. As long as SE is taken for granted, criteria for unity can only be deployed in a somewhat hypocritical and inconsistent fashion.

VIII. Rational Discovery

AOE provides physics with a fallible, non-mechanical but *rational* method for the discovery of revolutionary new physical theories. This involves modifying theories and metaphysical theses, at levels 2, 3 and perhaps 4, in an attempt to resolve conflicts between levels (and between theories at level 2), until it proves possible to formulate a thesis at level 3 with sufficient precision to become a new testable physical theory. It was essentially this rational method of discovery that was employed by Einstein in creating special and general relativity (Maxwell, 1993, pp. 275-305). Nothing like this is at all possible granted SE. From this standpoint, the only scientifically rational way to assess metaphysical ideas would be to favour those ideas that are compatible with existing well-confirmed physical theories. But good new fundamental physical theories, and their associated metaphysical ideas, almost always *contradict* pre-existing theories. SE leads us to favour precisely the wrong kind of metaphysical idea, from the standpoint of developing a good new scientific theory! No wonder such stalwart defenders of standard empiricism as Popper (1959, p. 31) and Reichenbach (1938, pp. 381-383) both deny the possibility of a rational method of discovery for science. No wonder that most physicists and philosophers of physics find the whole process of the discovery of fundamental new physical theories a profound mystery! (See, for example, Wigner, 1970, ch. 17.) Only the covert implementation of AOE in scientific practice, under the smokescreen of hypocritical allegiance to SE, has made it possible for some physicists to discover revolutionary new theories. Repudiation of SE, and the full, explicit implementation of AOE in scientific practice should enable physics to be even more successful than it has been up till now.

IX. Pessimistic Induction

Theoretical physics tends to advance from one false theory to another. Thus Galileo's laws of terrestrial motion and Kepler's laws of planetary motion are corrected by Newtonian theory, in turn corrected by general relativity. Classical physics is corrected by quantum theory, in turn corrected by quantum field theory. Philosophers of science, viewing this matter from a standard empiricist perspective, tend to regard the fact that physics advances from one false theory to another as having very negative implications for scientific progress. That physics will continue in this way has even been dubbed "the pessimistic induction" (Newton-Smith, 1981, p. 14). But viewed from the perspective of AOE, this manner of progression is actually to be expected, if physics really is making progress, and the cosmos really is physically comprehensible. For, if a theory, T_0, is precisely true throughout some restricted domain of phenomena D then, granted physicalism,[58] T_0 must specify precisely what does not change, U, throughout all phenomena in D, and the way U determines how things change in D. But, according to physicalism, U exists unchanged throughout all phenomena. Thus, if T_0 specifies the nature of U in D, it will be a straightforward matter to extend T_0 so that it specifies U for all physically possible phenomena, T_0 thus becoming the true theory of

[58] 'Physicalism' here means 'physicalism(n,1) with $8 \geq n \geq 5$'. We require a version of physicalism which asserts that there is an invariant U throughout all phenomena that are physically possible (according to that version of physicalism).

everything, T. Conversely, if T_o cannot be extended in this way to apply correctly to all phenomena, then T_o cannot be precisely true within D: T_o must be false. In brief, physicalism implies that a dynamical physical theory can only be precisely true of *anything* if it is (capable of being) precisely true of *everything*. Granted, then, that physics proceeds, not by attaining T in one bound, but rather by developing a succession of theories that apply, with ever increasing accuracy, to ever wider ranges of phenomena until eventually a theory of everything is attained, it is inevitable, granted physicalism, that physics will progress by the development of theories that are all *false* throughout their domains of application until the ultimate, unified true theory of everything is attained (which will be precisely true about everything).[59] Since physicalism predicts that physics will progress in this way, the fact that physics has so far thus progressed can only count in favour of physicalism: it cannot count against physicalism and AOE, as some have supposed.[60]

X. Natural Philosophy

Modern science began as natural philosophy – or "experimental philosophy" as it was sometimes called. In the time of Newton, in the 17th century, science was not only called "natural philosophy". It was conceived of, and pursued, as a development of philosophy. It brought together physics, chemistry and other branches of natural science as we know it today, with diverse branches of philosophy: metaphysics, epistemology, methodology, philosophy of science – even theology. Science and philosophy, which we see today as distinct, in those days interacted with one another and formed the integrated enterprise of natural philosophy.[61] But then the idea came to prevail – in part because of the impact of misunderstandings about Newton's views about scientific method – that evidence alone determines what is to be accepted and rejected in natural philosophy, or natural *science* as it came to be called.[62]

As a result of the widespread acceptance of this view – standard empiricism as I have called it here – empirical *science* came to be dissociated from non-empirical *philosophy*, and natural philosophy disintegrated (although the name still lingers on in some Scottish universities).

---

[59] Or rather, precisely true about that aspect of what exists which determines the way events evolve everywhere, at all times, throughout all phenomena.

[60] See, for example, Laudan (1980), Newton-Smith (1981).

[61] This point was well made long ago by Burtt, (1932).

[62] Newton came to be interpreted as advocating that evidence alone should determine what theories are to be accepted – even that theories should be arrived at by inductive reasoning from phenomena. But Newton formulated three of his four rules of reasoning in such a way that it is clear that these rules make assumptions about the nature of the universe. Thus rule 1 asserts: "We are to admit no more causes of natural things than such as are both true and sufficient to explain their appearances." And Newton adds: "To this purpose the philosophers say that nature does nothing in vain, and more is in vain when less will serve; for Nature is pleased with simplicity, and affects not the pomp of superfluous causes": see Newton (1962, vol. 2, p. 398). Newton understood that persistently preferring simple theories means that Nature herself is being persistently assumed to be simple (which violates standard empiricism). In so far as Newton's views about method played a major role in the general acceptance of standard empiricism after his death, it has to be said that it was a radical misunderstanding of these views that played this role. Newton's actual views on method are more sophisticated than all those 20th and 21st century scientists and philosophers of science who have taken one or other version of standard empiricism for granted.

The conception of science responsible for this disintegration of natural philosophy is however, as we have seen, untenable. Science cannot proceed without making highly problematic metaphysical assumptions concerning the knowability and comprehensibility of the cosmos. If science is to put scientific rationality fully into practice, and improve knowledge about how to improve knowledge in the light of improving knowledge, then it needs to explore and critically assess metaphysical assumptions and associated methods as an integral part of science itself.

The task of articulating and improving problematic aims and methods of science – the philosophy of science in other words – needs to be pursued as a part of science. We need to recreate natural philosophy.

XI. Philosophy of Science

Scientists and philosophers of science recognize that there are a number of baffling unsolved problems about the nature of scientific inquiry. There is the problem of induction, the problem of how theories can be verified by evidence.

There is the problem of specifying precisely what the methods of science are, especially in view of the fact that methods change over time, and from one scientific discipline to another. There is the problem of specifying non-empirical criteria theories must satisfy, such as simplicity, unity, explanatory power; and the problem of justifying the use of these criteria in science.

There is the problem of explicating what it can mean to say that physics makes progress, given that it advances from one false theory to another (the problem of verisimilitude). There is the problem of explaining how the discovery of revolutionary new theories is possible, given that such theories tend to conflict with pre-existing theories, and there are infinitely many possibilities to choose from.

Again and again, physicists have invented new theories which subsequently turn out to yield astonishingly accurate empirical predictions, often predictions of entirely new, unsuspected phenomena.[63] How is this possible?

There is the mystery of the applicability of mathematics to the physical world. Mathematicians on occasions develop new mathematical ideas in a context apparently remote from physics which subsequently turn out to be applicable to the physical world in a way no one initially anticipated.[64]

Again, how is this possible? Long standing attempts to solve these fundamental problems in the philosophy of science within the framework of standard empiricism have all failed.

[63] Especially notable in this respect are: Newtonian theory, Maxwellian electrodynamics, Einstein's special and general theories of relativity, Bohr's quantum theory of the atom, Heisenberg's and Schrödinger's non-relativistic quantum theory, Dirac's relativistic quantum theory of the electron, quantum electrodynamics of Tomonaga, Schwinger, Feynman and Dyson, and quantum electroweak dynamics and quantum chromodynamics developed by Yang, Mills, Salam, Weinberg, Gell-Mann and others.

[64] Especially noteworthy is Appollonius's characterization of conic sections over one and a half thousand years before Galileo and Kepler employed conic sections (the parabola and the ellipse) to describe terrestrial and planetary motion. Noteworthy, too, is Riemann's invention of Riemannian geometry some 60 years before Einstein depicted space-time as Riemannian in his general theory of relativity.

They are all readily solved, however, once AOE is accepted.[65] The long standing failure to solve these problems granted SE, and their straightforward resolution granted AOE, amounts to a devastating argument against SE and for AOE.

XII. Education

The transition from SE to AOE has implications for science education. Science needs to be taught as natural philosophy. University courses in physics need to include some discussion of metaphysical and methodological ideas, some exploration of problems of epistemology and philosophy of science.

XIII. Fundamental Problem

In moving from SE to AOE we are brought face to face with what is, in many ways, a rather grim vision of reality. Everything is just physics. Everything is made up of the all-powerful, invariant physical entity, U, present everywhere, throughout all possible phenomena, at all times and places, and the variable physical aspect of things, V, which change from instant to instant and place to place. What then becomes of everything that this seems to leave out? What becomes of the rich tapestry of the experiential world, the colours, sounds, smells and tactile qualities we experience? What becomes of our inner sensations, our inner states of consciousness? What becomes of free will, meaning and value, everything that seems precious in life? How can any of this exist if the cosmos really is physically comprehensible in the way that this has been explicated here?

One great merit of AOE, I suggest, is that it brings us face to face with this fundamental problem. It is, in my view, our fundamental problem of both thought and life. It can be formulated like this: How can our human world, imbued with perceptual qualities, consciousness, free will, meaning and value, exist and best flourish, embedded as it is in the

[65] How AOE solves the problem of induction is outlined above, in sections 7 and 8. There, I justified acceptance of theses at levels 7 to 4 as a part of theoretical scientific knowledge. Our current best fundamental theories in physics – general relativity and the standard model – are to be accepted because (a) they have met with sufficient empirical success, and (b) they accord sufficiently well with the thesis at level 4 in the hierarchy of theses of AOE. For earlier, more detailed accounts of how AOE solves the problem of induction, in each case adding to and improving earlier accounts, see Maxwell (2007, pp. 400-430; 2004a, pp. 205-220; 1998, ch. 5). See also Maxwell (2006a). It might be thought that AOE only specifies the nature of scientific method as far as theoretical physics is concerned. But, as I indicted in note 33, AOE can readily be generalized to apply to all branches of natural science, and to natural science as a whole: see Maxwell (2004a, ch. 2, especially pp. 41-47). Different sciences have somewhat different aims; and aims evolve as scientific knowledge evolves. Because of its meta-methodological character, AOE can do justice to the diversity of aims and methods, as one goes from one science to another, and from one time to another within one science; at the same time, AOE can do justice to the unity of method throughout this diversity of methods of natural science. How criteria of simplicity, unity and explanatory power can be explicated within the framework of AOE has been indicated in section 4 above. See also Maxwell (1998, chs. 3 and 4; 2004a, pp. 160-174). For my proposal as to how one can solve the problem of verisimilitude see Maxwell (2007, pp. 393-400 and 430-433). For my proposed solution to the mystery of the discovery of revolutionary new theories in physics, and the applicability of mathematics to the physical world, see Maxwell (1998, pp. 219-223). AOE does of course introduce a new problem: Why should the cosmos be physically comprehensible to us? "The eternal mystery of the universe is its comprehensibility", as Einstein (1973, p. 292) once put it. For my suggestions as to how this problem can be solved see Maxwell (2001, pp. 254-258).

physical universe? Elsewhere, I have argued that we need to interpret physics as being restricted to depicting what may be called the "causally efficacious" aspect of things – that aspect which determines (perhaps probabilistically) how events unfold. What things look like and feel like, what it is like to *be* a bit of the physical universe: these experiential aspects of things necessarily lie beyond the scope of physics, and any attempt to extend physics to incorporate the experiential would destroy the marvellously explanatory capacity of physics. Physics omits all mention of the experiential and the value-laden because it can without curtailing its predictive and explanatory task; and it must do this in order to retain the astonishing explanatory power of its theories.[66]

XIV. Wisdom-Inquiry

As I have argued at length elsewhere, AOE has revolutionary implications for rationality, for academic inquiry, and for the future of humanity.[67] It is not just science that has problematic aims; many other endeavours, individual, institutional and even global, have problematic goals as well. In this essay I have argued that the problematic character of the basic aim of physics means that we need to represent this aim in the form of a hierarchy of aims, thus providing physics with a fixed framework of aims and methods (assumptions and methods) within which much more problematic aims and methods can be improved as physics proceeds. These considerations can be generalized and applied to any human endeavour with problematic aims. We need a new conception of rationality – aim-oriented rationality (AOR) – a generalization of AOE, which emphasizes that, whatever we are doing, if our basic aim is problematic, we need to represent it in the form of a hierarchy of aims, thus providing ourselves with a framework of unproblematic aims and associated methods within which more specific and problematic aims and associated methods can be improved as we act. Applied to academic inquiry as a whole, AOR implies – I have argued – that we need a new kind of inquiry – *wisdom-inquiry* – which puts problems of living at the heart of the enterprise, and takes the basic aim to be to seek and promote wisdom, construed to be the capacity realize what is of value in life for oneself and others, wisdom thus including knowledge, technological know-how and understanding, but much else besides. AOR applied to our social world implies that much that we do – individually, institutionally and globally – needs to be reorganized so that problematic aims may be acknowledged and improved. Above all, AOR needs to be applied to the profoundly problematic endeavour to make progress towards as good a world as possible.

CONCLUSION

Granted the view that evidence alone decides what is to be accepted in science, it is clear that science cannot possibly be said to have established that the cosmos is physically comprehensible. But this view of science is untenable. Physics quite properly persistently only accepts theories that are (a) sufficiently successful empirically, and (b) sufficiently

[66] See works referred to in note 44.
[67] See Maxwell (2007). See also Maxwell (1976a; 2010a).

unified – even though endlessly many empirically more successful, disunified rivals can always be concocted. This means we must construe physics as accepting, as a part of scientific knowledge, a hierarchy of metaphysical theses concerning the knowability and comprehensibility of the cosmos, these theses becoming less and less substantial, and so more and more likely to be true, as we go up the hierarchy, and also more nearly such that their truth is required for science, or the pursuit of knowledge, to be possible at all. One of these theses is that the cosmos is physically comprehensible – that is, such that the true physical theory of everything is unified. It is, in other words, a part of current theoretical knowledge in physics that the cosmos is physically comprehensible.

This *aim-oriented empiricist* view of science has a range of fruitful implications for science itself, and for our understanding of science. It extends the scope of scientific knowledge and understanding about a matter of great significance: the ultimate nature of the cosmos. It provides us with explicit criteria for the assessment of the extent to which physical theories are *unified*. It provides physics with a rational if fallible method for the discovery of revolutionary new physical theories. It transforms science so that it becomes more nearly like natural philosophy of Newton's time – an enterprise that brings together and integrates empirical science, methodology, philosophy of science, metaphysics and epistemology. It facilitates the solution to a number of fundamental problems in the philosophy of science: problems of induction, verisimilitude, simplicity, and scientific discovery. This view of science helps to highlight the nature of the fundamental problem we face in thought and in life: How can our human world, imbued with all that we hold to be of value, exist and best flourish, embedded as it is in the physical cosmos? Finally, this view of science has implications for views about the nature of rationality, and the nature of rational inquiry quite generally.

REFERENCES

Aitchison, I. J. R., 1985, Nothing's Plenty: The vacuum in modern quantum field theory, *Contemporary Physics*, vol. 26, pp. 333-391.

Aitchison, I. J. R. and Hey, A. J. G., 1982, *Gauge Theories in Particle Physics*, Adam Hilger.

Atkins, P. W., 1983, *Molecular Quantum Mechanics*, Oxford University Press, Oxford.

Barnes, B., D. Bloor and J. Henry, 1996, *Scientific Knowledge: A Sociological Analysis*, University of Chicago Press, Chicago.

Bloor, D., 1991, *Knowledge and Social Imagery,* 2nd edn., University of Chicago Press, Chicago.

Bohr, N., 1949, Discussion with Einstein on Epistemological Problems in Atomic Physics. In: P. A. Schilpp (ed.) *Albert Einstein: Philosopher-Scientist*, vol. 1, Open Court, La Salle, Illinois, pp. 201-241.

Boscovich, R. J., 1966, *A Theory of Natural Philosophy*, MIT Press, Cambridge, Massachusetts (English language edition of 1763 edition).

Burtt, E. A., 1932, *The Metaphysical Foundations of Modern Science*, Routledge and Kegan Paul, London.

Chakravartty, A., 1999, Review, *Times Higher Education Supplement*, 24 September, p. 24.

Einstein, A., 1949, Autobiographical Notes. In: P. A. Schilpp (ed.) *Albert Einstein: Philosopher-Scientist*, vol. 1, Open Court, La Salle, Illinois, pp. 3-94.
Feyerabend, P., 1975, *Against Method: Outline of an Anarchistic Theory of Knowledge*, New Left Books, London.
Feynman, R., R. Leighton and M. Sands, 1965, *The Feynman Lectures on Physics* vol. II., Addison-Wesley, Reading, Mass.
Friedman, M., 1974, Explanation and Scientific Understanding, *Journal of Philosophy*, vol. 71, pp. 5-19.
Goldstein, H., 1980, *Classical Mechanics*, John Wiley, New York.
Goodman, N., 1954, *Fact, Fiction and Forecast*, Athlone Press, London.
Grebowicz, M., 2006, Where have the Philosophers been all this time? Reading Maxwell's Revolution, *Metascience*, vol. 15, pp. 141-144.
Griffiths, D., 1987, *Introduction to Elementary Particles*, John Wiley, New York.
Harris, M., 1980, *Cultural Materialism*, Vintage, New York.
Hume, D., 1959, *A Treatise of Human Nature*, Book 1, Everyman, London (first published 1738).
Iredale, M., 2005, Our Neurotic Friend, *The Philosopher's Magazine*, Issue 31, pp. 86-87.
Isham, C. J., 1989, *Lectures on Groups and Vector Spaces for Physicists*, World Scientific, London.
______, 1997, Structural Issues in Quantum Gravity. In: *General Relativity and Gravitation: GR 14*, World Scientific, Singapore, pp. 167-209.
Jeffreys H. and Wrinch, D., 1921, On Certain Fundamental Principles of Scientific Enquiry, *Philosophical Magazine*. Vol. 42, pp. 269-298.
Jones, H. F., 1990, *Groups, Representations and Physics*, Adam Hilger, Bristol.
Juhl, C. F., 2000, Review, *International Philosophical Quarterly*, vol. XL, pp. 517-518.
Kitcher, P., 1981, Explanatory Unification, *Philosophy of Science*, vol. 48, pp. 507-531.
Kneller, G. F., 1978, *Science as a Human Endeavor*, Columbia University Press, New York.
Koyré, A., 1965, *Newtonian Studies*, Chapman and Hall, London.
Laudan, L., 1980, A Confutation of Convergent Realism, *Philosophy of Science*, vol. 48, pp. 19-48.
Mandl, F. and Shaw, G., 1984, *Quantum Field Theory*, John Wiley, New York.
Maxwell, N., 1966, Physics and Common Sense, *British Journal for the Philosophy of Science*, vol. 16, pp. 295-311.
______, 1968a, Can there be Necessary Connections between Successive Events?, *British Journal for the Philosophy of Science*, vol. 19, pp. 1-25.
______, 1968b, Understanding Sensations, *Australasian Journal of Philosophy*, vol. 46, pp. 127-146.
______, 1972a, A Critique of Popper's Views on Scientific Method, *Philosophy of Science* 39, pp. 131-52.
______, 1972b, A New Look at the Quantum Mechanical Problem of Measurement, *American Journal of Physics*, vol. 40, pp. 1431-1435.
______, 1974, The Rationality of Scientific Discovery, *Philosophy of Science* 41, pp. 123-153 and 247-295.
______, 1976a, *What's Wrong With Science? Towards a People's Rational Science of Delight and Compassion*, Bran's Head Books, Frome, England.

______, 1976b, Towards a Micro Realistic Version of Quantum Mechanics, *Foundations of Physics*, vol. 6, 275-292 and 661-676.
______, 1982, Instead of Particles and Fields: A Micro Realistic Quantum "Smearon" Theory, *Foundations of Physics*, vol. 12, pp. 607-631.
______, 1984, *From Knowledge to Wisdom: A Revolution in the Aims and Methods of Science*, Basil Blackwell, Oxford.
______, 1988, Quantum Propensiton Theory: A Testable Resolution of the Wave/Particle Dilemma, *British Journal for the Philosophy of Science*, vol. 39, pp. 1-50.
______, 1993, Induction and Scientific Realism: Einstein versus van Fraassen, *British Journal for the Philosophy of Science* 44, pp, 61-79, 81-101 and 275-305.
______, 1994, Particle Creation as the Quantum Condition for Probabilistic Events to Occur, *Physics Letters A*, vol. 187, pp. 351-355.
______, 1998, *The Comprehensibility of the Universe*, Oxford University Press, Oxford.
______, 2000a, A new conception of science, *Physics World* 13, no. 8, pp. 17-18.
______, 2000b, The Mind-Body Problem and Explanatory Dualism, *Philosophy*, vol. 75, pp. 49-71.
______, 2001, *The Human World in the Physical Universe: Consciousness, Free Will and Evolution*, Rowman and Littlefield, Lanham, Maryland.
______, 2002, The Need for a Revolution in the Philosophy of Science, *Journal for General Philosophy of Science* 33, pp. 381-408.
______, 2004a, *Is Science Neurotic?*, Imperial College Press, London.
______, 2004b, Non-Empirical Requirements Scientific Theories Must Satisfy: Simplicity, Unification, Explanation, Beauty, http://philsci-archive.pitt.edu/1759/
______, 2004c, Does Probabilism Solve the Great Quantum Mystery?, *Theoria*, vol. 19/3, no. 51, pp. 321-336.
______, 2005, Popper, Kuhn, Lakatos and Aim-Oriented Empiricism, *Philosophia* 32, Nos. 1-4, 181-239.
______, 2006a, Practical Certainty and Cosmological Conjectures. In: *Is there Certain Knowledge?*, ed. Michael Rahnfeld, Leipziger Universitätsverlag, Leibzig, pp. 44-59.
______, 2006b, Aim-Oriented Empiricism: David Miller's Critique, PhilSci-Archive, http://philsci-archive.pitt.edu/3092/1/Miller's_Critique.doc.
______, 2007, *From Knowledge to Wisdom* (2nd ed., revised and enlarged, Pentire Press, London).
______, 2008, Do We Need a Scientific Revolution?, *Journal for Biological Physics and Chemistry*, vol. 8, no. 3, pp. 95-105.
______, 2009a, Muller's Critique of the Argument for Aim-Oriented Empiricism, *Journal for General Philosophy of Science*, vol. 40, pp. 103-114.
______, 2009b, How Can Life of Value Best Flourish in the Real World?. In: *Science and the Pursuit of Wisdom: Studies in the Philosophy of Nicholas Maxwell*, edited by Leemon McHenry, Ontos Verlag, Frankfurt, pp. 1-56.
______, 2010a, *Cutting God in Half – And Putting the Pieces Together Again: A New Approach to Philosophy*, Pentire Press, London.
______, 2010b, Reply to Comments on *Science and the Pursuit of Wisdom*, *Philosophia*, vol. 38, pp. 667-690.
______, 2011a, A Priori Conjectural Knowledge in Physics. In: *What Place for the A Priori?*, edited by Michael Shaffer and Michael Veber, Open Court, Chicago, pp. 211-240.

______, 2011b, Three Philosophical Problems about Consciousness and their Possible Resolution, *Open Journal of Philosophy*, vol. 1, no. 1, pages 1-10.
______, 2011c, Is the Quantum World Composed of Propensitons?. In: *Probabilities, Causes and Propensities in Physics,* edited by Mauricio Suárez, Synthese Library, Springer, Dordrecht, pp. 221-243.
______, 2012a, 'Arguing for Wisdom in the University: An Intellectual Autobiography', *Philosophia*, vol. 40, no. 4.
______, 2012b, In: Praise of Natural Philosophy: A Revolution for Thought and Life, *Philosophia*, vol. 40, no. 4.
______, 2012c, What's Wrong With Aim-Oriented Empiricism?, http://philpapers.org/archive/MAXWWW-3.1.doc.
______, 2013, Popper's Paradoxical Pursuit of Natural Philosophy. In: *Cambridge Companion toPopper*, edited by Jeremy Shearmur and Geoffrey Stokes, Cambridge University Press, Cambridge.
McHenry, L., 2000, Review, *Mind*, vol. 109, pp. 162-166.
McNiven, C., 2005, Review, *Journal of Consciousness Studies*, vol. 12, pp. 88-89.
Miller, D., 2006, *Out of Error*, Ashgate, London.
Moriyasu, K., 1983, *An Elementary Primer For Gauge Theory*, World Scientific, Singapore.
Muller, F. A., 2004, Review, *Studies in History and Philosophy of Modern Physics*, vol. 35, pp. 109-110 and 117.
______, 2008, In: Defence of Constructive Empiricism: Maxwell's Master Argument and Aberrant Theories. *Journal of General Philosophy of Science,* 39, pp. 131-156.
Newton, I., 1962, *Principia*, Berkeley, University of California Press, Berkeley, vol. 2, p. 398 (first published in 1687).
Newton-Smith, W. H., 1981, *The Rationality of Science*, Routledge and Kegan Paul, London.
Penrose, R., 1986, Gravity and State Reduction. In: *Quantum Concepts of Space and Time,* edited by R. Penrose and C. J. Isham, Oxford University Press, Oxford, pp. 129-146.
________, 2004, *The Road to Reality*, Jonathan Cape, London.
Poincaré, H., 1952, *Science and Hypothesis*, Dover, New York.
Popper, K. R., 1959, *The Logic of Scientific Discovery*, Hutchinson, London.
______, 1963, *Conjectures and Refutations*, Routledge and Kegan Paul, London.
______, 1998, *The World of Parmenides*, Routledge, London.
Reichenbach, H., 1938, *Experience and Prediction*, Chicago University Press, Chicago.
Roush, S., 2001, Review, *The Philosophical Review*, vol. 110, pp. 85-87
Salmon, W., 1989, *Four Decades of Scientific Explanation*, University of Minnesota Press, Minneapolis.
Smart, J. J. C., 2000, Review, *British Journal for the Philosophy of Science*, vol. 51, pp. 907-911.
Van Fraassen, B., 1985, Empiricism in the Philosophy of Science. In: P. M. Churchland and C. A. Hooker, (eds.) *Images of Science*, University of Chicago Press, Chicago, pp. 259-260.
Vicente, A., 2010, An Enlightened Revolt: On the Philosophy of Nicholas Maxwell, *Philosophia*, vol. 38, pp. 631-648.
Watkins, J. W. N., 1984, *Science and Scepticism*, Princeton University Press, Princeton.
Weinert, F., 2000, Review, *Philosophy*, vol. 75, pp. 296-309.
Wigner, E. P., 1970, *Symmetries and Reflections*, MIT, Cambridge, Mass.

In: Recent Advances in Cosmology
Editors: Anderson Travena and Brady Soren
ISBN: 978-1-62417-943-3

Chapter 2

MACH, EINSTEIN, DARK MATTER AND KNOWABILITY

Marián Ambrozy*
Department of Social Science, Faculty of Education,
Catholic University, Ruzomberok, Slovakia
Inštitút Štefana Náhalku, Pedagogická fakulta, Katolícka univerzita,
Poprad, Slovakia; Súkromné gymnázium, s.r.o. Spišská Teplica, Slovakia

ABSTRACT

At first sight it seems that the general theory of relativity is inspired by the so called Mach´s principle. To what extent does Einstein in his theory digress from the concepts of Ernst Mach? How can we apply Mach´s principle in physics today mainly in relation to the dark matter phenomenon?

Kurt Gödel, gravedigger of Hilbert´s program, created the model of rotating, homogeneous, stationary and anisotropic universe. To what extent is it compatible with Mach´s principle?

Dark matter is one of the basic problems of modern physics. The article attempts to look at this problem from the viewpoint of the theory of knowledge and uses the same viewpoint to compare it with the problem of knowability in mathematics.

Ernst Mach believes that physicists should study physical phenomena and not what is behind them, if there is anything at all. In his opinion, observation and experiment are the only suitable methods for physics. Physics deals with invariant, laws, phenomena but it does not tackle the issue of essence. It is not competent to express any standpoints to the question of determinism and indeterminism. Physics mentions power and energy, however, these are only fictions which help our imagination. They do not represent the real essence.

Ernst Mach´s original idea claiming that the inertia of a particle here and now occurs as a result of its interaction with other particles anywhere in the universe was adopted by Einstein.

* ambrozy.marian@gmail.com.

Einstein´s speculation was based on the attempt to solve one of Kant´s aporias – a cosmological idea. The father of the special theory of relativity and general theory of relativity wanted to prove the validity of a model of closed universe with a fixed number of integrating particles. Such model assumes that the distribution of matter is closely linked to geometry whereas inertial systems are defined only via distribution of matter and energy. This model interconnects perturbations and the general theory of relativity. John A. Wheeler expressed his positive attitude to the said model in one of his last interviews. He considered it pregnantly mathematically formulated and physically rational.

Mach´s principle which played a major role in formulation of the general theory of relativity is best captured in standard closed universes with standard spherical topology. As Herman Bondi accurately described it – it is the definition of local inertial systems via distribution of energy and mobility in the universe by means of averages. Einstein used Mach´s principle as a basis for the principles of general covariance and equivalence of inertial and gravitational forces. Dieter Brill believes that Mach´s principle is a factor giving physical meaning to quantities which depend on coordinates.[1] Despite the aforementioned many questions regarding the relation between Mach´s principle and general theory of relativity remain unanswered and standpoints often differ. Mach´s principle tries to answer the question what is the source of inertia. Mach assumes that the source of inertia are distant cosmic matters. Motion towards these matters differentiates between inertial and non-inertial frames. It is the interaction with distant matters that explains inertia. It is important to emphasize that Mach´s principle stands the test as it does not contradict the observed facts.[2]

Let us follow Einstein´s thought experiment: there are two fluid bodies very remote from all other bodies and from each other. At this distance only the gravitational forces apply. Parts of the two bodies with a constant distance between them rotate at a uniform angular velocity about the axis that passes through the centers of both bodies. Measurements show that one of the rotating bodies is a sphere and the other an ellipsoid. "Newtonian mechanics does not give a satisfactory answer to this question."[3] Mechanical properties of both bodies can be determined by properties that have not been taken into consideration so far. Dissimilar behavior of the two bodies is observable. The reason behind such dissimilar behavior can be the distant matters. There are inconsistencies with respect to relation between Mach´s principle and general theory of relativity. Argument refuting any relation between the two states that inertia forces and Coriolis forces would not change even if the Solar System co-existed with an empty space only. On the other hand, the fact that the universe is filled with matter and motions within local inertial frame are related to the motion of matter supports the existence of relation between Mach´s principle and general theory of relativity. Mach was, after all, a very strong critic of Newton.

Kurt Gödel created his own model of the universe. He agreed with the idea that the passage of time and conscience are interrelated. In special theory of relativity, different observers have different perception of time. According to Kant, passage of time is connected with a priori form of sense, i.e. a thing for itself not a thing in itself. Passage of time is thus a

---

[1] Barbour, J.; Pfister, H. (eds.); *Mach Principle: From Newton´s Bucket to Quantum Gravity*. Birkhauser, Boston, 1995.

[2] Bičák, Jiří; *Machův princip v současné kosmologii*, In: Dub, Peter; Musilová, Jana; (eds.), Ernst Mach, Fyzika filosofie vzdělávání, Brno, Muni Press, 2010, ISBN 978-80-210-48, p. 147.

[3] Novotný, Jan; *Ernst Mach, Kurt Gödel a vesmír*, In: XVI International Colloquium on the Management of Educational Process, Brno, 2008, ISBN 978-80-7231-511-6, p. 3.

matter of conscience. Friedman cosmology of open expanding universe preferred more objective perception of time than Kant and that is why Gödel inclined to homogeneous anisotropic and stationary model of the universe. In Gödel´s universe it is impossible to introduce the moving present. What is the relation between the Gödel´s universe and Mach´s principle? Gödel himself said that inertial field is to a large extent independent of the state of motion of the matter. As we do not have an absolute space we cannot speak of rotation of the universe. General theory of relativity works with the inertial field. Matter can rotate as determined against inertial field determined by the axis of a free flywheel. This rotation does not include the imaginary axis the world rotates around. If the distances between couples of physical particles remain the same in time, then the world rotates as a whole. As Gödel himself states it contradicts the Mach´s principle, however, it does not contradict the general theory of relativity. Nevertheless, does the Gödel´s standpoint have the form of total exclusion?

Is it possible to revise the general theory of relativity with the aim to prove that the source of the inertia forces is the distant dark matter? Einstein´s and Mach´s standpoints agree and disagree at the same time. Einstein believes that the gravitational forces have adapted to the inertia forces. The gravitational forces are determined by metric field of space – time which is connected with the field of energy tensor – impuls via Einstein´s equations, i.e. there is a connection with motion and distribution of matter. Einstein considers gravitational forces indistinguishable from inertial forces. This fact proves that Mach and Einstein had similar viewpoints. However, their views also differed. Mach believed that inertia is dependant on configuration of matter, dynamics is determined by kinetics and the two scientific disciplines differ from each other. On the other hand, Einstein in his general theory of relativity did not differentiate between kinematics and dynamics. It is important to emphasise that it is not possible to claim that Gödel´s rotating universe and Mach´s principle are totally incompatible. On the other hand, one cannot state that the nature of space-time is determined exclusively by matter. Einstein did not share Mach´s belief that the source of physics is observable matter creating a body. The outcome of Einstein´s general theory of relativity is space-time. As suggested by H. Poincaré it is necessary to add one more condition to the elementary physical laws. For example, in relativity we have to add the fourth coordinate – time to the law of momentum conservation.[4] This space-time has metric properties also without matter. Mach´s assertion that metric fields should not be given priority is thus no longer relevant.

The relation between Mach´s ideas and latest observations is very controversial. Views contradicting the general theory of relativity are given above. On the other hand, there is one significant phenomenon – the role of Mach´s principle as an ideological base for introduction of cosmological constant λ. Whereas the original solving of Einstein´s equations is not in accord with Mach´s principle, the introduction of λ was inspired by Mach. The results of current observations show that Eistein was right and prove that Mach´s intuition was very progressive.

Albert Einstein believed that a thinker´s goal should be turning astonishment into something graspable. Just like Newton, Einstein also tried to grasp nature using the language of mathematics – the program of achieving unity of physics via its geometrisation. “He shows that even the simplest mechanical motion – rectilinear and uniform – always results in

[4] Feynman, Richard P.; Leighton, Robert B.; Sands, Mathew; *Feynmanovy přednášky z fyziky 1*, Havlíčkův Brod, 2000, ISBN 80-7200-405-0, p. 245

changing the metric type of the space itself."[5] *(my translation)* In order to characterize gravitation states of the objects of physics one has to apply Riemannian spaces. Changes of motion of material points and changes of the metric space are the most general types of mathematical structures of physical theory. The above statement prevails in contemporary physics as well. Einstein wanted to go even further. Science started to talk about such aspect of physics geometrization which could be completed with axiomatic synthesis. None of Einstein´s attempts to build unified geometrized field theory ended successfully. The system of nature is richer and more sophisticated and such attempt was confronted with the problem of reduction of changes of space metrics to electric interaction and weak leptonic and mesonic reactions. Moreover, development of scientific terms did not stagnate either. Thanks to bourbakists mathematics discovered new better terms applicable in analysis of geometrical space such as limit transition, adjacency, and similarity. The above abstract structures are carefully implemented into metric structures in line with metric theorems of topology. However, there is a question one has to ask: "Is it possible to topologically characterize metric spaces? It seems so but general characteristics is quite complex and its proving is very laborious."[6] *(my translation)*

Einstein laid the foundations for principles of modern cosmology. It is known that Einstein considered the introduction of cosmological constant λ a mistake. Originally, the cosmological constant was supposed to help him elaborate a static description of the universe. Cosmological constant λ ensured that the universe did not expand or shrink. The aim was to decrease the gravitational force in order to prevent collapsing of galaxies in static universe. Russion cosmologist Alexander Friedman took a different approach. He introduced a model which admitted the idea of expanding universe thus enabling him to solve Einstein´s equations without using the constant. Einstein abandoned the cosmological constant and admitted its introduction was a mistake, however, contemporary science has a different opinion. Einstein introduced the constant on the basis of wrong assumption. He acknowledged the constant as a correct assumption until 1931. Some of Einstein´s adherents, e.g. Arthur Eddington, did not abandon the constant at all. On the contrary, Eddington considered the constant a source of universe expansion discovered by Edwin Hubble. From time to time, scientists took the cosmological constant out of Einstein´s thrash can and re-examined it. However, bad reputation it acquired led to its exlusion from most of the discussions of practical cosmology.[7]

Naturally, current situation is completely different. Constant progress and numerous physical discoveries favor the cosmological constant. Discoveries in the field of cosmology, made mainly in the last decade of the 20th century, led to significant changes in understanding of the universe. Seemingly paradoxically, the new scientific knowledge was created at studying of supernovas.

Supernovas are luminuous objects which are several billion times brighter than the sun. Metaphorically speaking, supernovas can be used as a taylor´s tape measure for measuring the dimension of the universe. Fritz Zwicky and Walter Baade were the first to study and name supernovas. Detection of supernovas was accompanied by numerous technical difficulties.

---

[5] Akčurin, I. A.; *Matematizácia ako princíp jednoty fyzikálnych teórií*, In: Metodologické princípy fyziky, Bratislava, 1984, p. 217.

[6] Adámek, Jiří; Koubek, Václav; Reiterman, Jan; *Základy obecné topologie*, Praha, 1977, p. 230.

[7] Kirshner, Robert, P.; *Výstřední vesmír*, Praha, 2005, ISBN 80-7185-729-7, p. 97.

For example, in 1970s the electronic detectors were too small therefore scientists had to use glass boards with a layer of photographic emulsion – silver salts dissolved in gelatin. The fact that the seeming brightness of the object decreases with the square of the distance was yet another technical problem. On the other hand, the 1970s witnessed certain progress in observation of changing brightness of supernovas. Only very bright objects, such as supernovas of the Ia type, can help in searching for reliable value of the Hubble constant and in specification of the age of the universe because they are all equally bright and can be thus used as etalon candles.[8] Bruno Leibundgut, supporter of the idea of invariant development of the above mentioned supernovas, also studied supernovas and in 1989 he significantly contributed to their reliable detection and observation in the dynamics of their development. In 1990s supernovas were successfully used for measuring distances of galaxies. Scientists discovered several extraordinary supernovas such as SN 1991T, SN 1991bg, SN 1994G, etc. and included the influence of cosmic dust on the color of supernovas into their measuring of the distances of supernovas. The results of measurings were based on analyses of several measurings via different filters which helped analyze the light of various colors.

In 1990s science made a very interesting discovery which had a great impact on cosmology and which is to a large extent connected with the phenomenon of dark matter. The first scientist who started to talk about dark matter and supernovas was Fritz Zwicky. Zwicky was an unconventional astronomer who discovered that some galaxies in Coma cluster move faster than originally expected which led him to a conclusion that the cluster was much more massive than he had thought. According to Zwicky´s own estimate the amount of dark matter in the observed cluster amounted to 90%. Jan Oort was the first to suggest the existence of a cloud full of comets which was named after him and is now known as the Oort cloud. Even though his detection was not confirmed, it has not been refuted either. Majority of astronomers have never seen the Oort cloud, nevertheless, they take it as a fact since its existence has never been refuted. Their approach is fully in line with Popper´s falsification. Dark matter had not been in the centre of scientific interest until 1970s when James Peebles and Jeremiah Ostriker introduced a theory according to which many spiral galaxies would cease to exist due to instability if the main part of their matter rested in disks. Such types of galaxies have to be stabilized by means of the so called galactic halo. Star matter is not sufficient to ensure conditions discovered by Peebeles and Ostriker and therefore the "halo" has to be as material as the matter of the galactic disk.[9] Vera Rubin, astronomer and doctorand of George Gamow, also studied the above issues and noticed that luminosity of the matter in spiral galaxy shows that the matter is concentrated in its centre. However, observations of its rotational velocity speak of an increasing amount of non-luminuous matter, i.e. visible picture of spiral galaxies thus does not correspond with measurings to a large extent.

In years 1997 - 1999 a group dominated by Brian Schmidt, Nick Suntzeff, Peter Garnavich, Adam Riess, Robert Kirshner and many others made several discoveries which led to a discovery of accelerated expansion of the universe. "Perspective of the future of the universe prevailing in cosmology until years 1998/1999 (which was also accepted by the general public) did not know dark energy and was therefore based on a simpler equation $\Omega m + \Omega k = 1$. According to this perspective the future was defined by a single parameter Ωk (or

[8] Ferris, Timothy; *Všetko o vesmíre*, Bratislava, 2005, ISBN 80-88993-85-7, p. 54.

[9] Ferris, Timothy; *Všetko o vesmíre*, Bratislava, 2005, ISBN 80-88993-85-7, p. 123.

equivalently *Ωm*)."[10] *(my translation)* Observations made by the Hubble telescope and evaluated in those years resulted in early exclusion of the possibility that ω=1, i.e. the proportion of the observed density of the universe and the so called critical density of the universe – separation of the eternal expansion of the universe and distant collapse of the universe which in practice contradicted the opinions on deceleration of the expansion of the universe. On the contrary, observations implied the universe would continue to expand forever. Moreover, further observations gave evidence of accelerated expansion of the universe. Supernovas proved to be the etalons. Data indicated that distant supernovas appear weaker than in case of lower density of the universe. This means the distances are greater and the expansion is faster. Observations had one direct consequence relating to determination of the value of ω_m. With cosmological constant equaling 0, ω_m reached negative values. In order to prevent such empirical impossibility, discoverers decided to reconsider the nonzero cosmological constatnt. The above assumption helped harmonize the results of observations in 1998.

The above case was not the first time the cosmological constant was used. Hubble´s estimate of his constant was 500km.s^{-1}.Mpc^{-1}. The best estimate of the age of the universe was 2 billion years. However, application of the above constant led to a very bizzare result – the universe is younger than the Earth. In order to correct the apparently incorrect result cosmologists reached for the cosmological constant λ again. When E. Hubble found out that he had disproportionately multiplied his constant, the cosmological constant λ was sent to the depository of scientific history and there it stayed until its revival in the late 1990s. Basic assumption, something like Gadamer´s "*Vorurteil*", that its value equals zero was still valid in the access paradigm. The need to explain the said phenomena started to turn this physical axiom upside down again.

Effects observed on supernovas SN 1999ff, SN 1999fv, SN 1997ff, etc. are not the only evidence of the accelerating expansion of the universe. Year 2003 brought a new discovery which confirmed the accelerating expansion of the universe by means of measuring microwave relict radiation by WMAP satellite. Fluctuations of relict radiation contain data on density of cosmic energy. The assumed existence of dark energy should provide the data for the inexplicable part which has been measured.

Naturally, science started to search for reasons behind acceleration of the universe. Universe is observed by means of luminuous i.e. baryonic matter which we see via electromagnetic waves. Current research shows that luminuous matter constitutes only about 4% of the total density of cosmic matter. Matter graspable through observation thus represents minority and rarer matter. The rest consists of an entity of unknown composition which is divided into dark matter and dark energy. Their compositions remain unknown though there are assumptions that the dark matter includes also a very weakly luminuous component - a baryonic part. Dark matter has not been directly observed, nevertheless, its existence is assumed deductively on the basis of empirically verified data. "Dark matter helps to change scattered interstellar gas into compact luminuous stars."[11] (*own translation*) Existence of the dark matter is also indicated by information obtained from measurements of circumferential velocities of stars in spiral galaxies which also explains evolving and moving

[10] Melo, Ivan; *Tmavá energia, zrýchlenie a plochosť vesmíru*, In: *Pokroky matematiky, fyziky a astronomie*, Vol. 46 (2001), No. 2, p. 91.

[11] Petrovič, Pavol; *Fyzika II.*, Košice, 2012, ISBN 978-80-8143-010-7, p. 95

of galaxies in clusters. "There have been attempts to avoid this postulate via application of suitable mystification of the Newton gravity law (the so called MOND theory – Modified Newton dynamics) introduced by physicist M.Milgrom in 1983, however, majority of physicists still prefers the existence of dark matter."[12] (*my translation*)

Dark energy is an extraordinarily remarkable topic of contemporary cosmology. With respect to dark energy it is important to emphasize the fact that "the existence of large amounts of dark energy with negative pressure causes acceleration of the expansion of the universe the ultimate fate of which is still hidden behind unknown properties of dark energy."[13] (*my translation*) New observations have shown that the universe is not dominated by dark matter but by dark energy. Supernovas of the Ia type are reliable indicators of universe expansion. The most probable factor responsible for accelerated expansion of the universe is dark energy. Cosmological constant λ is considered the simplest explanation of dark energy. Besides the cosmological constant, adherents of the string theory take into account also some of the additional particles of the string theory, however, this theory "does not offer any concrete answer to the question which of the number of candidates are the genuine ones."[14] (*my translation*) Carrier of the dark energy thus remains unknown. The simplest solution to the above question is to consider the cosmological constant the carrier of dark energy. Remarkably, Gennady Shipov, Russian theoretical physicist, points at parallel existence of positive and negative energy. "In accordance with the laws of symmetry, Ja. Terleckij emphasized the assumption that in every physical field with positive density of energy field $\rho^{+}>0$ there is a double of the field with negative density $\rho^{-}<0$. From the above assumption it emerges that particles of both positive and negative matter must originate at origination of particles from vacuum having zero average energy and zero average momentum."[15] (*my translation*)

CONCLUSION

There are many aspects in which Mach´s principle inspired Einstein to formulate the general theory of relativity. In relation to motion of material sources he speaks of inclusion of energy and gravitational waves momentum. "The Machian origin of the cosmological constant was fully enlightened by Yakov Zeľdovich, who showed that λ can be identified with the energy density of vacuum."[16] Ernst Mach and some of his physical ideas are thus still alive thanks to renaissance of the cosmological constant.

Ernst Mach looks at the issue of knowability from a viewpoint we can name skepticism. Mach believed a body is a complex of sensations. In his opinion, sensations are real whereas the reality of the carrier of sensations has not been proven. Reality rests in sensations. "The

[12] Krempaský, Július; *Čím žije súčasná fyzika?*, In: *Poznatky modernej fyziky a ich aplikácia do vyučovania fyziky*, Ružomberok, 2011, ISBN 978-80-8084-798-2, p. 14.

[13] Melo, Ivan; *Tmavá energia, zrýchlenie a plochosť vesmíru*, In: *Pokroky matematiky, fyziky a astronomie*, Vol. 46 (2001), No. 2, p. 89.

[14] Petrovič, Pavol; *Fyzika II.*, Košice, 2012, ISBN 978-80-8143-010-7, p. 99.

[15] Shipov, Genadii, I.; Theory of Physical Vacuum, Moskow, 1993, ISBN 5-7273-0011-8, p. 91.

[16] Stuchlík, Zdeněk, *Influence of the Dark Energy on Astrophysical Phenomena in Active Galactic Nuclei,* In: Dub, Peter; Musilová, Jana; (eds.), Ernst Mach, Fyzika filosofie vzdělávání, Brno, Muni Press, 2010, ISBN 978-80-210-48, p. 151.

question which is often asked, whether the world is real or whether we merely dream it, is devoid of all scientific meaning. Even the wildest dream is a fact as much as any other."[17] Mach did not differentiate between appearance and reality.

Investigative Mach does not consider reality of the world a scientific problem. What matters is the fact that dark matter, dark energy as well as cosmological constant λ provide satisfactory answers to scientific questions. In his work the issue of knowability of the world by science is dealt with in accordance with subjectivity of perception of the world´s reality. If complexes of sensations form a body, then bodies consist of mere elements of sensations and science is only a process of adjusting thoughts within relevant field of experience. Mach the solipsist would thus consider the reality of dark matter and dark energy a pseudoproblem because empirical field represented by physics has a specific number of reasons enabling the assumption of given entities which fact, after all, does not contradict the Mach´s principle.

REFERENCES

Adámek, Jiří; Koubek, Václav; Reiterman, Jan; *Základy obecné topologie*, Praha, 1977

Akčurin, I. A.; *Matematizácia ako princíp jednoty fyzikálnych teórií*, In: Metodologické princípy fyziky, Bratislava, 1984.

Barbour, J.; Pfister, H. (eds.); *Mach Principle: From Newton´s Bucket to Quantum Gravity*. Birkhauser, Boston, 1995.

Bičák, Jiří; *Machův princip v současné kosmologii*, In: Dub, Peter; Musilová, Jana; (eds.), Ernst Mach, Fyzika filosofie vzdělávání, Brno, Muni Press, 2010, ISBN 978-80-210-48.

Ferris, Timothy; *Všetko o vesmíre*, Bratislava, 2005, ISBN 80-88993-85-7.

Feynman, Richard P.; Leighton, Robert B.; Sands, Mathew; *Feynmanovy přednášky z fyziky 1*, Havlíčkův Brod, 2000, ISBN 80-7200-405-0.

Kirshner, Robert, P.; *Výstřední vesmír*, Praha, 2005, ISBN 80-7185-729-7.

Krempaský, Július; *Čím žije súčasná fyzika?*, In: *Poznatky modernej fyziky a ich aplikácia do vyučovania fyziky*, Ružomberok, 2011, ISBN 978-80-8084-798-2.

Mach, Ernst; *Analýza pocitov*, In: *Antológia z diel filozofov č. 7, Pozitivizmus, voluntarizmus, novokantovstvo*, Bratislava, 1967.

Melo, Ivan; *Tmavá energia, zrýchlenie a plochosť vesmíru*, In: *Pokroky matematiky, fyziky a astronomie*, Vol. 46 (2001), No. 2.

Novotný, Jan; *Ernst Mach, Kurt Gödel a vesmír*, In: XVI International Colloquium on the Management of Educational Process, Brno, 2008, ISBN 978-80-7231-511-6

Petrovič, Pavol; *Fyzika II.*, Košice, 2012, ISBN 978-80-8143-010-7.

Stuchlík, Zdeněk, *Influence of the Dark Energy on Astrophysical Phenomena in Active Galactic Nuclei,* In: Dub, Peter; Musilová, Jana; (eds.), Ernst Mach, Fyzika filosofie vzdělávání, Brno, Muni Press, 2010, ISBN 978-80-210-48.

Шипов, Генадий, И.; *Теория физического вакуума*, Москва, 1993, ISBN 5-7273-0011-8.

[17] Mach, Ernst; *Analýza pocitov*, In: *Antológia z diel filozofov č. 7, Pozitivizmus, voluntarizmus, novokantovstvo*, Bratislava, 1967, s. 204.

In: Recent Advances in Cosmology
Editors: Anderson Travena and Brady Soren
ISBN: 978-1-62417-943-3

Chapter 3

GEOMETRIZATION OF MATTER: A NOVEL PARADIGM IN GENERAL RELATIVITY*

Ram Gopal Vishwakarma†
Unidad Académica de Matemáticas
Universidad Autónoma de Zacatecas,
Zacatecas, ZAC, Mexico

Abstract

Theory of general relativity has been agreed upon by experts for almost a century and is believed to describe accurately all gravitational phenomena ranging from the solar system to the Universe. However, there is a price for this success which is often ignored. More than 95 percent of the content of the energy-stress tensor has to be dark, in the form of inflaton, dark matter and dark energy, which do not have any non-gravitational or laboratory evidence. Moreover, the dark energy poses a serious confrontation between fundamental physics and cosmology.

The present situation reminds us of Einstein's 'biggest blunder' when he forced his theory to predict a static Universe. It seems that we are making a similar blunder by forcing the energy-stress tensor into the field equations while the observations indicate that it is not needed. We seem to have a deep-rooted conviction that the spacetime will remain empty unless we 'fill' it by the energy-stress tensor. However, we have been ignoring numerous evidences telling earnestly that the so-called vacuum field equations $R^{\mu\nu} = 0$ do not necessarily represent an empty spacetime. It is shown that the energy, momenta and angular momenta of the gravitational and the material fields do exist in the spacetime of $R^{\mu\nu} = 0$, whose effects are revealed through the geometry.

This may appear orthogonal to the usual understanding. Nevertheless, it is in striking agreement with the theory and observations at all scales, without requiring the dark sectors of the standard approach. Moreover, the new finding evades the long-standing problems of the standard cosmology, besides providing natural explanations to some unexplained puzzles.

PACS 04.20.Cv, 04.20.-q, 95.30.Sf, 98.80.Jk

Keywords: GR - fundamental problems and general formalism - cosmological observations

---

*Based on an essay (arXiv:1206.2795) awarded 'HONORABLE MENTION' of the year 2012 by the Gravity Research Foundation.

†E-mail address:vishwa@matematicas.reduaz.mx

1 Introduction

Einstein's theory of general relativity (GR), which provides the current description of gravitation in modern physics, ranks as one of the crowning intellectual achievements in theoretical physics. It is a geometric theory of gravitation which describes gravity not as a 'force' in the usual sense but as a manifestation of the curvature of spacetime. In particular, the curvature of spacetime is directly related to the energy-stress tensor $T^{\mu\nu}$ through the Einstein field equations

$$R^{\mu\nu} - \frac{1}{2} R\, g^{\mu\nu} = -\frac{8\pi G}{c^4} T^{\mu\nu}, \quad (\mu, \nu = 0, 1, 2, 3). \tag{1}$$

It should be noted that equations (1) attribute the source of curvature entirely to matter, as the tensor $T^{\mu\nu}$ does not include the energy, momenta and stresses associated with the gravitational field itself, though it does incorporate all the candidates of material fields including the dark energy.

Although GR is not the only relativistic theory of gravitation, it is the simplest theory that has survived the tests of nearly a century of observational confirmation ranging from the solar system to the largest scales, the Universe itself. However, this success is achieved provided we admit three completely independent new components in the energy-stress tensor – inflaton, dark matter and dark energy, which are believed to play significant roles in the dynamics of the Universe during their turns. However, there is, until now, no direct observational or laboratory evidence for any of these dark sectors. Additionally, the mysterious dark energy, which has been conjured up primarily to fit the observations of supernovae of type Ia (SNeIa), poses a serious confrontation between fundamental physics and cosmology.

One may notice that all the above-mentioned dark ingredients are introduced in the standard cosmology through the energy-stress $T^{\mu\nu}$, eschewing which, the Einstein's field equations reduce to the vacuum field equations

$$R^{\mu\nu} = 0. \tag{2}$$

A curious proposal allowing a possible way out of the difficulties, would be then a theory wherein $T^{\mu\nu}$ is absent, for example, equations (2) in the framework of GR. In this context, it would be interesting to note that all the observations, ranging from the solar system to the Universe, can be explained successfully in the framework of equations (2), without taking refuge in the dark sectors of the standard cosmology (as we shall see in the following). However, this success gained by equations (2), would not have any value unless there is a possibility to represent matter in the framework of equations (2) without taking recourse to $T^{\mu\nu}$. In this connection, it would be important to note (as will be established in the following) that equations (2) do not represent an empty space and the sources of gravitation do exist there in the form of the geometry itself, implying that the energy-stress tensor $T^{\mu\nu}$ is redundant.

This reminds us of Einstein's 'biggest blunder' when he forced his equations to predict a static Universe (by imposing the cosmological constant) though a more natural implication of GR - the expanding Universe - was already known to him. While Einstein's blunder was perhaps motivated by his religious conviction that the Universe must be eternal and unchanging, in the present case it is one's deep-rooted conviction that the space would remain empty unless it is filled by $T^{\mu\nu}$. However, the rejection of $T^{\mu\nu}$ from equations (1) is further supported by a recent study [1] which discovers various inconsistencies and paradoxes in the basic formulation of the energy-stress tensor $T^{\mu\nu}$, concluding that like a proper energy-stress tensor of the gravitational field, a clean relativistic formulation of matter fields neither exists[1]. On the other hand, the new paradigm based on equations (2), is not only supported by various theoretical and observational evidences, but also circumvents the long-standing problems of the standard cosmology, besides resolving some unexplained puzzles (as we shall see in the following).

Then, let us explore this new paradigm starting from the study of the observational supports given to the 'vacuum' field equations, by asking the following question. What are the observations/experiments which have directly tested equations (1)? The classical tests of GR consider $T^{\mu\nu} = 0$. The same is true for the more precise tests of GR made through the observations of the radio pulsars, which are rapidly rotating strongly magnetized neutron stars. The pulsar tests assume the neutron stars as point-like objects and look for the relativistic corrections in the post-Keplerian parameters by measuring the pulsar timing. The tests do not even require to know the exact nature of the matter that pulsars and other neutron stars are made of. As $T^{\mu\nu} = 0$ implies $T = 0 = R$ in which case equations (1) reduce to (2), all we can claim is that it is only equations (2) which have been verified by the classical tests of GR. As these tests have been limited to our galaxy only, let us see how these equations fair against the cosmological observations. For this purpose, let us first solve equations (2) for a homogeneous and isotropic spacetime, as is revealed from the observations on a large enough scale. Obviously, the considered symmetry of homogeneity and isotropy requires the metric to be the Robertson-Walker one given by

$$ds^2 = c^2dt^2 - S^2\left(\frac{dr^2}{1-kr^2} + r^2d\theta^2 + r^2\sin^2\theta\, d\phi^2\right), \tag{3}$$

where $S(t)$ is the scale factor of the Universe. For the metric (3), equations (2) yield

$$R^0_{\ 0} = \frac{3}{c^2}\frac{\ddot{S}}{S} = 0, \tag{4}$$

$$R^1_{\ 1} = R^2_{\ 2} = R^3_{\ 3} = \frac{1}{c^2}\left(\frac{\ddot{S}}{S} + 2\frac{\dot{S}^2}{S^2} + 2kc^2\frac{1}{S^2}\right) = 0, \tag{5}$$

which uniquely determine

$$S = ct \text{ with } k = -1, \tag{6}$$

so that the final solution reduces to

$$ds^2 = c^2dt^2 - c^2t^2\left(\frac{dr^2}{1+r^2} + r^2d\theta^2 + r^2\sin^2\theta\, d\phi^2\right). \tag{7}$$

[1]The details of this study are discussed in section 4.

2 Support from the Cosmological Observations to $R^{\mu\nu} = 0$

2.1 Observations of supernovae of type Ia

A supernova of type Ia (SNIa) occurs when a carbon-oxygen white dwarf star in a binary system accretes enough mass from its companion to reach a critical mass and hence undergoes a thermonuclear explosion in its core. Because of the near uniformity of the mass of the white dwarf stars, the SNeIa produce nearly the same peak luminosity. This allows them to be used as standard candles to measure the distance to their host galaxies because the apparent magnitude of the SNeIa depends primarily on the (luminosity) distance. As the distances are model-based quantities in cosmology, one can use the high-redshift observations to test and compare the models.

The solution (6) is efficient enough to define uniquely, without requiring any inputs from the matter fields, the luminosity distance $d_{\rm L}$ of a source of redshift z by

$$d_{\rm L} = cH_0^{-1}(1+z)\sinh\{\ln(1+z)\}, \tag{8}$$

where H_0 represents the present value of the Hubble parameter $H = \dot{S}/S$.

The over-faintness of SNeIa, for their redshift, is usually explained in terms of an accelerating expansion caused by the dark energy. However, this is not the only explanation. For example, it is already known that the coasting model, based on solution (6), albeit non-accelerating (neither deceleration), is also consistent with the observations of SNeIa *without requiring any dark energy* (see, for example, [2]). In [3], Perlmutter et al. have found from the fitting of their data on the SNeIa that the performance of the empty model ($\Omega_{\rm m} = 0 = \Omega_\Lambda$) is practically identical to that of the best-fit unconstrained cosmology with a positive Λ. Let us consider a newer dataset, for example, the 'new gold sample' of 182 SNeIa [4][2], which is a reliable set of SNeIa with reduced calibration errors arising from the systematics. It can be checked that the model (8) provides an excellent fit to the data with a value of χ^2 per degrees of freedom (DoF) $= 174.29/181 = 0.96$ and a probability of the goodness of fit $Q = 63\%$. Obviously the standard ΛCDM model has even a better fit as it has more free parameters: χ^2/DoF $= 158.75/180 = 0.88$ and $Q = 87\%$ obtained for the values $\Omega_{\rm m} = 1 - \Omega_\Lambda = 0.34 \pm 0.04$. The best-fitting model (8) and the ΛCDM model have been compared with this sample of data in Fig. 1.

2.2 Observations of High-Redshift Quasars

Let us now consider the data on the angular size and redshift of 256 quasars with their redshift in the range 0.5–3.8 compiled by Jackson and Dodgson [6]. These data points were selected from a bigger sample of 337 ultracompact radio sources originally compiled by Gurvits [7]. These sources, of angular sizes of the order of a few milliarcseconds (ultracompact), were measured by the very long-baseline interferometry. The sample of Jackson and Dodgson is composed of short-lived quasars deeply embedded inside the galactic nuclei, which are expected to be free from evolution on a cosmological time scale and thus comprise a set of standard rods, at least in a statistical sense. These sources are distributed

[2]Although various newer SNeIa datasets are available, however, the way they are analyzed has left little scope for testing a theoretical model with them. This issue has been addressed in [5].

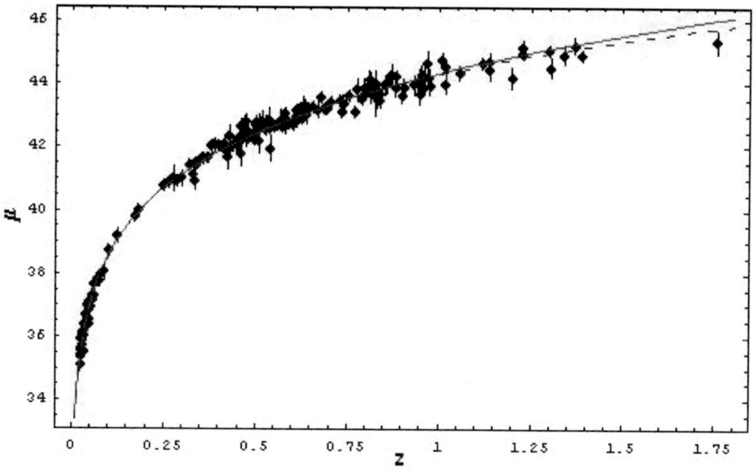

Figure 1: The 'new gold sample' of 182 SNeIa from Riess et al [4] is compared with some best-fitting models. The solid curve corresponds to the model (8) and the dashed curve corresponds to the spatially-flat ΛCDM model $\Omega_{\rm m} = 1 - \Omega_\Lambda = 0.34 \pm 0.04$.

into 16 redshift bins, each bin containing 16 sources. This compilation has recently been used by many authors to test different cosmological models [2, 8]. In order to fit the data to the model, we derive the $\Theta - z$ relation in the following.

The (apparent) angular size Θ of a source, of the proper diameter *d*, is given by

$$\Theta(z) = \frac{0.0688dh}{H_0 d_{\rm A}} \text{milliarcseconds}, \tag{9}$$

where *d* is measured in pc, h is the present value of the Hubble parameter in units of 100 km s^{-1} Mpc^{-1}, and the angular diameter distance $d_{\rm A} = d_{\rm L}/(1+z)^2$.

We find that the present model has a satisfactory fit to the data with χ^2/DoF$= 20.78/15 = 1.39$ and $Q = 14\%$. In order to compare, we find that the best-fitting ΛCDM model has a slightly better fit: χ^2/DoF $= 16.03/14 = 1.15$ and $Q = 31\%$ obtained for the values $\Omega_{\rm m} = 1 - \Omega_\Lambda = 0.21 \pm 0.08$. These models are shown in Fig. 2.

2.3 Observations of H_0 and the Oldest Objects

The age of the Universe t_0 is the time elapsed since the Big Bang. It depends on the expansion dynamics of the model and is given by

$$t_0 = \int_0^\infty \frac{H^{-1}(z)}{1+z} dz. \tag{10}$$

Hence, the Hubble parameter controls the age of the Universe, which in tern depends on the free parameters of the model. For example, by the use of the Friedmann equation, equation (10) reduces to the following in the standard cosmology:

$$t_0 = \frac{1}{H_0} \int_0^\infty \frac{dz}{(1+z)\sqrt{\Omega_{\rm m}(1+z)^3 + \Omega_\Lambda - \Omega_{\rm k}(1+z)^2}}. \tag{11}$$

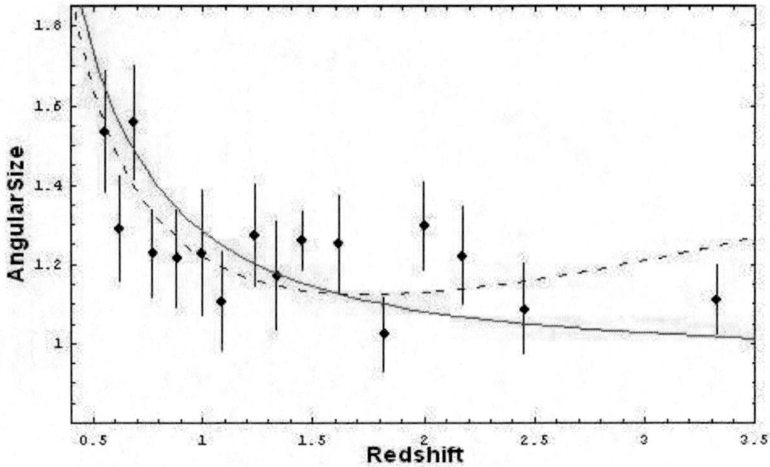

Figure 2: The data on the ultra compact radio sources [6] is compared with some best-fitting models. The solid curve corresponds to the model represented by equation (7) and the dashed curve corresponds to the spatially-flat ΛCDM model $\Omega_m = 1 - \Omega_\Lambda = 0.21 \pm 0.08$.

Although t_0 is a model-based parameter, a lower limit is put on it by requiring that the Universe must be at least as old as the oldest object in it. This is done through t_{GC}, the age of the globular clusters in the Milky Way which are among the oldest objects we so far know. The parameter H_0 can be estimated in a model-independent way, for example, from the observations of the low-redshift SNeIa, in which case the predicted magnitude does not depend on the model-parameters. One can use this value to calculate the age of the Universe in a particular theory which is to be compared with the age of the oldest objects. Thus the measurements of H_0 and t_{GC} provide a powerful tool to test the underlying theory.

In order to check the efficiency of this test, let us apply it to the Einstein-deSitter model ($k = \Lambda = 0$), for example. By using the current measurements of $H_0 = 71 \pm 6$ km s^{-1} Mpc^{-1} from the Hubble Space Telescope Key Project [9] in equation (11), we get t_0 in the Einstein-deSitter model as 9.18 Gyr. This cannot be reconciled with the age of the oldest globular cluster estimated to be $t_{GC} = 12.5 \pm 1.2$ Gyr [10] and the age of the Milky Way as 12.5 ± 3 Gyr coming from the latest uranium decay estimates [11]. However, for the concordance ΛCDM model with $\Omega_m = 1 - \Omega_\Lambda = 0.27$ (as estimated by the WMAP project [12]), equation (11) gives a satisfactory age of the Universe $t_0 = 13.67$ Gyr which is well above the age of the globular clusters. The age of the Universe in the present model is given by $t_0 = H_0^{-1}$, as can be checked from (6). For the above-mentioned value of H_0, this gives $t_0 = 13.77$ Gyr which is even higher than the concordance model value.

2.4 Observations of Starburst Galaxies

Let us now consider the data on the apparent magnitude and redshift of the starburst galaxies. Recent work has indicated that HII starbusrt galaxies might be considered as standard candles because of a correlation between their velocity dispersion, H luminosity, and metallicity (see, for example, [13] and the references therein). Siegel et al. [14] compiled

Table 1. Magnitudes, with uncertainties, of 13 HII starburst galaxies

Galaxy	z	$\mu \pm \sigma$
Q0201-B13	2.17	$47.49^{+2.10}_{-3.43}$
Q1623-BX432	2.18	$45.45^{+1.97}_{-3.07}$
Q1623-MD107	2.54	$44.82^{+0.31}_{-1.58}$
Q1700-BX717	2.44	$46.64^{+0.31}_{-1.58}$
CDFa C1	3.11	$45.77^{+0.31}_{-1.58}$
Q0347-383 C5	3.23	$47.12^{+0.44}_{-0.32}$
B2 0902+343 C12	3.39	$46.96^{+0.71}_{-0.81}$
Q1422+231 D81	3.10	$48.81^{+0.38}_{-0.40}$
SSA22a-MD46	3.09	$46.76^{+0.56}_{-0.51}$
SSA22a-D3	3.07	$49.71^{+0.43}_{-0.41}$
DSF2237+116a C2	3.32	$47.73^{+0.25}_{-0.25}$
B2 0902+343 C6	3.09	$45.22^{+1.38}_{-1.76}$
MS1512-CB58	2.73	$47.49^{+1.22}_{-1.57}$

a sample of 15 HII-like starburst galaxies with redshifts in the range 2.17 - 3.39 (by using the data available in the literature) in order to constrain $\Omega_{\rm m}$. Mania and Ratra [13] modified this sample by excluding two HII galaxies (Q1700-MD103 and SSA22a-MD41) that show signs of a considerable rotational velocity component and used the resulting sample (shown in Table 1) to constrain the cosmological models of dark energy.

For this sample, the present model provides the minimum value of χ^2/DoF= 53.54/12=4.46 with $Q = 3.4 \times 10^{-7}$, whereas the standard ΛCDM model gives the minimum χ^2/DoF= 53.32/10=5.33 with $Q = 6.5 \times 10^{-8}$ for the values $\Omega_{\rm m} = 0.19$, $\Omega_\Lambda = 0.98$. It would not be fair to claim that any of these models fits the data well. Perhaps the inherent scatter of the data is large (as is clear from Fig. 3), and a large sample size is required to perform the test and to get any meaningful constraint on the cosmological parameters. However, this much is clear from the fitting results that the performance of the present model vis-a-vis that of the standard model, is better. The fitting results are shown in Fig. 3.

2.5 Observations of CMB

Finally, let us study the compatibility of the present model with the observations of the cosmic microwave background radiation (CMB). This background radiation is formed from the photons decoupled from the baryon-photon plasma at the redshift $z_{\rm dec} = 1100$. As the photons were in thermal equilibrium with matter in the plasma before decoupling, the fingerprints of the inhomogeneity in the matter during the epoch of decoupling, is imprinted on the radiation in the form of small fluctuations in its temperature with respect to small variation in the directions in the sky. Since the Universe is optically thin after this epoch, this information remains frozen in the radiation and is observed today. Hence the observed anisotropy in the CMB temperature can be quantified in terms of the size of the structures on the surface of the 'last scatter'. For example, a region which has a proper size $L_{\rm dec}$ on

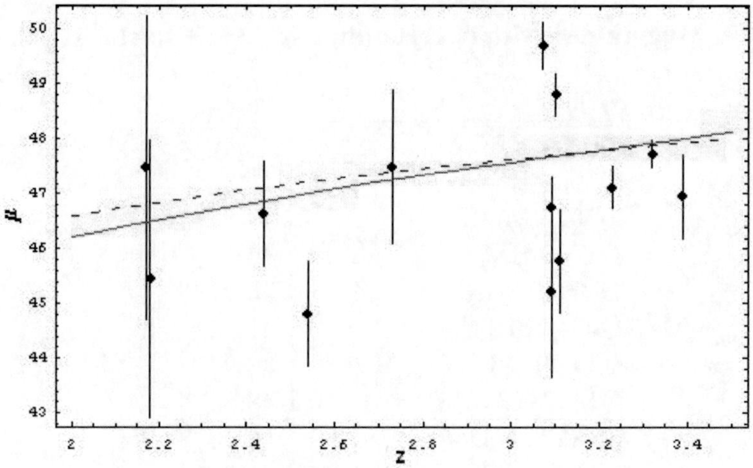

Figure 3: The HII-like starburst galaxy data [13] is compared with some best-fitting models. The solid curve corresponds to the model represented by equation (7) and the dashed curve corresponds to the ΛCDM model $\Omega_{\rm m} = 0.19$, $\Omega_\Lambda = 0.98$.

the surface of the last scatter, will subtend an angle θ at the observer today, given by

$$\theta = \frac{L_{\rm dec}}{d_{\rm A}(z_{\rm dec})}, \tag{12}$$

where $d_{\rm A}(z_{\rm dec})$ is the (angular diameter) distance to the surface of decoupling. If one considers $L_{\rm dec}$ to be equal to the Hubble radius $d_{\rm Hub}(t_{\rm dec}) = cH^{-1}(t_{\rm dec})$, equation (12) gives $\theta \approx 1^{\rm o}$ in the standard cosmology. As the CMB appears, in the observations, to be highly isotropic on all angular scales greater than $1^{\rm o}$, the length scale of the order of $d_{\rm Hub}(t_{\rm dec})$, is usually interpreted in terms of a horizon (in particular, the sound horizon) giving the largest coherent structure in the Universe at $t_{\rm dec}$, since this should be the largest distance a sound wave in the tightly coupled baryon-photon fluid could have traveled since the Big Bang until the epoch of decoupling.

A word of caution is needed here. It should be noted that the Hubble radius defined by $d_{\rm Hub}(t) = cH^{-1}(t)$ (which is an arbitrary definition) provides only a characteristic distance scale in the Universe at t and is *not* a horizon, as it does *not* have any causal significance (since it does not arise naturally from any light propagation formula). Although in the standard cosmology, the particle horizon (as well as the event horizon) have radii comparable to the Hubble radius $d_{\rm Hub}$, there are other cosmological models (for example, the present model) which do not have any horizon.

Another important point to be noted is that if we interpret $L_{\rm dec}$, appearing in (12), as the sound horizon, giving the size of the largest coherent region on the last scattering surface in which the homogenizing signals passed at sound speed, then the standard cosmology predicts large anisotropies (not isotropy!) for angular scales greater than $\approx 1^{\rm o}$ - a result in contradiction with what is observed, as has been noted by Weinberg [15] and Narlikar [16]. Hence, it seems that the isotropy of the CMB cannot be explained in terms of some physical process operating under the principle of causality in the standard paradigm [15]. It is generally believed that inflation made the Universe smooth and left the seeds of structures, on the surface of the last scatter, of the order of the Hubble radius at that time[3].

Thus all one can say, permitted by the present situation, is that the CMB observations fix the length scale of the inhomogeneity on the surface of last scatter, which can be estimated in terms of the Hubble radius $d_{\rm Hub}(t_{\rm dec})$. For example, this length scale $L_{\rm dec}$ can be written as

$$L_{\rm dec} = \eta d_{\rm Hub}(t_{\rm dec}) = \eta c H^{-1}(t_{\rm dec}), \tag{13}$$

where the constant parameter η can be estimated from the observations of the CMB. Particularly, this can be estimated accurately by using the angular scale of the first peak in the observed angular power spectrum of the CMB, which is supposed to give, with a high precision, the physical scale of the density contrast during the epoch of decoupling. In terms of the Legendre multipole ℓ, where $\ell = \pi/\theta$, the WMAP observations [12] give the location of the first peak at $\ell = 220$. This is equivalent to $\theta = 0.82^{\rm o}$. Hence, we have to solve equation (12), taken together with (13), for $\theta = 0.82^{\rm o}$, which would be equivalent to fitting equation (12) (taken together with (13)) to the first peak in the angular power spectrum of the CMB observed by WMAP.

This solution, in the concordance ΛCDM model ($\Omega_{\rm m} = 1 - \Omega_\Lambda = 0.27$), yields the value $\eta = 0.82$. By considering $H_0 = 71$ km s^{-1} Mpc^{-1}, this gives the size of the structures at $t_{\rm dec}$ as $L_{\rm dec} = 182.4$Kpc. In the framework of the present model, the solution yields $\eta = 7.86$, giving $L_{\rm dec} = 30.14$ Mpc. A larger $L_{\rm dec}$, in the present model, is a characteristic of a higher expansion rate, which also results in a larger distance to the surface of the last scatter $d_{\rm A}(z_{\rm dec}) = 2.11$ Gpc, compared with $d_{\rm A}(z_{\rm dec}) = 12.78$ Mpc in the concordance model.

The length scale $L_{\rm dec}$ is expected to grow, due to the cosmic expansion, to a proper length L_0 today, given by $L_0 = (1 + z_{\rm dec})L_{\rm dec}$. Thus the present size of our patch of homogeneity and isotropy, is 33.19 Gpc in the present model compared with 200.82 Mpc in the concordance model. It would be worthwhile to mention here that Grishchuk [18] has found out, by combining available observations with plausible statistical assumptions, that the present size of this patch should be significantly bigger than the present Hubble radius[4]. He evaluated it to be about 500 times bigger. This result is more favourable to the present model, which predicts a larger size of our patch of homogeneity and isotropy, than the standard cosmology.

We have noted that the overall isotropy of the CMB cannot be explained in the standard paradigm in terms of some homogenization process taken place in the baryon-photon plasma operating under the principle of causality. However, the present model can successfully explain this, as the whole Universe is always causally connected in this model (as we shall see later). Thus the presence of the seeds of structures at the last scattering surface, finds its origin in the random quantum fluctuations in the primordial density of the early Universe (in a more natural way than the standard paradigm) without getting perturbed by the speculative inflation.

---

[3]It may be a matter of debate that if inflation made everything else smooth, why it left this significant signature [16]. It may also be noted that the growth of density perturbations, in the standard cosmology, pose difficulties for the inflationary predictions when confronted with the CMB observations [17].

[4]As the size of the horizons in the standard cosmology is of the order of the Hubble radius, the size of the observable (causally connected) Universe is regarded of the order of the Hubble radius (≈ 4.2 Gpc).

Although the angular power spectrum of the CMB also carries informations contained in the other peaks (for example, the ratio of the odd peaks to the even peaks determines the baryon density while the third peak can be used to get information about the dark matter density), however, it would not be worthwhile to follow blindly the standard strategy (followed in the standard cosmology) to calculate the finer details of the matter composition when we do not know about its dynamics in the present model. Additionally, as no constraint is put on the matter sector by any observation discussed above (since matter does not play a direct role in the dynamics of the present model), the constraints, if any, on the matter sector derived from the other peaks, would not be in conflict with the other observations.

3 Gravity of the 'Vacuum' Field Equations $R^{\mu\nu} = 0$

The above-described success in explaining the various cosmological observations in the framework of equations (2), would not have any value until there is a scope to represent the real Universe with matter in the framework of these equations. In order to explore this possibility, let us scrutinize the source of curvature and its existence in Einstein's equations.

The source of curvature/gravitation in GR is regarded as the tensor $T^{\mu\nu}$, which includes in it all the sources of energy, momenta and stresses including the cosmological constant or other dark energy candidates (if any) *except the energy, momenta and stresses associated with the gravitational field itself*. However, the spacetime in GR is not a passive background but becomes a dynamical physical entity that may be deformed, stretched and even spread in the form of gravitational waves. So, it must contain energy, momenta and stresses. Realizing this fact, Einstein tried to include the energy-stress tensor of the gravitational field in the right hand side of equation (1) emphasizing that "*the gravitational field must have an energy-stress tensor, like all other physical fields*". However, failing to find a proper energy-stress tensor of the gravitational field, he rejected this requirement in the final version of GR (in order to have a generally covariant theory) by declaring *"there may very well be gravitational fields without stress and energy density"*.

It can be said safely that despite the century-long dedicated efforts of many luminaries, the attempts to discover a unanimous formulation of the gravitational field energy in GR, has failed[5] due to the following three reasons: (i) the non-tensorial character of the energy-stress 'complexes' (pseudotensors) of the gravitational field; (ii) the lack of a unique agreed-upon formula for these pseudotensors (for example, five different definitions of the 'complex', viz., from Tolman, Papapetrou, Landau-Lifshitz, Moller and Weinberg exist, which may lead to different distributions even in the same spacetime background); (iii) the inherent difficulty in the localization of the gravitational energy.

[5]Inspired by the pioneering works of Einstein and Tolman on the total energy of physical systems, there have been attempts (see, for example, [19]) to assign meaningful notions to the pseudotensor-representation of the gravitational field. However, they often lead to paradoxes and inconsistencies. For example, from equation (88.4) and the ensuing discussion in Tolman's book (page 227 of [20]), one can show that if the gravitating system is a perfect fluid, the energy density of the resulting gravitational field $= 3\times$ the pressure of the fluid. Apart from posing Tolman-like paradoxes, this leads to a funny situation that the gravitational field is due to the pressure of the gravitating matter and in the case of a pressureless source, there is no gravitational field, though the very origin of the field, the (pressureless) matter, is still present! If this was correct, the Schwarzschild spacetime (around a pressureless mass) would have been necessarily flat!

It is generally agreed that the local gravitational energy density has ‘no weight’ (as gravity can be made locally zero in a suitable coordinate system, it is believed that there should be no energy density associated with it) and so, it does not serve as a source term to curve the spacetime [21]. It can then be concluded that according to the conventional wisdom, it is only the matter energy-stress tensor $T^{\mu\nu}$ in (1), which derives its physical significance by curving spacetime and which is the very source of the curvature and hence of gravitation.

Hence in the absence of $T^{\mu\nu}$, in which case equations (1) reduce to the so-called ‘vacuum’ field equations (2), one should expect essentially a flat spacetime. However, it is already known that this is not true, and a space with dimensions four or more can have curvature even for $T^{\mu\nu} = 0$. Let us consider the following well-known examples.

Schwarzschild solution: Discovered by Karl Schwarzschild in 1915 immediately after GR was formulated, the solution forms the corner-stone of GR. It represents the spacetime structure outside an isotropic mass in an empty space [22]:

$$ds^2 = \left(1 + \frac{K}{r}\right) c^2dt^2 - \frac{dr^2}{(1 + K/r)} - r^2d\theta^2 - r^2 \sin^2\theta \, d\phi^2, \tag{14}$$

where K is a constant of integration. In fact, all the experiments which have so far been carried out to test GR (except the Gravity Probe B experiments, which are based on the predictions of the Kerr solution), are based on the predictions by this solution.

Kerr solution: Discovered by Roy Kerr in 1963, the solution describes the spacetime surrounding a spherical mass m spinning with angular momentum per unit mass = α (so that its total angular momentum $= m\alpha c$). In the Boyer-Lindquist coordinates [22], the solution takes the form

$$ds^2 = \left(1 - \frac{r_S r}{\rho^2}\right) c^2dt^2 - \frac{\rho^2}{\Delta}dr^2 - \rho^2 d\theta^2$$

$$- \left(r^2 + \alpha^2 + \frac{r_S r \alpha^2}{\rho^2}\sin^2\theta\right)\sin^2\theta \, d\phi^2 + \frac{2r_S r\alpha}{\rho^2}\sin^2\theta \, d\phi \, cdt, \tag{15}$$

where $\rho^2 = r^2 + \alpha^2\cos^2\theta$, $\Delta = r^2 - r_S r + \alpha^2$ and $r_S = 2Gm/c^2$ is the Schwarzschild radius. When $\alpha = 0$, the solution reduces to the Schwarzschild solution.

The solutions (14) and (15) are intrinsically curved and they cannot be transformed to the flat Minkowski metric by any possible coordinate transformation, as the Riemann-Chistoffel curvature tensor $R_{\mu\nu\delta\sigma} \neq 0$ in these cases. Since $T^{\mu\nu}$ is considered as the sole source of gravitation/curvature in equations (1), appearance of curved solutions in its absence, indicates that some (unexpected) source must be present inherently in equations (2) itself. Let us unveil it.

3.1 Gravitational Energy Appears Through Geometry!

Let us first consider the Schwarzschild solution (14). The mystery of the presence of curvature in this solution is related to the central singularity of the spherically symmetric space represented by (14) (which is associated, through a correspondence between the Newtonian and Einsteinian theories of gravitation in the case of a weak field, to an isotropic mass sitting at the centre $r = 0$). Thus, the source of the curvature in GR is either $T^{\mu\nu}$ or a singularity in $g_{\mu\nu}$. However, it should be noted that the metric (14) represents space exterior to the central mass at $r = 0$ and *not* the point $r = 0$ itself, where the metric breaks down. So, how can a mass situated at the point $r = 0$ (which is not even represented by the metric) curve the space of (14) at the points for which $r > 0$? Obviously, one cannot expect the Newtonian theory of action-at-a-distance to work in the framework of GR which is a local theory.

A little reflection suggests that the agent responsible for the curvature in (14) at the points for $r > 0$, must be the gravitational energy, which can definitely exist in an empty space.

However, if this is true, one should be able to calculate the gravitational energy from the metric (14). We can certainly do this by the following two simple observations:

1. The metric (14) departs from the flat spacetime in the term K/r, implying that this term must be having the source of curvature.

2. We have shown that source of curvature in (14), at the points $r > 0$, must be the gravitational energy.

Taken together, these two points imply that K/r must be the gravitational energy (in the units with c=1) in (14). This is in perfect agreement with the way the value of the constant K is determined. Let us recall that the constant K in equation (14) is specified in terms of the Newtonian gravitational potential energy (by requiring that in the case of a weak gravitational field, Newton's law should hold) giving $g_{00} = 1 + 2\psi/c^2$ where $\psi = -Gm/r$ is the gravitational energy (per unit mass) at a distance r from the central mass m producing the field. This fixes the constant K as

$$K = -\frac{2Gm}{c^2}. \tag{16}$$

It is thus established that the source of the curvature of spacetime in (14) is the energy of the gravitational field present at the points exterior to $r = 0$. This also establishes beyond doubts that the gravitational energy does gravitate and it already exists in equations (2) implicitly in the geometry, through the non-linearity of the field equations, and no additional incorporation thereof is needed. This fits very well in the story of the failure to discover a true energy-stress tensor of the gravitational field. The energy-stress tensor of the gravitational field does not exist simply because the tensor is not needed in the geometric framework of gravitation, it already exists there inherently in the geometry! We should note that in the weak-field approximation, the GR equations do reduce to the usual Newtonian dynamical equations; gravitational energy, force, etc. do emerge respectively from the metric, the Christoffel symbol, etc., *without adding any formulation of the gravitational field energy to the Einstein equations.*

The above-gained insight about the futility of the energy-stress tensor of the gravitational field is also corroborated by the Kerr metric (15) wherein the angular momentum also contributes to its curvature. This is an entirely new concept for an energy-stress tensor (in the framework of GR), which has no place for the angular momentum.

One would not show much inhibition to agree that the gravitational energy is inherently present in equations (2) *without* incorporating any formulation thereof. However, one would maintain that equations (2) represent otherwise empty spacetime *outside* the matter source. This interpretation is supported by the fact that the Schwarzschild and Kerr solutions (14) and (15) represent space outside the source mass, and not the point where the mass exists. However, the interpretation of the curvature in terms of the gravitational energy in empty space, does not appear satisfactory in the case of the Kasner solution, as we shall see in the following.

Kasner solution: This important cosmological solution, discovered by Edward Kasner [23] in 1921, is given by

$$ds^2 = c^2dt^2 - t^{2p_1}dx^2 - t^{2p_2}dy^2 - t^{2p_3}dz^2, \tag{17}$$

where the constants p_1, p_2 and p_3 satisfy

$$p_1 + p_2 + p_3 = 1, \quad p_1p_2 + p_2p_3 + p_3p_1 = 0.$$

This curved Bianchi type I metric is believed to describe an 'empty' homogeneous Universe which expands and contracts anisotropically at different rates in different directions (for example, for $p_1 = p_2 = 2/3$ and $p_3 = -1/3$, the space is expanding in two directions and contracting in the third). The metric was rediscovered, in a more useful form, by V. V. Narlikar and K. R. Karmarkar [24] in 1946[6]:

$$ds^2 = c^2dt^2 - (1+nt)^{2p_1}dx^2 - (1+nt)^{2p_2}dy^2 - (1+nt)^{2p_3}dz^2, \tag{19}$$

where n is an arbitrary constant. It may checked that by the use of the transformations $1 + nt = n\bar{t}$, $n^{p_1}x = \bar{x}$, $n^{p_2}y = \bar{y}$, $n^{p_3}z = \bar{z}$, the metric (19) takes the form (17) in the new coordinates.

3.2 Matter Too Registers Its Presence Through Geometry!

The conventional interpretation of the Kasner solution (17) is obscure and questionable. As the solution is interpreted in terms of an empty homogeneous Universe, the only possible source (in the absence of matter) can be a singularity. The solution (17) does contain a singularity at $t = 0$ (at $t = -1/n$ in the case (19), which is just a rescaling of time[7]), but *not* at any other time. However, the solution is curved at all times! A past singularity, which does not exist now, fueling the gravitational energy now without any other source, does not seem compatible with our understanding of the gravitational energy.

[6]It would be intersting to note that the Kasner metric has been rediscovered yet again in 1969 by L. K. Patel and P. C. Vaidya [25] in the form

$$ds^2 = (1+nt)^4[c^2dt^2 - dx^2 - dy^2] - (1+nt)^{-2}dz^2. \tag{18}$$

By the use of the transformation $(1 + nt) = (3n\ \bar{t})^{1/3}$, the metric (18) can be written in the form (17) for $p_1 = p_2 = 2/3, p_3 = -1/3$ (from an unpublished work of Professor Abdus Sattar).

[7]As the singularity in (19) now appears at $t = -1/n$, it seems that it can be avoided with a positive n. However, as the average scale factor (= spatial volume$^{1/3}$ $= 1 + nt$) has the same behaviour in the interval $-1/n \le t \le 0$ as it has for $0 \le t$ (i.e., no bouncing behaviour at $t = 0$), this feature to push the singularity back before the time $t = 0$, appears just a rescaling of time and does not have any bearing on the removal of the singularity.

A more apparent source of curvature in the Kasner metric (17) is a net non-zero momentum resulting from the anisotropic expansion/contraction of the homogeneous space, since the space expands and contracts at different rates in different directions in it. However, what actually expands/contracts is sharply controversial in this interpretation: How is it possible for space, which is utterly empty, to expand/contract? How can nothing expand/contract? It does not make sense to think of expanding/contracting space *without* matter. Hence, the conventional interpretation of this solution has no physics, or wrong physics, and the Kasner solution has remained a 'curious' solution. The puzzle cannot be overlooked by arguing that all the solutions of Einstein's theory are not physically meaningful, and the Kasner solution is one of them. This would raise doubts over the general validity of the theory.

An important point regarding the Kanser metric, which has not been paid attention is that, unlike the Schwarzschild and Kerr metrics, the Kasner metric represents a cosmological solution, which is not expected to have any 'outside'. Since the ultimate source of the gravitational field is matter, the (homogeneously distributed) matter source, present at the time of singularity in the Kanser metric, must be present at all other times as well, as it *must not* have been destroyed mysteriously! This simply means that the Kasner metric represents a homogeneous distribution of matter expanding/contracting anisotropically! However, if this is correct, we should be able to decipher the presence of this matter from its geometry, as we could do in the case of the Schwarzschild metric. This can similarly be done from the following two points.

1. For $n = 0$, the metric (19) reduces to the Minkowski metric at any time t[8]. This implies that this term must be having the source of curvature in terms of a dynamic quantity.

2. The source of curvature in metric (19) is a net non-zero momentum, resulting from an anisotropic motion of the homogeneously distributed matter (space) expanding and contracting at different rates in different directions.

If this is correct, we should expect the constant n to be expressed in terms of the momentum density, say, P. It is indeed possible to express n in terms of P, G and c in order to meet its natural unit (which is the dimension of the inverse of time). Written along the lines of (16), the expression for n finds a unique expression

$$n = \sqrt{\frac{GP}{c}}, \tag{20}$$

which is in agreement with the two points mentioned above. It is difficult to verify (20) with a classical analogue (as we did in the case of the Schwarzschild metric) since in the Newtonian gravitational theory, the field is independent of the motion of the source. This is similar to the lack of a classical analogue of the Kerr metric.

Does it then mean that like the energy, momentum and angular momentum of the gravitational field, those of the matter fields are also included in equations $R^{\mu\nu} = 0$ inherently whose effects are revealed through the geometry? We shall see in the next section that only an affirmative answer to this question, can provide a consistent explanation to the flatness of the homogeneous-isotropic solution (7) of $R^{\mu\nu} = 0$.

[8]The same is not true with vanishing p_1, p_2 and p_3, where there is no free t and the metric looses its dynamic character.

3.3 A New Paradigm in GR

The above-obtained revolutionary result, that the gravitational as well as the material fields are built-in ingredients of the geometry of equations $R^{\mu\nu} = 0$, puts forward a question. Why is the solution (7) flat while the solutions (14, 15), (17−19), of the same equations (2), are curved? (It may be noted that by the use of the transformations $\bar{t} = t\sqrt{1+r^2}$, $\bar{r} = ctr$, the metric (7) can be brought to manifestly Minkowskian form [26].)

One may argue that solution (7) is Minkowskian simply because $T^{\mu\nu}$ is vanishing in equations (2). However, if this is so, why do we get curved solutions (14, 15), (17−19) from the same equations? If equations (2) contain gravitational field sourcing the curvature present in solutions (14, 15), (17−19), they must do so in solution (7) also. Obviously, solutions (14, 15), (17−19) have singularities to fuel the gravitational field, while solution (7) does not. But, what is there to stop the singularity to occur in (7)? The only difference between the solutions (7) and (14, 15), (17−19) is that they have different types of symmetries in their spacetime structures. While the metric (7) is homogeneous and isotropic, the metrics (14, 15), (17−19) are either inhomogeneous or/and anisotropic. However, how a relaxation in the homogeneity and/or isotropy can result in a singularity, cannot be answered by the conventional wisdom. A possible explanation to the present situation leads us to the following two possibilities.

1. **Conventional Wisdom:** Equations $R^{\mu\nu} = 0$ represent empty spacetime structure and can support curved as well as flat spacetime solutions. However, they are unable to explain how a solution acquires curvature or flatness. For example, the solutions (19) and (7) represent the similar spacetime structures with the only difference that while the homogeneous space in (19) is expanding and contracting in different directions at different rates, the same space is expanding or contracting isotropically in (7). How this difference accounts for their curved and flat states and controls the appearance of the singularity, cannot be explained by equations $R^{\mu\nu} = 0$.

2. **New Paradigm:** The spacetime of equations $R^{\mu\nu} = 0$ is not really empty, the geometry of the equations does contain built-in impressions of the gravitational as well as the material fields, and no extra formulation thereof is required. Further, the structure of the geometry is determined by the net contribution from the material and the gravitational fields (of the chosen matter distribution) which may be manifested in the form of a singularity.

Though the second possibility appears new and orthogonal to the usual understanding, it provides not only a sensible meaning to the Kasner solution (in the absence of the energy-stress tensors of the gravitational or the matter fields), but also a reasonable explanation to the flatness of solution (7). Hence, the solution (7) is flat because it represents a homogeneously distributed matter throughout the space at all times, which is either expanding or contracting isotropically. As the positive energy of the matter field would be exactly balanced, point by point, by the negative energy of the resulting gravitational field (contrary to the case of the Schwarzschild solution where there is only the gravitational energy and no matter at the points represented by the metric), this would provide a net vanishing energy. Neither there would be any momentum contribution from the isotropic expansion or contraction of the material system (contrary to the case of the Kasner solution). Hence, in the absence of any net non-zero energy, momentum or angular momentum, the spacetime of (7) must not have any curvature.

This implies that it is the symmetry of the chosen spacetime structure, which determines whether a solution of $R^{\mu\nu} = 0$ will be curved or flat. This fact is also reflected in the singularities in the curved solutions (14, 15, 17): while the Schwarzschild solution (describing the spacetime structure exterior to a point mass) has a point singularity, the Kerr solution (describing the spacetime structure exterior to a rotating mass) has a ring singularity, the Kasner solution (in which the t = constant hypersurfaces are expanding and contracting at different rates in different directions) presents a singularity of the oscillating kind at $t = 0$.

The discovery, of the net vanishing energy-momentum-angular momentum in a homogeneous distribution of matter expanding/contracting isotropically, appears consistent with several investigations and results which indicate that the total energy of the Universe is zero. Thus, a flat spacetime solution, which has so far been a notion of special relativity (SR) only, can be achieved in the real Universe in the presence of matter, which originates dynamically from the field equations, and is not just put by hand as in SR. Further, the appearance of a flat spacetime in the presence of matter is also not impossible in the conventional approach. For example, it has been shown in [27] that a conformally coupled matter does not always curve spacetime.

It may be mentioned that solution (7) is generally considered as the Milne model, which is not quite correct when viewed with the traditional meaning (of the empty Universe). It should be noted that the Milne's Universe is not empty, albeit its evolution dynamics being the same as that of the empty FRW model. Milne's model of the Universe (based on kinematical relativity with foundations different from those of GR) considers matter filled in the Minkowskian spacetime wherein the matter does not interact with the geometry due to some unknown reasons. The present theory, governed by the field equations $R^{\mu\nu} = 0$, not only explains why the homogeneous, isotropic Universe is Minkowskian, but also predicts when the curved solutions are possible.

4 Why do we Need the New Paradigm?

We have seen above that the appearance of the source of curvature through the geometry of equations $R^{\mu\nu} = 0$ without adding any formulation of the energy-stress tensor of the gravitational field to these equations, is vindicated by the story of the failure to discover the energy-stress tensor of the gravitational field. Do we have any similar experience from the energy-stress tensor of the matter fields? Yes! Recently, it has been shown that a critical analysis of the relativistic formulation of matter given in terms of $T^{\mu\nu}$, reveals some surprising inconsistencies and paradoxes inherently present in its basic formulation [1, 28]. At this point it would be worthwhile to have a digression into the details of this study.

4.1 Non-conventional Sources of Gravitation in $T^{\mu\nu}$

In the following, we focus our attention on the relativistic formulation of matter given by the energy-stress tensor $T^{\mu\nu}$, *without considering any particular theory of gravitation*. After all, all the modern theories of gravitation use this tensor to represent the matter source.

Usually, this tensor may have contributions from, or may be supplemented by, other sources as well, for example, the scalar fields (representing generalized 'matter') or the geometric 'matter' resulting from the modification of the effective Lagrangian of the gravitational field by adding higher-order terms in the curvature invariants (as is done in the extended theories of gravitation [29]). To be more precise, we shall derive our results from the 'conservation' of $T^{\mu\nu}$, i.e., $T^{\mu\nu}_{;\nu} = 0$ which is identically satisfied, through the Bianchi identities, in the simple cases of the metric theories of gravitation, for example, GR. There are however, other theories, for example, the Brans-Dicke theory, wherein this conservation does not follow from the divergence of the left hand side of the field equations. However, it is a standard practice to assume $T^{\mu\nu}_{;\nu} = 0$ (as an extra degree of freedom) in this type of theories. Hence, the results we shall obtain in the following, apply to almost all the theories of gravitation.

Modelling matter by $T^{\mu\nu}$ revolutionizes the way we used to think about the source of gravity. As mass density is the source of gravity in Newtonian theory, we expect energy density to take over this role in the relativistic generalization of Poisson's equation. To our surprise however, all the ten (independent) components of $T^{\mu\nu}$ become contributing source of gravity, making a revolutionary prediction that not only the energy density but pressure and momentum too have gravitational effects. Though the momentum of the matter disappears from the scene once one uses a coordinate system comoving with the matter, however, it is not possible to avoid the pressure which becomes an integral part of the relativistic description of matter. In order to exemplify this, let us consider the simplest example of $T^{\mu\nu}$ which belongs to the case of a perfect fluid (which has no shearing stresses, viscosity or heat flow) and represents/approximates a great many macroscopic physical systems, including scalar fields, dark energy candidates and perhaps the Universe itself. In a coordinate system x^α, specifying the metric tensor $g_{\mu\nu}$, the $T^{\mu\nu}$ of the perfect fluid is given by

$$T^{\mu\nu} = (\rho + p)u^\mu u^\nu - pg^{\mu\nu}, \tag{21}$$

where ρ, the energy density of the fluid and p, its isotropic pressure, are measured by an observer in a locally inertial reference frame which happens to be moving with the fluid at the instant of measurement. In the considered coordinate system x^α, the fluid as a whole has 4-velocity $u^\alpha = dx^\alpha/ds$. *It is important to note that the term ρ, appearing in* (21), *includes not only the rest mass of the individual particles of the fluid but also their kinetic energy, internal energy (for example, the energy of compression, energy of nuclear binding, etc.) and all other sources of mass-energy* [21].

The mystery of the gravitational effects of the non-conventional sources (i.e., the components of $T^{\mu\nu}$ other than the energy density), is seldom addressed at a sufficiently foundational level. It has not been realized, for example, that the formulation of matter in terms of $T^{\mu\nu}$, modifies at the deepest level its well-known properties, so that the intuitive Newtonian knowledge needs revision upon closer investigation. Let us consider, for example, the gravitational effects of pressure, which is the focal point of this section. Since pressure enters $T^{\mu\nu}$ as a scalar, just like the energy density, and since the dimensions of the pressure are those of the energy density, it implies that the pressure (somehow) *contains energy density*. (To what degree it contains energy density, depends on the equation of state.) This is an entirely new physics! Albeit counter-intuitive, this is perfectly consistent with equation (21) wherein the pressure appears as being added to the energy density. Further, this new physics is corroborated by many examples wherein pressure contributes to the energy density of the fluid. One may mention, for example, the pressure contributing to the active gravitational mass density $(\rho + 3p)/c^2$ in the Raychaudhuri equation

$$\frac{3\ddot{\ell}}{\ell} = 2(\omega^2 - \sigma^2) + \dot{u}^{\alpha}_{;\alpha} - \frac{4\pi G}{c^2}(\rho + 3p), \tag{22}$$

which is the relativistic analogue of Poisson's equation and determines the average contraction/expansion of a self-gravitating fluid in GR. Here ℓ represents the volume behaviour of the fluid and the kinematical quantities ω, σ and $\dot{u}^{\alpha}$ respectively measure rotation, shear and acceleration of the fluid flow. As another example, one can mention the pressure contributing to the passive gravitational mass density $(\rho + p)/c^2$ of the fluid in the Tolman-Oppenheimer-Volkoff equation

$$\frac{dp}{dr} + (\rho + p)\frac{d}{dr}\ln(|g_{00}|)^{1/2} = 0, \tag{23}$$

which represents the relativistic analogue of the classical hydrostatic equilibrium of a star. Here r is the radial coordinate and g_{00} is the time-time component of the static spherically symmetric metric modelling the isotropic material of the star. Let us note that the active and passive gravitational masses are not equal in the modern theories of gravitation. The binding energy of the gravitational field is believed to be responsible for this. However, why the contributions from the gravitational energy to the different masses are not equal, has remained a mystery in these theories. Similar examples can be given from the alternative theories of gravity as well (one can see [29], for a review on the extended theories of gravity).

Notwithstanding strongly supported by various examples, the prediction that pressure contains energy, faces paradoxes and inconsistencies. There is already a longstanding Tolman paradox [30] which can be described as follows.

4.2 Tolman Paradox

Tolman [20] has derived a formula for the total energy E, including gravitational energy, of a static system:

$$E = mc^2 = \int (\mathcal{T}_0^0 - \mathcal{T}_1^1 - \mathcal{T}_2^2 - \mathcal{T}_3^3)\, d^3x,$$

where $\mathcal{T}_\nu^\mu = T_\nu^\mu \sqrt{-g}$ is the tensor-density corresponding to the energy-stress tensor. For the energy-stress tensor (21), this formula reduces to

$$E = mc^2 = \int (\rho + 3p)\sqrt{|g_{00}|}\, dV, \tag{24}$$

which applies to a wide class of cases including the one in which the matter is confined to some limited region. Here g_{00} is the time-time component of the metric in an asymptotically flat coordinate system representing the spacetime at a sufficient distance from the material system of interest lying in the neighbourhood of the origin of the coordinate system, and dV is the proper spatial volume element of the fluid sphere. In modern terminology, m is the ADM mass measured by a faraway inertial observer.

The Tolman paradox (related with the consequences of the gravitational effect of pressure) can be described as the following. A pressure-less matter at rest in a container exhibits the total mass m. However, converting the matter inside the container into disordered radiation ($p = \rho/3$) would double the total mass. Or alternatively, a conversion of gamma rays of total energy $E = mc^2$, enclosed in the container, into electron-positron pairs would lead to a total mass half[9] of m. In both cases, the conservation of E is seriously violated.

The Tolman paradox has not received much attention, as the formula (24) involves contributions from the pseudo tensor of the gravitational energy (ascertained by the presence of $\sqrt{-g}$ in (24)) which is notorious for violating its (local) conservation, as has been mentioned before. Thus one can readily claim that the gravitational energy is responsible for the above-mentioned violation of the conservation of E in the Tolman paradox (though it is apparent that the paradox results from the term p, and not from $\sqrt{-g}$ in equation (24)). In addition to the Tolman paradox, there have also been claims for the violation of energy in the cosmological context (see, for example, [31]). However, like the Tolman paradox, these claims have also not attracted much attention, as these usually take refuge, for a possible resolution, in the gravitational energy, which has been of an obscure nature and a controversial history.

So, is it possible to ascertain, beyond doubts, if there is really some fundamental problem in the relativistic formulation of matter, for example, given by the energy-stress tensor (21)? If so, this must be done without implicating gravitational energy. Let us try to do this in the following.

4.3 New Paradoxes

As the central point of our analysis in the following will be the energy-stress tensor of a perfect fluid given by equation (21), it would not be out of place to appreciate its derivation. The standard way to derive the tensor (21) is the following. First, it is defined in the absence of gravity, i.e., in special relativity (SR). Then, the general expression for the tensor in the presence of gravity, in coordinates, say x^σ (in which the metric tensor is identified as $g_{\mu\nu}$), is imported from SR through a coordinate transformation. The bridge between the ideal case of SR and the actual case in the presence of gravity, is provided by a locally inertial coordinate system (LICS) which can always be found at any point of spacetime (thanks to the principle of equivalence). Let us consider an LICS $x^\sigma_\star$ which, by definition, provides an infinitesimal spacetime neighbourhood around the selected point wherein the metric tensor $g_{\mu\nu}$ does not change significantly (and reduces to the Lorentzian metric $\eta_{\mu\nu}$). Hence $\partial g_{\mu\nu}/\partial x^\sigma_\star = 0$ and the effects of gravity disappear in the neighbourhood (but not in a finite region). This LICS is so chosen that a perfect fluid element is at rest in the neighbourhood of the position and time of interest. As an observer at the selected point sees the fluid around him isotropic, the energy-stress tensor $T^{\mu\nu}_\star$ (in the coordinates $x^\sigma_\star$)

[9] As energy and momentum are conserved when two photons collide to give rise to an electron-positron pair, so, if all the photons pairs do not meet this requirement, only those will contribute to the paradox, which meet the requirement. While their momentum will balance each others', their energy will be converted to the rest mass of the created electron-positron pairs. However, the term $3p$ appearing in the Tolman integral (24) would remain unaccounted, and hence would contribute to the paradox.

takes the form characteristic of spherical symmetry:

$$T_\star^{00} = \rho, \; T_\star^{11} = T_\star^{22} = T_\star^{33} = p \text{ and } T_\star^{\sigma\rho} = 0 \text{ for } \sigma \neq \rho, \tag{25}$$

where ρ and p are, respectively, the energy density and pressure of the fluid measured by the considered observer. As the fluid element is at rest in the above-mentioned neighbourhood, the spatial components of the 4-velocity vector then vanish, i.e.

$$u_x \equiv \frac{dx}{d\tau} = 0, \; u_y \equiv \frac{dy}{d\tau} = 0, \; u_z \equiv \frac{dz}{d\tau} = 0 \text{ and } \frac{dt}{d\tau} = 1, \tag{26}$$

where we have specified $x_\star^0 \equiv ct$, $x_\star^1 \equiv x$, $x_\star^2 \equiv y$, $x_\star^3 \equiv z$; and the proper time interval $d\tau = ds/c$. In the general coordinate system x^i, the energy-stress tensor $T^{\mu\nu}$ is then given by

$$T^{\mu\nu} = \frac{\partial x^\mu}{\partial x_\star^\alpha} \frac{\partial x^\nu}{\partial x_\star^\beta} T_\star^{\alpha\beta}, \tag{27}$$

which reduces to (21) by noting that $g^{\mu\nu} = \eta^{\alpha\beta}(\partial x^\mu/\partial x_\star^\alpha)(\partial x^\nu/\partial x_\star^\beta)$ and $dx^\mu/ds = \partial x^\mu/\partial x_\star^0$, owing to the relations in (26). Equation (21) thus derived, is valid at all points, as an LICS can always be found at any point of spacetime so that the fluid element is at rest in an infinitesimal spacetime neighbourhood around the point.

After establishing the general expression for the energy-stress tensor of the perfect fluid, let us study its divergence which is famous for describing the mechanical behaviour of the fluid. As our main motive is to understand the mysterious gravitational effects of the pressure of the fluid without implicating the gravitational energy, let us consider the same LICS which has been used to derive (21) so that the subtleties of gravitation and the gravitational energy disappear locally in this coordinate system. Let us reconsider the same fluid element, which is at rest in the chosen spacetime neighbourhood. Though the spatial components of the 4-velocity vector vanish in this neighbourhood, the derivatives of the velocity will not be zero in general, except for its temporal component, which will vanish, as we see in the following: the general formula for the interval, $ds^2 = g_{\mu\nu}\, dx_\star^\mu\, dx_\star^\nu$ implies that

$$g_{00}\left(c\frac{dt}{ds}\right)^2 + g_{11}\left(\frac{dx}{ds}\right)^2 + ... + 2g_{01}c\frac{dt}{ds}\frac{dx}{ds} + ...$$

$$+g_{33}\left(\frac{dz}{ds}\right)^2 = 1,$$

which on differentiation gives

$$\frac{\partial}{\partial x_\star^\alpha}\left(\frac{dt}{ds}\right) = 0, \tag{28}$$

by virtue of the relations in (26) and by recalling that $\partial g_{\mu\nu}/\partial x_\star^\alpha = 0$, $g_{\mu\nu} = \eta_{\mu\nu}$ in the chosen neighbourhood. As has been mentioned earlier, we now consider a vanishing divergence of (21)

$$T^{\mu\nu}_{\;;\nu} = 0, \tag{29}$$

which, in the chosen coordinates, is given by

$$\frac{\partial T^{\mu\nu}}{\partial x_\star^\nu} = 0. \tag{30}$$

The temporal component of equation (30) for the case $i = 0$, can be written in accordance with (26) and (28) as

$$\frac{d}{dt}(\rho\delta v) + p\frac{d}{dt}(\delta v) = 0, \tag{31}$$

which can also be written as

$$\delta v\frac{d\rho}{dt} + (\rho + p)\frac{d(\delta v)}{dt} = 0, \tag{32}$$

where $\delta v = \delta x\delta y\delta z$ is the proper volume of the fluid element. The usual interpretation of equation (31) is that the rate of change in the energy of the fluid element is given in terms of the work done against the external pressure. This seems reasonable at the first sight, but cracks appear in it after a little reflection. The first concern, as also noticed by Tolman [20], is that the fluid of a finite system can be divided into similar fluid elements and the same equation (31) can be applied to each of these elements, meaning that proper energy $(\rho\delta v)$ of every element is decreasing when the fluid is expanding or increasing when the fluid is contracting. This leads to a paradoxical result that the sum of the proper energies of the fluid elements which make up an isolated system, is not constant! Tolman overlooked this problem by assuming a possible role of the gravitational energy in it. We however remind that there cannot be any role of the gravitational energy in equation (31) which has been derived in a LICS.

Another more important issue related with equation (31) is the following. We have already mentioned that the term ρ includes in it all the possible sources of mass and energy (excluding the gravitational energy). Hence, so is included in it the energy equivalent to the work done against the external pressure. If we already know that energy (in the form of work done) is being supplied to the system or getting released from it (which can be calculated by checking by what amount δv is increasing or decreasing), why can't this too be taken care of by the term ρ? There is no natural law which dictates that ρ cannot include particular types of energies. So, if ρ includes the energy equivalent to the work done against the external pressure, the additional work contained in the second term of equation (31) violates the conservation of energy!

Yet another paradox has been discovered recently [28] which demonstrates, more vividly, subtle inconsistencies in (21). This can be derived for the spatial components of equation (30). By the use of (26) and (28) in (30), we can write for the case $i = 1$, the following equation

$$\frac{\partial p}{\partial x} + \frac{(\rho + p)}{c^2}\frac{du_x}{dt} = 0, \tag{33}$$

which has also been obtained by Tolman in some other context. Here $du_x/dt = du_x/d\tau = \partial u_x/\partial t$ (in the chosen coordinates) is the acceleration of the fluid element in the x-direction. As any role of gravity is absent in this equation, it can be interpreted as the relativistic analogue of the Newtonian law of motion: the fluid element of unit volume, which moves under the action of the force $\partial p/\partial x$, has got the inertial mass $(\rho+p)/c^2$. But it is surprising that the inertial mass of the fluid element has got an additional contribution from p, though without any apparent source! Equation (33), taken at the face value, reveals that p should be carrying some kind of energy (density) as p/c^2 contributes to the inertial mass (density) of the fluid element. However, as the term ρ includes in it all the sources of energy-mass, so if p 'somehow' contains energy, it must be at the cost of violating the celebrated law of the conservation of energy! Though equation (33) is not an energy conservation equation, but that does not allow it to defy the law of conservation of energy.

Finally, we derive the Tolman integral (24) in an LICS wherein it reduces to

$$E = mc^2 = \int (T^0_0 - T^1_1 - T^2_2 - T^3_3)\, d^3x = \int (\rho + 3p)\, dV, \tag{34}$$

which does not contain any gravitational energy, but still pronounces the Tolman paradox, though the formula may be valid for a sufficiently small volume of the fluid. From equations (32), (33) and (34), we notice something unexpected: we still encounter different (unequal) mass densities $(\rho + 3p)/c^2$ and $(\rho + p)/c^2$ of the same fluid element though there remains no possibility, in the present situation, of any contribution from the gravitational energy to these masses to make them equal, as they are derived in an LICS!

Although the equality of the inertial and gravitational masses is the starting point of any metric theory of gravitation, however, the resulting theories have not achieved their own defining feature unequivocally. The two gravitational masses, the active and the passive ones, have remained unequal and the equality of the inertial and the active gravitational masses has remained a dream. The usual interpretation thereof seeks refuge in the gravitational energy. However, the present analysis ensures that there is no possibility of any role of the gravitational energy in making the different masses equal. Perhaps the origin of this problem is elsewhere.

It is thus established that the relativistic description of matter given by (21) suffers from some subtle inherent inconsistencies in its basic formulation. The point to note is that there is no role of the notorious (pseudo) energy of the gravitational field in these problems. It is perhaps not correct to formulate the tensor from two basically distinct kinds of quantities – energy density, a scalar quantity and stress, a directional quantity. The tensor treats them at equal footing by recognizing a component $T^{\mu\nu}$ as a scalar (irrespective of the values of μ and ν) linked with the surface specified by μ and ν in the *hypothetical* 4-dimensional fluid, in the same way as the component $G^{\mu\nu}$ is linked with the curvature of the same surface. This leads to a sound mathematics and we do not notice any inconsistency until we relate the tensor $T^{\mu\nu}$ with the *real* fluid which is 3-dimensional and *not 4-dimensional*.

4.4 Corrections and their Consequences

There already exists the resolution of the Tolman paradox. Let us take tips from there to resolve the other paradoxes described above and study the consequences. The standard resolution of the Tolman paradox, first presented by Misner and Putnam [32] in 1959, is provided through the following prescription: "If the pressure is confined by non-gravitational means, there must be tension (negative pressure) in the walls (to keep the fluid inside and the field static) which will compensate the additional energy causing the paradox." In fact, the integrated value of the negative pressure ($-p = -T^i_i$, $i = 1, 2, 3$, no summation over i) will contribute to (24) a term which just counterbalances the increased mass arising from the $3p = (T^1_1 + T^2_2 + T^3_3)$ term. More recently this resolution has been confirmed by Ehlers et al. [33] in a more refined version by giving a tensorial treatment to the surface tension of the walls and the trapped fluid sphere.

It should be noted that the key point of these resolutions – the counterbalancing the pressure of the fluid by stresses in the wall – does not emanate either from the derivation of $T^{\mu\nu}$ or from the underlying gravitational theory. Rather it has been *put by hand* as a correction. After getting corrected, equation (24) reduces to the classical result

$$E = \int T_0^0 \, d^3x = \int \rho\sqrt{|g_{00}|} \, dV. \tag{35}$$

Let us apply this clue to the paradoxes represented by equations (31) and (33), for the case when the system is held by non-gravitational means[10]. As the tension of the wall is transmitted to the fluid element, each of the T_x^x, T_y^y and T_z^z present in the considered fluid element, is balanced by a -T_i^i from the wall. Hence the term $pd(\delta v)/dt$ in equation (31) [which can also be written as $pd(\delta v)/dt = \delta y \delta z \, T_x^x d(\delta x)/dt + \delta z \delta x \, T_y^y d(\delta y)/dt + \delta x \delta y \, T_z^z d(\delta z)/dt$] is balanced by $-\delta y \delta z \, T_x^x d(\delta x)/dt - \delta z \delta x \, T_y^y d(\delta y)/dt - \delta x \delta y \, T_z^z d(\delta z)/dt = -pd(\delta v)/dt$, modifying equation (31) to

$$\frac{d}{dt}(\rho \delta v) = 0, \tag{36}$$

which is now free from the paradoxes. We notice something very remarkable in these corrections. The corrections made in equations (24) and (31) are effectively equivalent to replacing the active gravitational mass density $(\rho + 3p)/c^2$ as well as the inertial mass density $(\rho + p)/c^2$ (which is also the passive gravitational mass density) by the simple (total) Newtonian mass density ρ/c^2. *Hence the long-sought-after validity of the principle of equivalence is achieved naturally here!* This modifies equation (33) to

$$\frac{\partial p}{\partial x} + \frac{\rho}{c^2}\frac{du_x}{dt} = 0, \tag{37}$$

We notice that there is only one contribution from T_i^is, viz. from T_x^x, to equation (33) whereas all the three T_i^is contribute to equations (24) and (31). The reason is that equation (33) represents a balance of forces along the x-direction only. Similar contributions from T_y^y and T_z^z would appear to the other two similar equations derived from equation (30) for the cases $i = 2$ and $i = 3$ respectively. Obviously, equation (34) gets corrected to

$$E = \int T_0^0 \, d^3x = \int \rho \, dV. \tag{38}$$

The corrected equations are convincing and free from the conceptual problems. Additionally, they ascertain the equality of the gravitational and inertial masses, which is remarkable. However, replacing the different general-relativistic energy densities, appearing

[10] The nature of the corrections is not clear when the system is held by gravitation. Should one consider the systems held by the gravitation and those by non-gravitational means, equivalent? Even in the situation when the system is held by gravitation, one would not get any contribution from the gravitational energy in equations (31)–(34) (as they are derived in a LICS), but still sufferring from the paradoxes. This shows the equivalence of the cases held by gravitation and non-gravitational means, so far as the paradoxes are concerned.

in the equations of gravitation and cosmology, by the total energy density ρ, would be revolutionary. Although these obligatory corrections (which must be performed in order to rectify the relativistic formulation of matter) are important in their own right, however, the corrected equations do not seem consistent with many cosmological observations.

For example, equation (36) would predict that matter density and radiation density both scale as $(\delta v)^{-1}$ which would not be compatible with the standard theory of the thermal evolution of the Universe. The standard cosmology (and other theories) with the corrected equations, would not be compatible with an accelerating Universe (with any sensible matter source with $\rho > 0$), as the equations would not have a bearing on the pressure of the matter. This will render these theories inconsistent with many observations. This indicates that the relativistic formulation of matter by $T^{\mu\nu}$ is not compatible with the geometric description of the metric theories. This view is vindicated by the following points indicating additional inconsistencies in the very basic assumption on the formulation of $T^{\mu\nu}$ which have not been realized earlier.

1. As we have seen earlier, the energy-stress tensor $T^{\mu\nu}$ of a fluid is first formulated in SR, in a small neighbourhood of an inertial observer. Then the expression of the tensor in the presence of gravity is imported, from SR to GR, through a coordinate transformation. However, the simple point which has not been noticed, which makes this derivation questionable, is that an inertial coordinate system is valid only at a point, and *not* in a neighbourhood, however small it is (Christoffel symbols can be made vanishing only at a point, and *not* in a neighbourhood). But a fluid element cannot be defined at a point, we do need a neighbourhood. Hence, at the best, the tensor $T^{\mu\nu}$ may approximate a dust (with vanishing pressure) but cannot represent a fluid (with non-zero pressure) which requires more than one particle to generate pressure.

2. As the derivation of $T^{\mu\nu}$ assumes its validity in the absence of gravitation (in the flat spacetime), this goes contradictory to the very notion of $T^{\mu\nu}$ being the source of curvature. To exemplify this, let us note that in a flat spacetime, the left hand side of equation (1) vanishes automatically, but not the right hand side which has to be put equal to zero by hand. That is, the source of curvature can exist there without producing any curvature! Although equation (1), being a geometric formulation of gravitation, is expected to be valid in a curved spacetime, it must also reduce to the no-gravitation case consistently. This provides another reason why $T^{\mu\nu}$ should not appear in the field equations of gravitation.

3. Angular momentum is a fundamental and unavoidable characteristic of matter, as is witnessed from the subatomic to the highest scales. As $T^{\mu\nu}$ fails to include the angular momentum, it cannot be regarded a true representative of the matter. We have already noticed that the angular momentum (which registers its inherent presence through the geometry) does appear as a source of curvature/gravitation in the Kerr solution.

This indicates that, like a proper energy-stress tensor of the gravitational field, a correct relativistic formulation of the matter fields neither exists. A similar view was expressed about four decades ago by J. L. Synge, one of the most distinguished mathematical physicists of the 20th Century: *"The concept of energy-momentum-stress is simply incompatible with general relativity"*.

4.5 Any Observational Support for $T^{\mu\nu}$?

It should be noted that the relativistic description of the matter, in the form of the energy-stress tensor (21), has never been tested in any direct experiment. The experimental tests, which have verified GR, have been carried out in empty space by considering $T^{\mu\nu} = 0$, as has been mentioned earlier. Hence, the celebrated tests of GR have remained limited to test only the geometric aspect of the theory.

The observed deflection of starlight (when it passes past the Sun), twice as much as predicted by a heuristic argument made in Newtonian gravity, appears roughly consistent with Tolman's formula (24) for the disordered radiation [20], which is often considered as an observational evidence for the correctness of $T^{\mu\nu}$. Ironically, a result derived by Tolman himself (page 250 of [20]) appears contradictory to this. In a weak gravitational field, like that of the sun, where Newtonian gravitation can be regarded as a satisfactory approximation, equation (24) can be written as $E = \int \rho dV + (1/2c^2) \int \rho\psi dV$, where ψ is the Newtonian gravitational potential. As ψ is negative, we note that the general relativistic active gravitational mass E/c^2 of the gravitating body is obviously less than its proper mass $(1/c^2) \int \rho dV$ and is expected to give a lower value for the gravitational deflection of light! It should be noted that the correct interpretation of the observations of the bending of light comes from the correct geometry resulting from the Schwarzschild solution of the vacuum field equations $R^{\mu\nu} = 0$, and *not* from formula (24).

It can then be said safely that the relativistic formulation of matter given by the energy-stress tensor $T^{\mu\nu}$, does not have any direct observational support. The only support one can name, comes through the consistency between the formulations of the disordered radiation derived from the energy-stress tensor of the perfect fluid and that resulting from Maxwell's theory of classical electrodynamics. However, in view of the paradoxes and inconsistencies with $T^{\mu\nu}$ discussed in the preceding sections, one should be cautious in this matter and investigate the role of the hidden symmetries of the disordered radiation causing the illusionary consistency.

It should be noted that being a geometric theory of gravitation, GR eliminates any possibility to represent gravitation in terms of a force. Rather the theory replaces the effects of the force as through geometry. Similarly, the effects of stresses, momenta, angular momenta and energy density too are revealed through geometry. In this view, the theoretical crisis in the formulation of $T^{\mu\nu}$, mentioned above, acquires a new meaning in the present context: the tensor is not compatible with the geometric formulation of gravitation simply because, like the gravitational field, the effects of the matter field are already present inherently in the geometry!

Einstein argued that *"On the basis of the general theory of relativity, space as opposed to 'what fills space', has no separate existence"*. If this is true, considering a spacetime structure (conditioned by the equations $R^{\mu\nu} = 0$) must be equivalent to considering the accompanying fields (material and gravitational) as well, and there should be no need to add any extra formulation thereof to the field equations.

This, in fact, leaves equation $R^{\mu\nu} = 0$ as the only possibility for a consistent field equation of gravitation, wherein the matter fields as well as the gravitational field appear through the geometry, i.e., they are represented by the metric field. One may argue that a consistent field equation of gravitation is expected to reduce to Poisson's equation $\nabla^2\psi = 4\pi G\rho$ in a weak stationary gravitational field. However, this requirement no longer seems mandatory, as it has already been compromised in the concordance ΛCDM cosmology. It should be noted that the Einstein field equations with a non-zero Λ do *not* fulfill this requirement [34]. While, there is no scope in the standard paradigm to mend this shortcoming, it would not be correct to compare the new theory (being a fundamentally different theory wherein matter does not appear explicitly) with the Poisson equation (wherein matter appears explicitly).

We shall see, in the following, that the new theory governed by equation $R^{\mu\nu} = 0$, has some attractive features, as it can obviate the long-standing problems of the standard paradigm.

5 Features of the New Paradigm

The standard cosmology has well-known shortcomings, such as the problem of the initial singularity, the cosmological constant problem and the problems of the special initial conditions, appearing in the horizon problem and the flatness problem. Interestingly, the new theory circumvents all these problems, as we shall see in the following.

5.1 No Singularity

The success of the standard Big Bang cosmological model is marred by another unwanted feature, the singularities, besides the speculative dark sectors. There can be no question that the singularity is an unphysical feature of any theory. Hence, there have been attempts to discover singularity-free cosmological solutions of Einstein's equations. Generally this is achieved by violating the energy conditions, though there are also examples of the singularity-free cosmological solutions which do not violate the energy and causality conditions [35].

As we have already noted, the cosmological model (stemming from the present theory), represented by solution (7), is singularity free, since the singularity present in it is merely a coordinate singularity which can be removed in its Minkowskian form by a redefinition of the coordinates.

5.2 Machian

Although the Mach principle has been one of Einstein's guiding principles and has played a central role in the development of GR, the resulting theory is not completely Machian. Solutions of the field equations discovered by, for example, de Sitter and Godel, disobey Mach's principle.

In the new paradigm, however, the metric tensor is determined uniquely by the overall distribution of the gravitational and the material fields. This is what Einstein had envisioned (though could not accomplish)! So, there is no usual problem of the 'occurrence of curved solutions (non-inertial worldlines) without the presence of matter' (if matter is solely identified with $T^{\mu\nu}$), since the so-called 'vacuum' solutions do not really describe a spacetime devoid of matter; what they describe is a spacetime in which the matter field is represented by the geometry (the metric field). So Mach's Principle does hold in general in the new theory, provided we characterize it by demanding the determination of the metric tensor from the overall distribution of the energy-momentum-angular momentum resulting from the accompanying gravitational and material fields, and *not* from the matter alone.

5.3 No Horizons

The horizon problem in the standard (ΛDCM) model arises from the observed uniformity of the CMB, which has the same temperature everywhere (except for tiny, stochastic fluctuations), even in regions on opposite sides of the sky, which appear to lie outside of each other's causal horizon. Since no physical process propagating at or below light speed could have brought them into thermal equilibrium, it appears that the Universe required special initial conditions, which are supposed to be provided by inflation. Hence the problem, in essence, appears because a (particle) horizon exists in the standard cosmology. This can be checked from the following. The particle horizon is defined by

$$d_{\rm PH}(t) = cS(t)\int_0^t \frac{dt'}{S(t')}, \tag{39}$$

which measures the largest (comoving) proper distance light could have traveled in the age of the Universe at a time t. As a finite value exists for $d_{\rm PH}$ in the standard cosmology, this means that the Universe has a horizon in the theory. However, this problem does not exist in the new paradigm as $d_{\rm PH} = \infty$ at any time, as can be easily checked from (6) and (39). Hence, the whole Universe is always causally connected, which explains the observed uniformity of CMB without invoking the hypothetical inflaton field.

Neither exists the event horizon in the new paradigm. The event horizon, defined by

$$d_{\rm EH} = cS(t_0)\int_{t_0}^{t_{\rm max}} \frac{dt'}{S(t')}, \tag{40}$$

where $t_{\rm max}$ is the time-coordinate at the end of the Universe, which would be infinite in the case of an ever-expanding Universe, like the one in the present model. Clearly, $d_{\rm EH} = \infty$ in the present model and the event horizon neither exists in this model.

5.4 No Cosmological Constant or Flatness problems

The origin of the cosmological constant problem lies in a conflict between the energy-stress tensor $T^{\mu\nu}$ in equation (1) and the energy density of vacuum in the quantum field theory (QFT). The vacuum energy, according to the QFT, results from the quantum vacuum fluctuations which provide an energy contribution of the order of the Planck mass. In GR, the vacuum energy can be represented by the cosmological constant (through a particular equation of state $p_\Lambda/c^2 = \rho_\Lambda$ of $T^{\mu\nu}$). Friedman solution of equation (1) then provides an estimate of the vacuum energy in terms of H_0^2. This is, however, smaller than the QFT-value by a factor of $\approx 10^{120}$! This discrepancy has been called 'the worst theoretical prediction in the history of physics!'

Interestingly, this problem is evaded in the new paradigm owing to the fact that the energy-stress tensor is absent and the matter, which is represented through the geometry, does not appear explicitly in the dynamical equations. Neither exists the coincidence problem (associated with the cosmological constant problem).

The flatness problem of the standard cosmology requires the initial density of matter (in the energy-stress tensor) to have a value extremely fine-tuned to its critical value (corresponding to a flat Universe). Even a tiny deviation from this value would have had drastic effects on the nature of the present Universe. However, the problem is circumvented in the new paradigm due to the absence of the energy-stress tensor, as has been explained above.

5.5 Limitations of the New Theory

The field equation of gravitation, in the new theory, is $R^{\mu\nu} = 0$, which is simpler and free from any conceptual problems. Nevertheless, as matter does not appear explicitly in it, it provides limited scope for various studies which can be done in the framework of the standard paradigm. For example, it would be difficult to study the evolution of the structures resulting from the perturbation of matter, computation of anisotropies in the CMB resulting from the perturbation of matter, and gravitational collapse of matter. It would also not be possible to explain the observations of the baryon acoustic oscillations (BAO) in the framework of equations $R^{\mu\nu} = 0$, as the mechanism of interaction of matter is not specified. However, these should be considered as a limitation of the geometric theory of gravitation.

6 Predictions and Evidences

A good theory is expected to make accurate predictions, so that it can be falsified. The above-obtained revolutionary result, that the gravitational as well as material fields are built-in ingredients of the geometry of equations $R^{\mu\nu} = 0$, has a powerful prediction. If the field equations $R^{\mu\nu} = 0$ represent real spacetime with matter and energy, they are expected to represent the interior solutions to Schwarzschild and Kerr solutions as well.

In this context, it would be worthwhile to mention that the so-called Schwarzschild interior solution (which is alternatively expressed as the Tolman-Oppenheimer-Volkoff equation, providing the standard representation of the interior of a static spherically symmetric star) turns out to be unphysical. The speed of sound $= c\sqrt{dp/d\rho}$ becomes infinite in a constant density-fluid. Neither could be found so far, any solution (representing the interior of a rotating star) which could be matched to the Kerr solution. It seems that we have been searching for the interior solutions in a wrong place. They should, in fact, be searched in the framework of the field equations $R^{\mu\nu} = 0$, as noted above.

Hence the structure of spacetime inside a spherically symmetric matter distribution should also be represented by (14)[11] but, with $K = 0$ leading to a flat[12] spacetime. K

[11] A spherical spacetime structure will also result inside an isotrpic distribution of matter, besides outside an isotropic mass.

[12] Recently it has been shown [36] that for a strictly homogeneous fluid, ρ must be vanishing leading to a flat metric.

must be zero in this case, since otherwise the metric would have a singularity at $r = 0$ without any proper reason. This appears logical as an isotropic matter distribution contains an isotropic distribution of the matter energy, in addition to the gravitational energy, and since the positive energy of the matter fields is exactly balanced by the negative energy of the resulting gravitational field at each point, the net energy must be zero everywhere, leading to a flat spacetime. This also appears consistent with the homogeneously distributed matter expanding or contracting isotropically leading to the flat metric (7).

On the usual matching of the interior and exterior solutions: It is usually expected that the exterior and the interior solutions (in the case of Schwarzschild solutions, for example) must match at the boundary. Hence, the interior Minkowskian metric with exterior curved Schwarzschild metric (14) will raise an eyebrow. However, at or just outside the boundary of a finite volume of matter (not a point mass), the spacetime *will not* be spherically symmetric (unless we go infinitely away from the boundary). Hence, the points at or just outside the boundary, *cannot* be represented by the isotropic Schwarzschild exterior solution (14), we need another (possibly more complicated) metric to represent them. So, the usual matching of the Schwarzschild interior and exterior solutions does not seem correct.

This also predicts that the interior of the Kerr metric should be a curved rotating spacetime solution of equations $R^{\mu\nu} = 0$ which should reduce to Minkowskian form (Schwarzschild interior solution) for vanishing rotation, since the Kerr solution reduces to the Schwarzschild exterior solution for a vanishing rotation.

Conclusion

GR provides an accurate description of a wide range of gravitational phenomena. One point has, however, marred this success in describing the physical world - the dark ingredients of $T^{\mu\nu}$ in the form of inaton, dark matter and dark energy - which do not have any laboratory evidence and remain unidentified.

Motivated by this fact and guided by the strong observational support to the so called 'vacuum' field equations $R^{\mu\nu} = 0$, we discover that energy-stress tensor is a redundant part of the Einstein field equations and the source of gravitation is the geometry itself. A critical analysis of the different solutions of equations $R^{\mu\nu} = 0$ supports this view indicating that equations $R^{\mu\nu} = 0$ do not represent an empty spacetime and impressions of the gravitational as well as the material fields are inherently present in the geometry of these equations. This can be called the geometrization of matter, which appears as an intrinsic characteristic of a geometric theory of gravitation (like GR). That is, the existence of matter should be understood in terms of the geometrical quantities of spacetime. This result rests on the following logical ingredients.

1a. No formulation of the energy-stress tensor of the gravitational field, is included in equations $R^{\mu\nu} = 0$. In fact, a proper energy-stress tensor of the gravitational field does not exist.

1b. It is shown that the gravitational energy appears, *through geometry*, as a source of curvature in the solutions (Schwarzschild, Kerr) of equations $R^{\mu\nu} = 0$.

2a. Recently it has been discovered that the relativistic formulation of matter, given by the energy-stress tensor $T^{\mu\nu}$, is plagued with paradoxes and inconsistencies, implying that, like the energy-stress tensor of the gravitational field, an indisputable energy-stress tensor of matter neither exists.

2b. It is shown that the momentum density appears, *through geometry*, as a source of curvature in a cosmological solution (Kasner) of equations $R^{\mu\nu} = 0$.

3a. There is no scope, in the framework of GR, to include the angular momentum in $T^{\mu\nu}$.

3b. It is already known that the angular momentum too contributes to the curvature of spacetime, and it also appears, *through geometry*, for example, in Kerr solution of equations $R^{\mu\nu} = 0$.

This leaves equation $R^{\mu\nu} = 0$ as the only possibility for a consistent field equation of gravitation, wherein the matter fields as well as the gravitational field appear through the geometry (the metric field), without adding any formulation thereof to the field equations. The resultant non-zero energy-momentum-angular momentum of the two fields acts as the source of curvature.

Einstein believed that *"On the basis of the general theory of relativity, space as opposed to 'what fills space', has no separate existence"*. If this is true, considering a spacetime structure must be equivalent to considering the accompanying fields (material and gravitational) as well, and there should be no need to add any extra formulation thereof to the field equations. Though it may appear orthogonal to the usual understanding, is in striking agreement with the observations at all scales, without requiring the usual epicycles of the standard paradigm, such as inflaton, non-baryonic dark matter and dark energy. Moreover, it provides a consistent explanation to the curvature of Kasner solution and the flatness of the homogeneous-isotropic solution of $R^{\mu\nu} = 0$, which have remained unexplained puzzles so far. Additionally, the resulting cosmological model, in the new paradigm, has many attractive features. For example, it is Machian and free from the problems of the standard cosmology - the singularity, horizon, flatness and the cosmological constant problems.

It appears that the quest for a better understanding of gravity is far from over! It was Einstein's obsession that the vibrant geometrical part of GR is 'marble' and matter is 'wood', and that all attempts should be directed to turn wood into marble. It, finally turns out that the 'wood' is a redundant part of the theory, shunning which, enhances the beauty of the 'marble' in the extreme simplicity of the field equations $R^{\mu\nu} = 0$!

"A clever man commits no minor blunders." – Johann Wolfgang Goethe

Acknowledgment

The author benefited from valuable discussions with Professor Abdus Sattar and Professor Jayant Narlikar. Thanks are due to Vijay Rai for sending some important literature.

References

[1] R. G. Vishwakarma, Astrophys. Space Sci., 340, 373, 2012, (preprint arXiv:1204.1553).

[2] R. G. Vishwakarma, Nuovo Cim. B, 122, 113, 2007.

[3] S. Perlmutter, et al., Astrophys. J. 517, 565, 1999.

[4] A. Riess, et al., Astrophys. J. 659, 98, 2007.

[5] R. G. Vishwakarma and J. V. Narlikar, Res. Astron. Astrophys. 10, 1195, 2010 (preprint arXiv: 1010.5272 [astro-ph.CO]).

[6] J. C. Jackson and M. Dodgson, Mon. Not. R. Astron. Soc. 285, 806, 1997.

[7] L. I. Gurvits, *Astrophys. J.* 425, 442, 1994.

[8] S. K. Banerjee and J. V. Narlikar, Mon. Not. R. Astron. Soc. 307, 73, 1999;
R. G. Vishwakarma, Class. Quantum Grav. 17, 3833, 2000;
R. G. Vishwakarma and Parampreet Singh, Class. Quantum Grav. 20, 2033, 2003.

[9] J. R. Mould, et al., Astrophys. J. 529, 786, 2000.

[10] O. Y. Gnedin, O. Lahav and M. J. Rees, preprint: astro-ph/0108034.

[11] R. Cayrel, et al., Nat. 409, 691, 2001.

[12] D. Larson,et al., Astrophys. J. Suppl. 192, 16, 2011.

[13] D Mania and B. Ratra, preprint arXiv: 1110.5626 [astro-ph.CO].

[14] R. R. Siegel, et al., 2005, Mon. Not. R. Astron. Soc. 356, 1117, 2005.

[15] S. Weinberg, *Gravitation and Cosmology: Principles and Applications of the General Theory of Relativity* (John Wiley & Sons, 1972), p 525.

[16] J. V. Narlikar, *An Introduction to Cosmology* (Cambridge University Press, 2002), p 251, 253.

[17] L. P. Grishchuk, Space Sci. Rev., 148, 315, 2009.

[18] L. P. Grishchuk, Phys. Rev. D, 45, 4717, 1992.

[19] A. Mitra, *Gen. Rel. Grav.*, 42, 443, 2010.

[20] R. C. Tolman, *Relativity, Thermodynamics and Cosmology* (Oxford University Press, 1934).

[21] C. W. Misner, K. S. Thorn and J. A. Wheeler, *Gravitation* (W. H. Freeman and Company, New York, 1970).

[22] S. W. Hawking and G. F. R. Ellis, *The Large Scale Structure of Spacetime* (Cambridge University Press, 1973).

[23] E. Kasner, Am. J. Math. 43, 217, 1921;
A. Taub, Ann. Math. 53, 472, 1951.

[24] V. V. Narlikar, K. R. Karmarkar, Curr. Sci. 3, 69, 1946.

[25] L. K. Patel and P. C. Vaidya, Prog. Maths., 3, 158, 1969.

[26] J. V. Narlikar, *An Introduction to Cosmology* (Cambridge University Press, 2002).

[27] E. Ayon-Beato, C. Martinez, R. Tronoso and J. Zanelli, Phys. Rev. D 71, 104037, 2005 (preprint hep-th/0505086).

[28] R. G. Vishwakarma, Astrophys. Space Sci., 321, 151, 2009 (preprint arXiv:0705.0825).

[29] S. Capozziello and M. De Laurentis, *Phys. Rep.* 509, 167, 2011.

[30] J. Ehlers, I. Ozsvath and E. L. Schucking, Am. J. Phys. 74, 607, 2006 (arXiv: gr-qc/0505040).

[31] E. R.Harrison, Astrophys. J., 446, 63, 1995.

[32] C. W. Misner and P. Putnam, Phys. Rev., 116, 1054, 1959.

[33] J. Ehlers, I. Ozsvath, E. L. Schucking and Y. Shang, Phys. Rev. D, 72, 124003, 2005 (arXiv: gr-qc/0510041).

[34] Weinberg, S., *Gravitation and Cosmology: Principles and Applications of the General Theory of Relativity* (John Wiley & Sons, 1972), p155.

[35] J. M. M. Senovilla, *Phys. Rev. Lett.* 64, 2219, 1990;
N. Dadhich, L. K. Patel and R. Tikekar, *Pramana - J. Phys.* 44, 303, 1995.

[36] A. Mitra, Astrophys. Space Sci., 333, 169, 2010.

In: Recent Advances in Cosmology
Editors: Anderson Travena and Brady Soren
ISBN: 978-1-62417-943-3

Chapter 4

Relativistic Viscous Universe Models

***Iver Brevik*[1] *and Øyvind Grøn*[2]**
[1] Department of Energy and Process Engineering,
Norwegian University of Science and Engineering, Trondheim, Norway
[2] Oslo and Akershus University College of Applied Sciences,
Faculty of Technology, Art and Design, Oslo, Norway

Abstract

The research on relativistic universe models with viscous fluids is reviewed. Viscosity may have been of significance during the early inflationary era, and may also be of importance for the late time evolution of the Universe. Bulk viscosity and shear viscosity cause exponential decay of anisotropy, while nonlinear viscosity causes power-law decay of anisotropy. We consider also the influence from turbulence, in connection with future singularities of the universe (Big Rip and Little Rip). Finally, we review some recent developments of causal cosmology theories.

1. Viscous Universe Models

Misner [1967] [1] noted that the 'measurement of the isotropy of the cosmic background radiation represents the most accurate observational datum in cosmology, which is even more true today with the WMAP- and Planck measurements. An explanation of this isotropy was provided by showing that in a large class of homogeneous but anisotropic universes, the anisotropy dies away rapidly. It was found that the most important mechanism in reducing the anisotropy is neutrino viscosity at temperatures just above 10^{10} K (when the Universe was about 1 s old: cf. Zel'dovich and Novikov [1983] [2]). The first theory of relativistic viscous fluid was presented in Eckart [1940] [3]. Eckart's theory deals with first order deviation from equilibrium, while neglected second order terms are necessary to prevent non-causal behavior. Israel and Stewart [1976] [4] have developed a second order theory. Grøn [1990] [5] and Maartens [1995, 1996] [6, 7] have presented exhaustive reviews of research on cosmological models with non-causal and causal theories of viscous fluids, respectively.

Bulk viscosity driven cosmic expansion with the Israel-Stewart theory have been investigated by Zimdahl [1996] [8], Mak and Harko [1998] [9], Paul et al. [1998] [10] and by

Arbab and Beesham [2000] [11]. As noted by Lepe et al. [2008] [12], although Eckart's theory presents some causality problems, it is the simplest alternative and has been widely considered in cosmology, as documented in Grøn [1990] [5], which we refer to for works on these topics up to 1990. We will here review papers from 1990 and onwards.

Many types of observations favor that our universe is homogeneous and isotropic on scales above a billion light years. The observations of the temperature fluctuations in the cosmic microwave radiation favor that the universe is flat, i. e. that the total density of the matter and energy contained in the universe is equal to the critical density.

The discovery that the expansion of the universe accelerates could most simply be explained by repulsive gravity due to a cosmic vacuum energy with a density equal to about 70% of the critical density. The observations also favor a special type of vacuum energy which may be represented by a cosmological constant in Einstein's field equations. The energy-momentum tensor of this energy is proportional to the metric tensor. One may show that this means that every component of the energy-momentum tensor is Lorentz invariant [Zel'dovich 1968] [13] and Grøn [1986] [14]. Hence it is not possible to measure velocity with respect to this type of energy. It may therefore be called a Lorentz invariant vacuum energy, LIVE.

Furthermore a large amount of cold dark matter is needed to keep the galaxies and the hoops of galaxies together because of the rapid motions of the stars in the galaxies and of the galaxies in the hoops. Hence, about 30 % of the contents of the universe seem to be in the form of cold dark matter. The cosmologists therefore introduced a standard model of the universe dominated by two fluids, a Lorentz invariant vacuum energy, LIVE, and a cold fluid. The vacuum energy is usually called *dark energy* and the cold fluid is called *dark matter*. Since the observations of the temperature fluctuations in the cosmic microwave radiation favor that the universe is flat, we shall only consider flat universe models.

In this review we will focus upon universe models based upon the general theory of relativity. This means among other things that we shall only review works in which the gravitational parameter is constant. It should be mentioned, though, that some researchers have investigated universe models with variable gravitational parameter (see, for instance, Belinchón, Harko and Mak [2002] [15]).

2. The Standard Model of the Universe

For later comparison we shall first briefly summarize the main properties the standard model which has vanishing viscosity. In this model the total pressure and density are given by

$$\rho = \rho_M + \rho_\Lambda, \quad p = p_M + p_\Lambda = -\rho_\Lambda, \tag{1}$$

where ρ_M is the density of cold dark matter in the form of dust, $p_M = 0$, and ρ_Λ is the density of a Lorentz Invariant Vacuum Energy, LIVE, which has $p_\Lambda = -\rho_\Lambda$ (using units so that $c = 1$), and may thus be represented by a cosmological constant in Einstein's field equations.

The line-element has the form (using units so that the velocity of light in empty space is equal to 1),

$$ds^2 = -dt^2 + a(t)^2(dr^2 + r^2 d\Omega^2), \quad d\Omega^2 = d\theta^2 + \sin^2\theta d\phi^2. \tag{2}$$

The scale factor of this model is [Grøn 2002] [16]

$$a = K_s^{1/3} \sinh^{2/3}(\frac{t}{t_\Lambda}), \quad t_\Lambda = \frac{2}{\sqrt{3\kappa\rho_\Lambda}} = \frac{2}{3H_0\sqrt{\Omega_{\Lambda 0}}}, \quad K_s = \frac{1-\Omega_{\Lambda 0}}{\Omega_{\Lambda 0}}, \tag{3}$$

where $\kappa = 8\pi G$ is Einstein's gravitational constant. Here H_0 and $\Omega_{\Lambda 0}$ are the present values of the Hubble parameter and the density parameter of LIVE. The scale factor represents the distance between two galaxy clusters relative to their present distance. Hence $a(t_0) = 1$ where the present age of the universe is

$$t_0 = t_\Lambda \,\mathrm{arctanh}\sqrt{\Omega_{\Lambda 0}}. \tag{4}$$

Inserting the presently favored values 13.7×10^9 years and $\Omega_{\Lambda 0} = 0.7$ leads to $t_\Lambda = 11.4 \times 10^9$ years.

The Hubble parameter is

$$H = (2/3t_\Lambda)\coth(t/t_\lambda). \tag{5}$$

The deceleration parameter is

$$q = -\frac{a\ddot{a}}{\dot{a}^2} = -1 - \frac{\dot{H}}{H^2}, \tag{6}$$

giving

$$q = (1/2)[1 - 3\tanh^2(t/t_\Lambda)]. \tag{7}$$

The point of time t_1 when deceleration turns into acceleration is given by $q(t_1) = 0$ which leads to

$$t_1 = t_\Lambda \mathrm{arctanh}(1/\sqrt{3}). \tag{8}$$

The corresponding redshift is

$$z = \frac{1}{a(t_1)} - 1 = \left(\frac{2\Omega_{\Lambda 0}}{1-\Omega_{\Lambda 0}}\right)^{1/3} - 1, \tag{9}$$

which gives $t_1 = 7.4 \times 10^9$ years and $z(t_1) = 0.67$. In this model the age-redshift relationship is

$$t_e = t_0 \frac{\mathrm{arcsinh}\frac{\sqrt{\Omega_{\Lambda 0}}}{\sqrt{\Omega_{M0}(1+z)^{3/2}}}}{\mathrm{arcsinh}\sqrt{\Omega_{\Lambda 0}/\Omega_{M0}}}. \tag{10}$$

Here t_e is the emitter time of a signal arriving at the instant t_0. With a unit of time equal to 10^9 years and by inserting $t_0 = 13.7, \Omega_{\Lambda 0} = 0.7$ we can write equation (10) as

$$t_e = 11.3 \,\mathrm{arcsinh}\left[1.53(1+z)^{-1.5}\right]. \tag{11}$$

This is shown graphically in Figure 1.

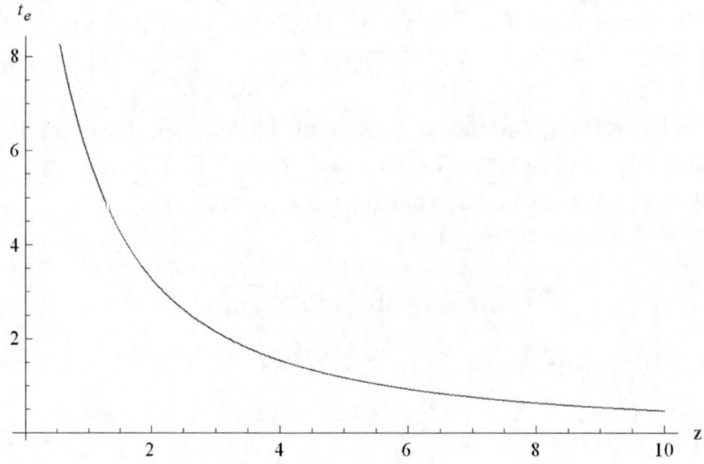

Figure 1. Emitter time in billion years as a function of cosmic redshift.

3. Viscous Fluid in an Expanding Universe

Let u^μ be components of the 4-velocity of a fluid element. The projection tensor onto a 3-space orthogonal to the world line of a fluid element is defined by

$$h_{\alpha\beta} = g_{\alpha\beta} + u_\alpha u_\beta. \tag{1}$$

The covariant derivative of the velocity field of the fluid can be written as a 3 times 3 matrix which can be separated into the antisymmetric part representing the *vorticity*,

$$\omega_{\alpha\beta} = \frac{1}{2}(u_{\mu;\nu} - u_{\nu;\mu})h^\mu_\alpha h^\nu_\beta, \tag{2}$$

the trace free, symmetrical part which represents the *shear*,

$$\sigma_{\alpha\beta} = \left[\frac{1}{2}(u_{\mu;\nu} + u_{\nu;\mu}) - \frac{1}{3}u^\lambda_{;\lambda}h_{\mu\nu}\right] h^\mu_\alpha h^\nu_\beta, \tag{3}$$

and the trace, which represents the *expansion*,

$$\theta = u^\lambda_{;\lambda}. \tag{4}$$

The four-acceleration of the fluid element, which is non-vanishing only for non-geodesic flow, is defined by

$$a_\alpha = a_{\alpha;\mu}u^\mu. \tag{5}$$

We then have

$$u_{\alpha;\beta} = \omega_{\alpha;\beta} + \sigma_{\alpha;\beta} + \frac{1}{3}\theta h_{\alpha\beta} - \sigma_\alpha u_\beta. \tag{6}$$

The energy-momentum tensor of a viscous fluid with proper density ρ and pressure p is

$$T_{\alpha\beta} = \rho u_\alpha u_\beta + (p - \xi\theta)h_{\alpha\beta} - 2\eta\sigma_{\alpha\beta}, \tag{7}$$

where η and ξ are the coefficients of shear and bulk viscosity, respectively. Einstein's field equations imply that the divergence of this tensor vanishes. From this one may deduce the equation of continuity of the fluid in the form [Kohli 2012] [17]

$$\dot{\rho} + (\rho + p)\theta - 4\eta\sigma^2 - \xi\theta^2 = 0. \tag{8}$$

The evolution of the divergence with time is given by the Raychaudhuri equation which may be written as [Ellis 2007] [18]

$$\dot{\theta} = a^{\lambda}_{;\lambda} + \frac{\kappa}{2}(3\xi\theta - \rho - 3p) + 2\omega^2 - 2\sigma^2 - \frac{1}{3}\theta^2 + \Lambda. \tag{9}$$

For a universe model containing a fluid with equation of state $p_M = w\rho_M$ and a LIVE interacting with each other, the interaction is usually represented by a term $Q(t)$ so that the equations of continuity of the fluids take the form [Jamil and Farooq [2010] [19]

$$\dot{\rho}_M + (1 + w)\rho_M\theta = -Q, \quad \dot{\rho}_\Lambda = Q. \tag{10}$$

4. Isotropic, Viscous Generalization of the Standard Universe Model

We now consider a homogeneous and isotropic universe with geodesic fluid flow. In this case $a^{\lambda}_{;\lambda} = \omega = \sigma = 0$ and eqs. (8) and (9) in Sect. 3 reduce to

$$\dot{\rho} + (\rho + p)\theta - \xi\theta^2 = 0, \tag{1}$$

and

$$\dot{\theta} = \frac{\kappa}{2}(3\xi\theta - \rho - 3p) - \frac{1}{3}\theta^2 + \Lambda. \tag{2}$$

Assuming that the 3-space is Euclidean the line element takes the form

$$ds^2 = -dt^2 + a(t)^2(dr^2 + r^2 d\theta^2 + r^2 \sin^2\theta d\phi^2). \tag{3}$$

With co-moving coordinates in the cosmic fluid the expansion of the fluid and the Hubble parameter are related by $\theta = 3H$. We shall assume that the universe contains two non-interacting fluids, Lorentz invariant vacuum energy, LIVE, that may be represented by a cosmological constant, $\kappa\rho_\Lambda = \Lambda$, and a fluid with density ρ_M and pressure $p_M = w\rho_M$. In this case Einstein's field equations take the form

$$\kappa\rho_M = 3H^2 - \Lambda, \tag{4}$$

and

$$\frac{\ddot{a}}{a} = -\frac{\kappa}{6}(1 + 3w)\rho_M + \frac{\Lambda}{3} + \frac{3\kappa\xi}{2}H. \tag{5}$$

Using that

$$\frac{\ddot{a}}{a} = \dot{H} + H^2, \tag{6}$$

these equations lead to

$$\dot{H} = -\frac{3}{2}(1+w)H^2 + \frac{3\kappa\xi}{2}H + \frac{1}{2}(1+w)\Lambda, \tag{7}$$

in accordance with eq.(2). The critical density is given by $\kappa\rho_{cr} = 3H^2$. The density parameters of the fluid and the LIVE are $\Omega_M = \rho_M/\rho_{cr}$, $\Omega_\Lambda = \Lambda/\kappa\rho_{cr}$. Defining a corresponding viscosity parameter $\Omega_{\xi 0} = \kappa\xi_0/H_0$ and using the definition (6) in Sect. 2 of the deceleration parameter equation (7) for the present time can be written as

$$q_0 = \frac{1}{2}(1+3w) - \frac{3}{2}[\Omega_{\xi 0} + (1+w)\Omega_{\Lambda 0}]. \tag{8}$$

Hence, the present value of the viscosity parameter of the mixture of a cosmic fluid and LIVE can be given in terms of the present values of the Hubble parameter, the deceleration parameter and the density parameter of LIVE. If the cosmic fluid is cold we may put $w = 0$, which leads to

$$\Omega_{\xi 0} = \frac{1}{3}(1 - 2q_0) - \Omega_{\Lambda 0}. \tag{9}$$

Pavon and Zimdahl [1993] [20] have used a corresponding formula to estimate the value of $\Omega_{\xi 0}$ that is necessary in order that the viscosity of the cosmic fluid shall be of significance for the evolution of the expansion history of the universe. A recent determination of the present value of the deceleration parameter from observational data give large uncertainties [Giostri et al 2012] [21], $-0.66 < q_0 < -0.20$. Furthermore, $0.69 < \Omega_{\Lambda 0} < 0.77$. Inserting these values into eq.(9) gives a lower boundary on $\Omega_{\xi 0}$ which is negative, which means that these data do not require any viscosity at all. Hence the motivation for studying cosmological models with viscous fluids does not come from observational data, but from the possibility of the existence of physical mechanisms that can produce viscosity. One such possibility that has been studied by Mathews et al [2008] [22], is production of viscosity by decay of dark matter particles to relativistic particles at a recent epoch with redshift $z < 1$.

Before reviewing more general models we shall consider the most simple viscous universe model. It is dominate by dust with constant viscosity coefficient, and has been studied by Padmanabhan and Chitre [1987] [23]. This model is reviewed not because it is physically realistic, but because it gives us an opportunity to demonstrate in a simple way typical features of universe models dominated by viscous fluids. For this model eq. (7) reduces to

$$\dot{H} = -\frac{3}{2}H^2 + \frac{3}{2}\Omega_{\xi 0}H_0 H. \tag{10}$$

Integrating this equation two times with the boundary conditions $H(t_0) = H_0, a(0) = 0, a(t_0) = 1$ leads to

$$H = \frac{1}{2}\Omega_{\xi 0}H_0\left[1 + \coth\left(\frac{3}{4}\Omega_{\xi 0}H_0 t\right)\right], a = \left[\frac{1-\Omega_{\xi 0}}{\Omega_{\xi 0}}\right]^{2/3}\left(e^{\frac{3}{2}\Omega_{\xi 0}H_0 t} - 1\right)^{2/3}, \tag{11}$$

where the age of the universe is given in terms of the present value of the Hubble parameter as

$$t_0 = \frac{4}{3\Omega_{\xi 0}H_0}\text{arctanh}\frac{\Omega_{\xi 0}}{2 - \Omega_{\xi 0}}. \tag{12}$$

If $\Omega_{\xi 0} H_0 t \ll 1$ the scale factor takes the approximate form $a \approx [(3/2)H_0 t]^{2/3}$ corresponding to the evolution of the universe dominated by viscosity free dust. Hence in this universe model the viscosity can be neglected at early times. On the other hand at late times so that $\Omega_{\xi 0} H_0 t \gg 1$ the expansion becomes exponential with $H = \kappa\xi_0$, $a \propto \exp(\kappa\xi_0)$, $\rho = 3\kappa\xi_0^2$, and the universe enters a late inflationary era with accelerated expansion. This was pointed out by Panmanabhan and Chitre [1987] [23] more than ten years before the late accelerated expansion of the universe was discovered. However, the time t_c after the viscosity becomes dominant may be be extremely late if there is no effective mechanism for creating viscosity in the cosmic fluids,

$$t_c = \frac{2}{3\Omega_{\xi 0} H_0} = \frac{c^2}{12\pi G \xi_0} = 1.5 \times 10^{10} \text{years} \left(\frac{10^9 \text{gcm}^{-1}\text{s}^{-1}}{\xi_0} \right). \tag{13}$$

Ren and Meng [2006] [24], Hu and Meng [2006] [25], and Mostafapoor and Grøn [2011] [26] have studied universe models in which the bulk viscosity coefficient of the viscous fluid has the following form

$$\xi = \xi_0 + \xi_1 \frac{\dot{a}}{a} + \xi_2 \frac{\ddot{a}}{\dot{a}}. \tag{14}$$

The motivation for considering this form for the coefficient of bulk viscosity is that from fluid mechanics we know that the viscosity is related to the motion of the fluid, i.e. to $\dot{a}$ and $\ddot{a}$. Inserting eq.(4.7) into eq.(4.6) we obtain

$$a\dot{H} = -bH^2 + cH + d, \tag{15}$$

where

$$a = 1 - \frac{3\kappa\xi_2}{2}, \quad b = \frac{3}{2}[1 + w - \kappa(\xi_1 + \xi_2)], \quad c = \frac{3\kappa\xi_0}{2}, \quad d = \frac{1}{2}(1+w)\Lambda. \tag{16}$$

Integration with $a(0) = 0$, $a(t_0) = 1$ and assuming $\kappa(\xi_1 + \xi_2) < 1$ and $w \geq 0$ so that $b > 0$ and $4bd + c^2 > 0$ gives

$$H(t) = \frac{c}{2b} + \frac{a}{b}\hat{H}\coth(\hat{H}t), \quad \hat{H}^2 = \frac{bd}{a^2} + \frac{c^2}{4a^2}, \tag{17}$$

and

$$a(t) = e^{\frac{c}{2b}(t-t_0)} \left[\frac{\sinh(\hat{H}t)}{\sinh(\hat{H}t_0)} \right]^{a/b}, \tag{18}$$

or

$$a(t) = (K_\xi)^{\frac{a}{2b}} e^{\frac{c}{2b}(t-t_0)} \sinh^{\frac{a}{b}}(\hat{H}t), \quad K_\xi = \frac{4b(bH_0^2 - cH_0 - d)}{4bd + c^2}, \tag{19}$$

where

$$t_0 = \frac{1}{\hat{H}} \text{arctanh} \frac{2a\hat{H}}{2bH_0 - c} \tag{20}$$

is the age of the universe. The corresponding scale factor and age of the universe if the viscosity vanishes are given by equations (3) and (4) in Sect. 2. It follows that for a given present value of the Hubble parameter the viscosity increases the age of the universe. Assuming that $\kappa\xi_0 \ll H_0$ the increase of the age due to the viscosity is of an order of magnitude

$$t_0 - t_{00} \approx \Omega_{\xi 0}^2 \, t_0. \tag{21}$$

Brevik and Heen [1994] [27] have used that in the plasma era of the universe the bulk viscosity derived from kinetic theory of gases has order of magnitude so that $(\kappa\xi_0)^{-1} \sim 10^{21}$ years. Since $(H_0)^{-1} \sim 10^{10}$ years this estimate of the magnitude of the bulk viscosity gives $\Omega_{\xi 0} \sim 10^{-11}$. During most of the evolution of the universe the viscosity is smaller than this. Hence, this form of viscosity is totally insignificant for the age of the universe.

Brevik and Stokkan [1996] [28] have, however, pointed out that impulsive processes at the end of the inflationary era may have produced great viscosity. In this extremely brief period the viscosity may have given significant contributions to the production of entropy in the universe, able to explain why the number of photons per baryon is so large, $\sim 10^9$, in our universe.

With vanishing cosmological constant, i.e. $d = 0$, the relationship between the Hubble parameter and the cosmological redshift $1 + z = a^{-1}$ is

$$H(z) = \frac{c}{b} + \left(H_0 - \frac{c}{b}\right)(1 + z^{b/a}). \tag{22}$$

Dou and Meng [2011] [29] have considered a universe model with $w = \xi_1 = \xi_2 = 0$. Then $c/b = \Omega_{\xi 0} H_0$, $b/a = 3/2$ and eq.(22) reduces to

$$H(z) = H_0 \left[\Omega_{\xi 0} + (1 - \Omega_{\xi 0})(1 + z)^{3/2}\right]. \tag{23}$$

Avelino and Nucamendi [2008] [30] have used supernova data to constrain the dimensionless viscosity parameter $\Omega_{\xi 1} = \kappa\xi_1$ by considering a matter-dominated universe model with bulk viscosity proportional to the Hubble parameter. In this case $\xi_0 = \xi_2 = w = \Lambda = 0$ giving $a = 1$, $c = d = 0$, and eq. (15) reduces to

$$\dot{H} = -bH^2. \tag{24}$$

The Hubble parameter, scale factor and density then are

$$H = \frac{H_0}{1 + bH_0(t - t_0)}, \quad a = \frac{1}{[1 + bH_0(t - t_0)]^{1/b}}, \quad \rho = \frac{\rho_0}{[1 + bH_0(t - t_0)]^2}, \tag{25}$$

with $b = (3/2)(1 - \Omega_{\xi 1})$. In terms of the redshift z the Hubble parameter can in this case be expressed as

$$H(z) = H_0(1 + z)^{(3/2)(1 - \Omega_{\xi 1})}. \tag{26}$$

Using a combination of observational data including emitter points of time at the early universe, i.e. with $z \gg 2$, Avelino and Nucamendi [2008] [31] found that $\Omega_{\xi 1}$ had to be negative in order that the properties of this universe model should not be in conflict with the data. But using only supernova data with $z < 2$ they found that agreement with the observations requires $\Omega_{\xi 1} = 0.48 \pm 0.04$, which is not physically impossible. Hence, they concluded that in order that a matter-dominated universe model with bulk viscosity proportional to the Hubble parameter shall be in agreement with observations, the viscosity must be due to a mechanism that produces viscosity mainly in the recent history of the universe. This is in agreement with the conclusion of Mathews et al [2008] [22].

From a general form of the solution corresponding to eq.(17) applied to the case of a universe with no dark matter, only dominated by dark energy with $w = -1$ with linear viscosity, i.e. $\xi_1 = \xi_2 = 0$ so that $b = 0$, Cataldo, Cruz and Lope [2005] [32] concluded that the scale factor is not defined by the field equations. However, this form of the solution is only valid for $b \neq 0$. The field equations determine the scale factor also in this case where eq.(7) reduces to

$$\dot{H} = \frac{3\kappa\xi_0}{2} H. \tag{27}$$

Integration with $a(t_0) = 0$ gives

$$H(t) = H_0 \exp\left[\frac{3\Omega_{\xi 0} H_0}{2}(t - t_0)\right],$$

$$a(t) = \exp\left\{\frac{2}{3\Omega_{\xi 0}}\left[e^{\frac{3\Omega_{\xi 0} H_0}{2}(t-t_0)} - 1\right]\right\}. \tag{28}$$

Hence a universe dominated by viscous dark energy with constant viscosity coefficient expands exponentially faster than a corresponding universe model without viscosity.

The form of the solution when $4bd + c^2 < 0$ is

$$H(t) = \frac{c}{2b} + \frac{a}{b}\hat{H}\cot(\hat{H}t), \quad \hat{H}^2 = -\left(\frac{bd}{a^2} + \frac{c^2}{4a^2}\right), \tag{29}$$

and

$$a(t) = e^{\frac{c}{2b}(t-t_0)}\left[\frac{\sin(\hat{H}t)}{\sin(\hat{H}t_0)}\right]^{a/b}. \tag{30}$$

Note that this solution is only valid for $d \neq 0$. The scale factor blows up to infinity, i.e. there is a Big Rip, at a point of time t_R given by $\hat{H}t_R = \pi$, or

$$t_R = \frac{2a\pi}{\sqrt{-(4bd + c^2)}}. \tag{31}$$

5. Viscosity and the Accelerated Expansion of the Universe

The question whether a matter dominated universe with bulk viscosity can drive the accelerated expansion of the universe has been discussed by Kremer and Devecchi [2003] [33], Fabris, Goncalves and Ribeiro [2006] [34], Avelino and Nucamendi [2009] [31]. In the universe model of Avelino and Nucamendi $\xi_1 = \xi_2 = 0$, $w = 0$, $\Omega_M = 1$, $\Omega_\Lambda = 0$, and the expression (12) in Sect. 4 for the scale factor reduces to

$$a(t) = \left[\frac{4(1 - \Omega_{\xi 0})}{\Omega_{\xi 0}^2}\right]^{1/3} \exp\left[(\Omega_{\xi 0}/2)H_0(t - t_0)\right] \sinh^{\frac{2}{3}}\left(\frac{3}{4}\Omega_{\xi 0} H_0 t\right), \tag{1}$$

where the age of the universe is

$$t_0 = \frac{4}{3\Omega_{\xi 0} H_0} \text{arctanh} \frac{\Omega_{\xi 0}}{2 - \Omega_{\xi 0}} = -\frac{2}{3\Omega_{\xi 0} H_0} \ln(1 - \Omega_{\xi 0}). \tag{2}$$

This form of the solution satisfies the boundary conditions $a(0) = 0, a(t_0) = 1$. The first of these are not satisfied by the form of the solution given by Avelino and Nucamendi. The solution (1) may be written as

$$a(t) = \left(\frac{1 - \Omega_{\xi 0}}{\Omega_{\xi 0}} \right)^{2/3} \left(e^{\frac{3}{2}\Omega_{\xi 0} H_0 t} - 1 \right)^{2/3}. \tag{3}$$

This universe model has earlier been considered by Brevik and Gorbunova [2005] [35] and by Grøn [2010] [36], and is also identical to the model in eq. (11) in Sect. 4. The Hubble parameter is

$$H(t) = \frac{\Omega_{\xi 0} H_0}{1 - e^{-(3/2)\Omega_{\xi 0} H_0 t}}. \tag{4}$$

It approaches the De Sitter model for $t \gg 1/\Omega_{\xi 0} H_0$ with a constant Hubble parameter equal to $\Omega_{\xi 0} H_0$. The deceleration parameter is

$$q = \frac{3}{2 \exp[(3/2)\Omega_{\xi 0} H_0 t]} - 1, \tag{5}$$

with present value

$$q(t_0) = (1/2)(1 - 3\Omega_{\xi 0}). \tag{6}$$

These expressions show that the expansion starts from a Big Bang with an infinitely great expansion velocity, but decelerates to a finite value. At the instant t_1 given by $q(t_1) = 0$ there is a transition to accelerated expansion, which will last for ever. The transition happens at

$$t_1 = \frac{2 \ln(3/2)}{3\Omega_{\xi 0} H_0}. \tag{7}$$

At this instant the scale factor has the value

$$a(t_1) = \left(\frac{1 - \Omega_{\xi 0}}{2\Omega_{\xi 0}} \right)^{2/3}. \tag{8}$$

The corresponding redshift is

$$z_1 = \left(\frac{2\Omega_{\xi 0}}{1 - \Omega_{\xi 0}} \right)^{2/3} - 1. \tag{9}$$

In order that the transition shall have happened at a past time, $a(t_1) < 1$, the bulk viscosity must be sufficiently large, $\Omega_{\xi 0} > 1/3$.

For this universe model, with Euclidean spatial geometry, the matter density is equal to the critical density,

$$\rho_M = \frac{3H^2}{\kappa} = \frac{3\Omega_{\xi 0}^2 H_0^2}{\kappa \left(1 - e^{-(3/2)\Omega_{\xi 0} H_0 t}\right)^2}. \tag{10}$$

Hence, the matter density approaches a constant value, $\rho_M \to (3/\kappa)\Omega_{\xi 0}^2 H_0^2$.

Avelino and Nucamendi [2009] [31] have used the most comprehensive supernova data to estimate the value of $\Omega_{\xi 0}$ that gives the best fit with observed data for a universe model containing dust with constant viscosity coefficient. The result was $\Omega_{\xi 0} = 0.64$, which is eleven orders of magnitude greater than the value coming from kinetic gas theory [27]. However, a mechanism for producing greater viscosity may be a bulk viscosity generation due to decay of dark matter particles into relativistic products; cf. Singh [2008] [37].

The final fate of a universe dominated by a viscous fluid has been discussed by Brevik and Gorbunova [2005] [35] and by Cataldo, Cruz and Lepe [2005] [32]. Consider first a universe without viscosity and dark energy, containing only a non-viscous fluid with $p = w\rho$. In this case eq. (15) in Sect. 4 reduces to

$$\dot{H} = -bH^2, \quad b = \frac{3}{2}(1+w). \tag{11}$$

The Hubble parameter, scale factor and density are given by eq. (15) in Sect. 4 with b given by eq. (11). For this universe model there is a Big Rip at an instant

$$t_{R0} = t_0 + \frac{2}{3(1+w)H_0}. \tag{12}$$

Cataldo et al. then considered a universe model with a fluid having $w < -1$ and constant viscosity coefficient ξ_0. Then $b < 0$ and $d = 0$ so neither of the above solutions are valid. In this case eq. (15) in Sect. 4 reduces to

$$\dot{H} = -\frac{3}{2}(1+w)H^2 + \frac{3}{2}\Omega_{\xi 0} H_0 H. \tag{13}$$

The Hubble parameter, scale factor and density for this universe model are

$$H = \frac{H_0}{\frac{1+w}{\Omega_{\xi 0}} + \left(1 - \frac{1+w}{\Omega_{\xi 0}}\right) e^{-\frac{3}{2}\Omega_{\xi 0}(t-t_0)}}, \tag{14}$$

$$a = \left[1 - \frac{1+w}{\Omega_{\xi 0}} + \frac{1+w}{\Omega_{\xi_0}} e^{\frac{3}{2}\Omega_{\xi 0} H_0 (t-t_0)}\right]^{\frac{2}{3(1+w)}}, \tag{15}$$

and

$$\rho = \frac{\rho_0}{\left[\frac{1+w}{\Omega_{\xi 0}} + \left(1 - \frac{1+w}{\Omega_{\xi 0}}\right) e^{-\frac{3}{2}\Omega_{\xi 0}(t-t_0)}\right]^2}. \tag{16}$$

In this case there is a Big Rip at the instant

$$t_R = t_0 + \frac{2}{3\Omega_{\xi 0} H_0} \ln\left(1 - \frac{\Omega_{\xi 0}}{1+w}\right). \tag{17}$$

To second order in $\Omega_{\xi 0}/(1+w)$ this formula gives

$$t_R \approx t_{R0} - \frac{\Omega_{\xi 0}}{3(1+w)^2 H_0}, \tag{18}$$

where t_{R0} is given in eq. (12). Hence the viscosity makes the Big Rip come earlier.

Cataldo et al. [2005] [32] have also considered a universe model in which the viscosity coefficient of the fluid is proportional to the square root of the density, $\xi = \alpha\rho^{1/2}$. Since the mass density is equal to the critical density, $\kappa\rho = 3H^2$, this corresponds to the case (19) with $b = (3/2)(1 + w - \sqrt{3}\alpha)$, where κ has been absorbed into α. This again is equivalent to a universe model with only one viscous fluid where the coefficient of viscosity is proportional to the Hubble parameter.

Similar universe models with variable gravitational and cosmological 'constants' have been investigated by Singh, Beesham and Mbokazi [1998] [38] and by Singh, Kumar and Pradhan [2007] [39].

6. Viscous Bianchi Type-I Universe Models

The influence of viscosity on Bianchi type-I universe models, which are the anisotropic generalizations of the flat Friedmann-Robertson-Walker universe models, has been investigated by Belinski and Khalatnikov [1975] [40]. They found that for large times such universes models with constant coefficients of bulk viscosity, will approach an isotropic steady-state universe model with de Sitter spacetime which expands exponentially. For asymptotically early times they found that there exists a Kasner era in which the effect of matter, radiation and viscosity are negligible. Heller [1978] [41] has concluded that such universe models have in general a stage near an unavoidable initial singularity in which the energy-momentum tensor has no influence on the cosmic evolution. However, Grøn [1990] [5] has found that in a Bianchi type-I universe filled with viscous Zel'dovich fluid the bulk viscosity may remove the initial singularity.

In this section we will review articles where the influence of viscosity on the evolution of anisotropic Bianchi universe models filled with viscous fluids has been studied, following Mostafapoor and Grøn [2012] [42].

The line element of a Bianchi type-I universe can be written in the form

$$ds^2 = dt^2 - R_i^2 \cdot (dx^i)^2, \tag{1}$$

where $R_1 = a(t), R_2 = b(t), R_3 = c(t)$ are directional scale factors. The energy-momentum tensor of the viscous fluid has the non-vanishing components

$$T_0^0 = \rho, \quad T_i^i = -p + 2\eta H_i + (3\xi - 2\eta)H - 9\alpha H\Delta H, \tag{2}$$

where $H_i = \dot{R}_i/R_i$ are the directional Hubble parameters, $H = (1/3)\sum_{i=1}^{3} H_i$, $\Delta H_i = H_i - H$, $p = w\rho$, and α is a non-linear viscosity coefficient. For these universe models the Raychaudhuri equation takes the form

$$\dot{H} = -3H^2 + \frac{\kappa}{2}(1 - w)\rho + \frac{3}{2}\kappa\xi H + \Lambda. \tag{3}$$

The anisotropy parameter is defined as [14]

$$A = \frac{1}{3}\sum_{i=1}^{3}\left(\frac{\Delta H_i}{H}\right)^2 = \frac{1}{9}\sum_{i<j}\left(\frac{H_i - H_j}{H}\right)^2. \tag{4}$$

From Einstein's equations then follows

$$\kappa\rho = (3/2)(2 - A)H^2 - \Lambda, \tag{5}$$

and

$$A = C\frac{\tau^{2(3\alpha-1)}e^{-2\Phi}}{H^2}, \quad \tau = abc, \quad \Phi = 2\kappa\int \eta dt, \tag{6}$$

where C is an integration constant which is given by the present values as $C = A_0 H_0^2 e^{2\Phi_0}$ with the normalization $\tau(t_0) = 1$. Insertion of eq. (6) into eq. (5) leads to

$$\kappa\rho = 3H^2 - (3/2)C\tau^{2(3\alpha-1)}e^{-2\Phi} - \Lambda. \tag{7}$$

A further insertion of eq. (7) into eq. (3) gives

$$\dot{H} = -\frac{3}{2}(1+w)H^2 + \frac{3}{2}\kappa\xi H - \frac{3C}{4}(1-w)\tau^{-2(1-3\alpha)}e^{-2\Phi} + \frac{1}{2}(1+w)\Lambda. \tag{8}$$

Saha [2008] [43] has investigated Bianchi type-I universe models containing fluids with different kinds of viscosity. As a first case he considered a model with vanishing shear viscosity. Then $\Phi = 0$. Assuming also that there is no non-linear viscosity so that $\alpha = 0$, eqs. (6) and (7) reduce to

$$A = C/(\tau H)^2 = 9C/\dot{\tau}^2, \tag{9}$$

and

$$\kappa\rho = 3H^2 - \Lambda - (3/2)C/\tau^2. \tag{10}$$

Saha [2008] [43] has found a similar equation, but with positive sign before Λ and claims that during the expansion of the universe τ increases and ρ decreases. Assuming that at some stage $\tau \to \infty$, $\rho \to 0$ it follows that (with Saha's sign) $3H^2 + \Lambda \to 0$. From this Saha concludes that the universe may be infinitely large only if $\Lambda \leq 0$. It seems, however, that Saha has got the cosmological constant with wrong sign in his field equations (2.22), so the correct conclusion is that a positive cosmological constant permits the scale factor to evolve towards an infinitly large value.

6.1. Bianchi Type-I Universe with Viscous Zel'dovich Fluid and LIVE

Saha also considered the case where the coefficient of viscosity is inversely proportional to the Hubble parameter, $\xi H = \text{constant} = 2C_\xi/3\kappa$. From eq. (8) is seen that in this case the viscosity acts like a cosmological constant. Assuming further that the cosmic fluid is a Zel'dovich fluid, eq. (8) takes the form (15) in Sect. 4 with $a = 1, b = 3, c = 0, d = C_\xi + \Lambda$. With the boundary conditions $H(t_0) = H_0, \tau(0) = 0, \tau(t_0) = 1$ it is found that the Hubble parameter, the volume scale factor and the anisotropy parameter are

$$H = \frac{\hat{H}}{3}\coth(\hat{H}t), \quad \tau = \left[\left(\frac{3H_0}{\hat{H}}\right)^2 - 1\right]^{1/2} \sinh(\hat{H}t), \quad A = A_0\left(\frac{H_0}{\tau H}\right)^2, \tag{11}$$

where $\hat{H} = 3(C_\xi + \Lambda)$ and A_0 is the present value of the anisotropy parameter. The age of the universe and the present values of the Hubble parameter are

$$t_0 = \frac{1}{\hat{H}}\text{arctanh}(\hat{H}t), \quad H_0 = \sqrt{\frac{2\kappa(\rho_0 + \rho_{\Lambda 0})}{3(2 - A_0)}}. \tag{12}$$

Furthermore, Saha considered a universe model with both bulk and shear viscosity. He assumed that the coefficient of shear viscosity is negative and proportional to the Hubble parameter, so that

$$(2\kappa\eta/3) = -H. \tag{13}$$

This gives

$$e^{-\Phi} = \tau,$$

so that eq. (7) with $\alpha = 0$ reduces to

$$\kappa\rho = 3H^2 - (3/2)C - \Lambda. \tag{14}$$

Hence the constant C is given in terms of present quantities as $C = 2H_0^2 - (2\kappa/3)(\rho_0 + \rho_{\Lambda 0})$. Then eq. (8) takes the form (15) in Sect. 4 with $a = 1$, $b = (3/2)(1+w)$, $c = (3/2)\kappa\rho$, $d = [H_0^2 - (\kappa/3)\rho_0](1-w) + \Lambda$, and the solution is given by eqs. (17)-(20) in Sect. 4. In the case of a Zel'dovich fluid, $b = 3$ and $d = \Lambda$.

A similar example was considered by Mak, Harko and Fung [1997] [44], who investigated a Bianchi type-I universe model with a viscous fluid consisting of a mixture of a Zel'dovich fluid (also called stiff matter because the velocity of sound is equal to the velocity of light) with $w = 1$ and LIVE with $w = -1$. In this model the only type of viscosity was a constant bulk viscosity. Then eq. (3) reduces to

$$\dot{H} = -3H^2 + \frac{3}{2}\kappa\xi H + \Lambda, \tag{15}$$

which has the same form as eq. (11) in Sect. 4 with $a = 1$, $b = 3$, $c = (3/2)\kappa\xi_0 = (3/2)\Omega_{\xi 0}H_0$, $d = \Lambda = 3\Omega_{\Lambda 0}H_0^2$. From eq. (13) in Sect. 4 the Hubble parameter then is

$$H = H_0\left[\Omega_{\xi 0}/4 + K_{\lambda\xi}\coth(3K_{\lambda\xi}H_0 t)\right], \quad K_{\lambda\xi} = \sqrt{\Omega_{\Lambda 0} + (\Omega_{\xi 0}/4)^2}. \tag{16}$$

According to eq. (15) in Sect. 4 the volume scale factor normalized to unity at present time is

$$\tau = K_\xi^{1/2} e^{(3/4)\Omega_{\xi 0}H_0(t-t_0)}\sinh(3K_{\lambda\xi}H_0 t), \quad K_\xi = \frac{1 - \Omega_{\Lambda 0} - \Omega_{\xi 0}/2}{K_{\lambda\xi}^2}. \tag{17}$$

Equation (16) in Sect. 4 gives for the age of this universe model

$$t_0 = \frac{1}{3K_{\Lambda\xi}H_0}\operatorname{arctanh}\frac{K_{\Lambda\xi}}{1 - \Omega_{\xi 0}/4}. \tag{18}$$

With $A(0) = 2$, which is equal to the anisotropy of a Kasner universe [14], eq. (6) gives for the anisotropy parameter

$$A = \frac{2}{e^{(1/2)(3\Omega_{\xi 0} + 8\Omega_{\eta 0})H_0 t}\left[\cosh(3K_{\lambda\xi}H_0 t) + \frac{\sinh(3K_{\lambda\xi}H_0 t)}{\sqrt{1 + 16\Omega_{\lambda 0}/\Omega_{\xi 0}^2}}\right]^2}, \tag{19}$$

where $\Omega_{\eta 0} = \kappa\eta_0/H_0$.

The Hubble parameter and hence, from eq. (7) also the energy density of the fluid, are both infinitely large at the beginning of the cosmic evolution. As t increases, the Hubble parameter and the energy density decrease and approach finite values. The universe starts from a Big Bang with a vanishing value of the volume scale factor.

6.2. Bianchi Type I Universe with Variable Shear and Bulk Viscosity

In order to consider a simple example of anisotropic universe models with variable viscosity, Mak, Harko and Fung [1997] [44] also investigated a universe model dominated by a Zel'dovich fluid with vanishing shear viscosity and bulk viscosity proportional to the Hubble parameter H averaged over the different directions, $\xi = \xi_1 H$. In this case eq. (8) reduces to

$$\dot{H} = -3\left(1 - \frac{1}{2}\kappa\xi_1\right)H^2. \tag{20}$$

The Hubble parameter and the scale factor are

$$H = \frac{H_0}{1 + K_{\xi 1}H_0(t - t_0)}, \quad \tau = [1 + K_{\xi 1}H_0(t - t_0)]^{3K_{\xi 1}},$$

$$K_{\xi 1} = 3[1 - (1/2)\kappa\xi_1]. \tag{21}$$

With the condition $A(t_0) = 2$ the anisotropy parameter of this universe model is

$$A = \frac{2}{[1 + K_{\xi 1}H_0(t - t_0)]^{8 - \frac{9}{2}\kappa\xi_1}}. \tag{22}$$

In order that this shall represent a Big Bang universe model the volume scale factor must obey $\tau(0) = 0$. Then the age of this universe model is

$$t_0 = \frac{1}{K_{\xi 1}H_0}. \tag{23}$$

In this case the Hubble parameter, volume scale factor and anisotropy parameters are

$$H = \frac{1}{K_{\xi 1}t}, \quad \tau = (K_{\xi 1}H_0 t)^{3K_{\xi 1}}, \quad A = \frac{2}{(K_{\xi 1}H_0 t)^{8-(9/2)\kappa\xi_1}}. \tag{24}$$

The effect of shear viscosity in such a universe may be investigated analytically by introducing a carefully chosen proportionality constant between the coefficient of shear viscosity and the Hubble parameter,

$$\eta = -\frac{3}{2}(1 - 3\alpha)H, \tag{25}$$

so that $\tau^{2(2\alpha-1)}e^{-2\Phi} = 1$. Furthermore we assume that the coefficient of bulk viscosity depends linearly upon H, $\xi = \xi_0 + \xi_1 H$. Then eq. (8) reduces to

$$\dot{H} = -\frac{3}{2}(1 + w - \kappa\xi_1)H^2 + \frac{3}{2}\kappa\xi_0 H - \frac{3}{4}C(1 - w) + \frac{1}{2}(1 + w)\Lambda. \tag{26}$$

Integration with $\tau(0) = 0$, $\tau(t_0) = 1$ gives

$$H(t) = \frac{\kappa\xi_0}{2(1 + w - \kappa\xi_1)} + \hat{H}\coth\left[\frac{3}{2}\hat{H}(1 + w - \kappa\xi_1)t\right], \tag{27}$$

where

$$\hat{H}^2 = \left(\frac{\kappa\xi_0}{2(1 + w - \kappa\xi_1)}\right)^2 + \frac{(1 + w)(\Lambda/3) + (1 - w)(C/2)}{1 + w - \kappa\xi_1}, \tag{28}$$

and the volume factor is

$$\tau(t) = e^{\frac{3\kappa\xi_0(t-t_0)}{2(1+w-\kappa\xi_1)}} \left(\frac{\sinh\left[\frac{3}{2}\hat{H}(1+w-\kappa\xi_1)t\right]}{\sinh\left[\frac{3}{2}\hat{H}(1+w-\kappa\xi_1)t_0\right]} \right)^{\frac{2}{1+w-\xi_1}}, \tag{29}$$

where

$$t_0 = \frac{2}{3\hat{H}(1+w-\kappa\xi_1)} \mathrm{arccoth}\left[\frac{H_0 - \frac{\kappa\xi_0}{2(1+w-\kappa\xi_1)}}{\hat{H}} \right]. \tag{30}$$

In this case the anisotropy parameter is given by

$$A = C/H^2, \tag{31}$$

and the density is

$$\kappa\rho = 3H^2 - \Lambda - 3C/2. \tag{32}$$

Here $C = (2/3)(3H_0^2 - \kappa\rho_0 - \Lambda)$, giving

$$\kappa\rho = \kappa\rho_0 + 3(H^2 - H_0^2). \tag{33}$$

7. Anisotropic Universe Model with Decaying Vacuum Energy

Bali, Singh and Singh [2012] [45] have considered a related universe model with Zel'dovich fluid, $w = 1$, and a decaying vacuum energy with density proportional to the Hubble parameter, $\Lambda = \alpha H$, where α is a positive constant. Then eq. (13) in Sect. 6 reduces to

$$\dot{H} = -\frac{3}{2}(2-\kappa\xi_1)H^2 + \left(\alpha + \frac{3}{2}\kappa\xi_0\right) H. \tag{1}$$

The general solution is

$$H = \frac{\beta H_0}{H_0\gamma - (H_0\gamma - \beta)e^{-\beta(t-t_0)}}, \tag{2}$$

where $\beta = \alpha + (3/2)\kappa\xi_0$, $\gamma = 3 - (3/2)\kappa\xi_1$ and $H_0 = H(t_0)$. The initial value of the Hubble parameter is

$$H(0) = \frac{\beta H_0}{H_0\gamma - (H_0\gamma - \beta)e^{\beta t_0}}. \tag{3}$$

Considering a universe model with an initial Big Bang having $H(0) = \infty$ gives the age of the universe model in terms of the present value of the Hubble parameter,

$$t_0 = -\frac{1}{\beta}\ln\left(1 - \frac{\beta}{H_0\gamma}\right). \tag{4}$$

For this universe model the expression for the Hubble parameter reduces to

$$H = \frac{\beta}{\gamma(1-e^{-\beta t})}. \tag{5}$$

Introducing an average scale factor $R = (R_1R_2R_3)^{1/3}$ so that $H = \dot{R}/R$ and integrating with the normalization $R(t_0) = 1$ we obtain

$$R = \left(\frac{e^{\beta t} - 1}{e^{\beta t_0} - 1}\right)^{1/\gamma}. \tag{6}$$

The decay of the density of the vacuum energy is given by

$$\Lambda = \frac{\alpha\beta}{\gamma(1 - e^{-\beta t})}. \tag{7}$$

The deceleration parameter as given in eq. (6) in Sect. 2 is

$$q = \gamma e^{-\beta t} - 1. \tag{8}$$

At early times $q \approx 2 - (3/2)\kappa\xi_1 > 0$ and the expansion decelerates. The deceleration is reduced by the component of the bulk viscosity proportional to the Hubble parameter. At the instant t_1 given by $q(t_1) = 0$, i.e. at $t_1 = (1/\beta)\ln\gamma$, there is a transition from decelerated to accelerated expansion. In the limit of large times this universe model enters a de Sitter era with constant Hubble parameter

$$H(t \to \infty) = \frac{\alpha + (3/2)\kappa\xi_0}{3 - (3/2)\kappa\xi_1}, \tag{9}$$

constant density of vacuum energy, and constant deceleration parameter $q(t \to \infty) = -1$.

8. Other Viscous Universe Models

Pradhan and Srivastava [2007] [46] have investigated viscous universe models of Bianchi type V with constant viscosity coefficient, and with viscosity coefficient proportional to the density, i.e. with $\xi = \xi_0$ and $\xi = \xi_2 H^2$. The line elements of these universe models have the form

$$ds^2 = -\frac{1}{2}dT^2 + n^2T^2dX^2 + n^{\frac{2}{n}}T^{\frac{2}{n}}e^{2X}(dY^2 + dZ^2). \tag{1}$$

With this line element as a point of departure the authors calculated the time dependence of the density and the cosmological 'constant', $\Lambda(T)$. Similar universe models have been investigated by Singh and Baghel [2009] [47].

Ram, Singh and Verma [2012] [48] have investigated universe models of Bianchi type II with variable G and Λ. Their solution of the field equations seem rather ad hoc, however, since they assume a specific form of the metric and calculate the time dependence of G, Λ and ξ from the field equations.

9. Viscosity, Turbulence, and Big Rip/Little Rip Cosmology

Let us return to the observed accelerated expansion of the universe. Current observations indicate that the equation-of-state parameter w lies in the region $w = -1.04^{+0.09}_{-0.10}$; cf. Nakamura et al. [2010] [49] and Amanullah et al. [2010] [50]. The case $w < -1$ (the phantom case) is the least understood region, as all the four energy conditions are violated. Although the theory is unstable from a quantum field theoretical viewpoint, it could be stable in classical cosmology. Observations indicate; cf. Caldwell [2001] [51] and Caldwell et al. [2003] [52], that the crossing of the cosmological constant/phantom divide took place in the near past, or will occur in the near future. An essential property of most of the phantom dark energy models is the Big Rip future singularity [51, 52] where the scale factor becomes infinite at a finite time in the future. A softer future singularity caused by phantom or quintessence dark energy is the so-called Type II singularity; cf. Barrow [2004] [53], Nojiri and Odintsov [2004] [54], where the scale factor is finite at the Rip time. Recently, an attempt to resolve the finite-time future singularities has been proposed by introducing mild phantom models where w asymptotically tends to -1 and where the energy density increases with time or remains constant, but where the singularity occurs in the infinite future; cf. Frampton et al. [2011, 2012] [55, 56, 57]. The key point here is that if w approaches -1 sufficiently fast, the time required for occurrence of the singularity is infinite. That is, in effect the singularity never happens.

The pioneering works on the future singularity [51, 52] considered the cosmic fluid to be nonviscous. Viscous Little Rip cosmology in an isotropic fluid (only bulk viscosity present) was recently worked out by Brevik et al. [2011] [58]. Let us give one of the characteristic results from that investigation: if the effective pressure p_{eff} is assumed to have the explicit form

$$p_{\text{eff}} = -\rho - A\sqrt{\rho} - 3\xi H \tag{1}$$

with A a positive constant, and if moreover the bulk viscosity ξ is assumed to satisfy the condition

$$3\xi H \equiv C = \text{constant}\,, \tag{2}$$

then the following expression is found for the time dependent energy density

$$\rho(t) = \left[\left(\frac{C}{A} + \sqrt{\rho_0}\right)\exp\left(\frac{1}{2}\sqrt{3\kappa}\,At\right) - \frac{C}{A}\right]^2 \tag{3}$$

(recall that $\kappa = 8\pi G$). Subscript zero refers to the present time. Thus an infinite time is required to reach the infinite energy density state. This is just the characteristic property of the Little Rip 'singularity'.

9.1. A Turbulent Approach

The simple description above, in terms of macroscopic bulk viscosity in the fluid, cannot be considered to be satisfactory in the later stages of the development of the universe where it approaches the future singularity. The reason is that the motion is then violent, and a transition into *turbulent* motion seems to be inevitable. The local Reynolds number must be expected to be very high. That brings in fact the *shear* viscosity back into the analysis (it is usually neglected in viscous cosmology due to the assumed spatial isotropy), now not in a macroscopic but in a local sense, causing the distribution of local eddies over the wave number spectrum. In this section we will review the cosmological turbulence theory as formulated by Brevik et al. [2012] [59].

What kind of turbulence should we expect? The natural choice is that of isotropic turbulence, which is a topic reasonably well understood. Thus, we should expect a Loitziankii region for low wave numbers where the energy density varies proportionally to k^4; for higher k we should expect an inertial subrange characterized by the formula

$$E(k) = \alpha\epsilon^{2/3}k^{-5/3} \tag{4}$$

with α the Kolmogorov constant and ϵ the mean energy dissipation per unit time and unit mass; and finally when the values of k become as high as the inverse Kolmogorov length η_L,

$$k \to k_L = \frac{1}{\eta_L} = \left(\frac{\epsilon}{\nu^3}\right)^{1/4} \tag{5}$$

with ν the kinematic viscosity, we enter the dissipative region where the local Reynolds number is of order unity and heat dissipation occurs. In accordance with common usage we shall consider the fluid system as quasi-stationary, and omit the production of heat energy. In practical cases it may be useful to combine these elements into the useful von Kármán interpolation formula which covers the whole wave number spectrum (cf., for instance, Panchev [1971] [60], Brevik [1992] [61], Carhart and Kostinski [1988] [62].

However, the full spectral theory of isotropic turbulence will not be needed in our first approach to the problem. Rather, we shall in the following focus attention on how the turbulent part of the energy density, called ρ_{turb}, can be estimated to vary from present time t_0 onwards. First, we write the effective energy density as a sum of two terms,

$$\rho_{\text{eff}} = \rho + \rho_{\text{turb}} \,, \tag{6}$$

where ρ denotes the conventional macroscopic energy density in the local rest inertial system of the fluid. It is natural to assume that ρ_{turb} is proportional to ρ itself. Further, we shall assume that ρ_{turb} is proportional to the scalar expansion $\theta = u^\mu{}_{;\mu} = 3H$. This because physically speaking the transition to turbulence is expected to be more pronounced in the violent later stages, and a proportionality to the scalar expansion is mathematically the most simple way in which to represent the effect. Calling the proportionality factor τ, we can thus write the effective energy density as

$$\rho_{\text{eff}} = \rho(1 + 3\tau H) \,. \tag{7}$$

Consider next the effective pressure p_{eff}. We split it into two terms,

$$p_{\text{eff}} = p + p_{\text{turb}} \,, \tag{8}$$

analogously as above. For the conventional non-turbulent quantities p and ρ we assume the standard relationship

$$p = w\rho \,, \tag{9}$$

where $-1 < w < -1/3$ in the quintessence region and $w < -1$ in the phantom region. The question now is: How does p_{turb} depend on ρ_{turb}? There seems to be no definite physical guidance to that problem, so we shall make the simplest possible choice in the following, namely write

$$p_{\text{turb}} = w_{\text{turb}}\,\rho_{\text{turb}} \,, \tag{10}$$

with $w_{\rm turb}$ a constant.

We shall consider two different possibilities for the value of $w_{\rm turb}$. The first is to put $w_{\rm turb}$ equal to w in Eq. (9), meaning that the turbulent matter behaves in the same way as the non-turbulent matter as far as the equation of state is concerned. This option is straightforward and natural, and is not quite trivial since $\rho_{\rm turb}$ and ρ behave differently, in view of Eq. (7). Our second option will be to assume that $w_{\rm turb}$ takes another, prescribed value. In view of the expected violent conditions near the future singularity, it might even be natural here to chose the value $w_{\rm turb} = +1$, i.e., the Zel'dovich fluid option.

The first and the second Friedmann equations can now be written

$$H^2 = \frac{1}{3}\kappa\rho(1+3\tau H)\,, \tag{11}$$

$$\frac{2\ddot{a}}{a} + H^2 = -\kappa\rho(w + 3\tau H w_{\rm turb})\,. \tag{12}$$

This may be compared with an earlier attempt of Brevik et al. [2011] [63] to introduce the turbulence in dark energy.

Equations (11) and (12) determine our physical model. Recall that its input parameters are $\{w, w_{\rm turb}, \tau\}$, all assumed constant. From these equations we can now describe the development of the Hubble parameter. For convenience we introduce the quantities γ and $\gamma_{\rm turb}$, defined as

$$\gamma = 1 + w, \quad \gamma_{\rm turb} = 1 + w_{\rm turb}\,. \tag{13}$$

We can then write the governing equation for H as

$$(1+3\tau H)\dot{H} + \frac{3}{2}\gamma H^2 + \frac{9}{2}\tau\gamma_{\rm turb}H^3 = 0\,. \tag{14}$$

This equation is in principle to be integrated from present time $t = t_0 = 0$ onwards, with initial value $H = H_0 = \dot{a}_0/a_0$.

If $T^{\mu\nu}_{\rm tot}$ denotes the total energy-momentum tensor for the cosmic fluid, we must have

$$T^{\mu\nu}_{{\rm tot};\nu} = 0\,, \tag{15}$$

as a consequence of Einstein's equation.

In most cases studied, the expression for $T^{\mu\nu}_{\rm tot}$ can be written down explicitly; this is so for non-viscous fluids as well as with macroscopic viscous fluids. In the present case this no longer true, however, since the turbulent energy is produced by shear stresses on a *small scale*, much less than the scale of the macroscopic fluid equations. That is, we are dealing with a non-closed physical system, of essentially the same kind as encountered in phenomenological electrodynamics in a continuous medium in special relativity. It implies that the source term in the energy balance equation has to be put in by hand.

Let henceforth $T^{\mu\nu}$ refer to the non-viscous part of the fluid. We may express the energy balance as

$$\dot{\rho} + 3H(\rho + p) = -Q\,, \tag{16}$$

where the source term Q is positive, corresponding to an energy sink for the non-viscous fluid. We shall put Q equal to $\epsilon\rho$, where the specific energy dissipation ϵ however shall be taken to involve the large Hubble parameter H in the later stage of the development. Let us assume the form

$$\epsilon = \epsilon_0(1 + 3\tau H)\,, \tag{17}$$

ϵ_0 being the specific energy dissipation at present time. This equation is seen to contain the same kind of development as assumed before; cf. the analogous Eq. (7). Thus, our ansatz for the energy balance reads

$$\dot{\rho} + 3H(\rho + p) = -\rho\epsilon_0(1 + 3\tau H)\,. \tag{18}$$

We now consider in some detail the mentioned two different options for the value of w_{turb}.

9.1.1. The Case $w_{\text{turb}} = w < -1$

This case means that the turbulent component of the fluid is regarded as a passive ingredient as far as the parameter w is concerned. The time development of ρ and ρ_{turb} will however be different. Equation (14) reduces to

$$\dot{H} + \frac{3}{2}\gamma H^2 = 0\,, \tag{19}$$

leading to

$$H = \frac{H_0}{Z}\,, \tag{20}$$

where we have defined

$$Z = 1 + \frac{3}{2}\gamma H_0 t \tag{21}$$

(note that $w < -1$ implies $\gamma < 0$). Thus we have a Big Rip cosmology, where the future singularity time t_s is given by

$$t_s = \frac{2}{3|\gamma|H_0}\,. \tag{22}$$

The scale factor becomes correspondingly

$$a = a_0 Z^{2/3\gamma}\,, \tag{23}$$

and from the first Friedmann equation (11) we get the non-turbulent energy density as

$$\rho = \frac{3H_0^2}{\kappa^2}\frac{1}{Z}\frac{1}{Z + 3\tau H_0}\,. \tag{24}$$

The ratio between turbulent and non-turbulent energy becomes

$$\frac{\rho_{\text{turb}}}{\rho} = 3\tau H = \frac{3\tau H_0}{Z}\,. \tag{25}$$

It is of main interest to consider the behavior near t_s. As $Z = 1 - t/t_s$ we see that

$$H \sim \frac{1}{t_s - t}\,, \qquad a \sim \frac{1}{(t_s - t)^{2/3|\gamma|}}\,, \tag{26}$$

$$\rho \sim \frac{1}{t_s - t}, \qquad \frac{\rho_{\text{turb}}}{\rho} \sim \frac{1}{t_s - t}. \tag{27}$$

Notice the difference from conventional cosmology: the behavior of H and a near the singularity is as usual, while the singularity of ρ has become *weakened*. The reason for this is, of course, the non-vanishing value of the parameter τ. Moreover, as $t \to t_s$ all the non-turbulent energy has been converted into turbulent energy. From a physical point of view, this is just as we would expect.

9.1.2. The Case $w < -1, w_{\text{turb}} > -1$

This case is thermodynamically quite different from the preceding one as the turbulent component of the fluid is no longer a passive ingredient. We have now $\gamma_{\text{turb}} = 1 + w_{\text{turb}} > 0$, which means that we cover the region $-1 < w_{\text{turb}} < 0$ also. In the latter region, the turbulent contribution to the pressure is still negative as above, while if $w_{\text{turb}} > 0$ the turbulent pressure becomes *positive*, just as in ordinary hydrodynamical turbulence.

The governing equation (14), written as

$$(1 + 3\tau H)\dot{H} = \frac{3}{2}H^2(|\gamma| - 3\tau\gamma_{\text{turb}}H), \tag{28}$$

tells us that at the present time $t = 0$ the condition

$$|\gamma| > 3\tau\gamma_{\text{turb}}H_0 \tag{29}$$

must hold, because at the present time the turbulent part is regarded as unimportant. This corresponds to the inequality $\dot{H} > 0$.

Equation (28) can be integrated to give t as a function of H,

$$t = \frac{2}{3|\gamma|}\left(\frac{1}{H_0} - \frac{1}{H}\right) - \frac{2\tau}{|\gamma|}\left(1 + \frac{\gamma_{\text{turb}}}{|\gamma|}\right)\ln\left[\frac{|\gamma| - 3\tau\gamma_{\text{turb}}H}{|\gamma| - 3\tau\gamma_{\text{turb}}H_0}\frac{H_0}{H}\right]. \tag{30}$$

A striking property of this expression is that it describes a *Little Rip* scenario. As $t \to \infty$, the Hubble parameter reaches a finite critical value

$$H_{\text{crit}} = \frac{1}{3\tau}\frac{|\gamma|}{\gamma_{\text{turb}}}. \tag{31}$$

The physical role of γ_{turb} is thus to postpone and weaken the development towards the future singularity.

A natural choice for the EoS parameter w_{turb} in the vicinity of the singularity, in view of the violent motions expected, would be

$$w_{\text{turb}} = +1, \tag{32}$$

that means, a Zel'dovich fluid. This is an extreme case, where the velocity of sound equals the velocity of light.

9.2. A One-Component Dark Fluid

We now turn to an approach that is quite different from the one above, namely to consider the cosmic fluid as a *one-component* fluid. Thus the distinction between a non-turbulent and a turbulent fluid component is avoided altogether. This new approach is actually more close to the usual picture in hydrodynamics, where a fluid is known to shift suddenly from a laminar to a turbulent state.

Consider the following picture: the universe starts from present time $t = 0$ as an ordinary viscous fluid with a bulk viscosity ξ, and develops according to the Friedmann equations. We assume as before that the EoS parameter $w < -1$, meaning that the universe develops in the viscous era towards a future singularity. Before this happens, however, at some instant $t = t_*$, we assume that there is a sudden transition of the whole fluid into a turbulent state, after which the EoS parameter is $w_{\rm turb}$ and the pressure accordingly $p_{\rm turb} = w_{\rm turb}\,\rho_{\rm turb}$. As before we assume that $w_{\rm turb} > -1$, and for simplicity we take ξ, as well as w and $w_{\rm turb}$, to be constants. One may ask: What is the resulting behavior of the fluid, especially at later stages?

The problem can easily be solved, making use of the condition that the density of the fluid has to be continuous at $t = t_*$. One gets; cf. Brevik and Gorbunova [2005] [35], Brevik et al. [2010] [64]

$$H = \frac{H_0\, e^{t/t_c}}{1 - \frac{3}{2}|\gamma| H_0 t_c (e^{t/t_c} - 1)}\,, \tag{33}$$

$$a = \frac{a_0}{\left[1 - \frac{3}{2}|\gamma| H_0 t_c (e^{t/t_c} - 1)\right]^{2/3|\gamma|}}\,, \tag{34}$$

$$\rho = \frac{\rho_0\, e^{2t/t_c}}{\left[1 - \frac{3}{2}|\gamma| H_0 t_c (e^{t/t_c} - 1)\right]^{2}}\,, \tag{35}$$

where t_c is the 'viscosity time'

$$t_c = \left(\frac{3}{2}\kappa\xi\right)^{-1}. \tag{36}$$

The values H_*, a_*, ρ_* at $t = t_*$ are thereby known.

In the turbulent era $t > t_*$ we can make use of the same expressions (33) - (35) as above, only with substitutions $t_c \to \infty$ ($\xi \to 0$), $t \to t - t_*$, $w \to w_{\rm turb}$, $H_0 \to H_*$, $a_0 \to a_*$, $\rho_0 \to \rho_*$. Thus

$$H = \frac{H_*}{1 + \frac{3}{2}\gamma_{\rm turb} H_* (t - t_*)}\,, \tag{37}$$

$$a = \frac{a_*}{\left[1 + \frac{3}{2}\gamma_{\rm turb} H_* (t - t_*)\right]^{2/3\gamma_{\rm turb}}}\,, \tag{38}$$

$$\rho = \frac{\rho_*}{\left[1 + \frac{3}{2}\gamma_{\rm turb} H_* (t - t_*)\right]^{2}} \tag{39}$$

(recall that $\gamma_{\rm turb} > 0$). Thus the density ρ, at first increasing with increasing t according to Eq. (35), decreases again once the turbulent era has been entered, and goes smoothly to zero as t^{-2} when $t \to \infty$. In this way the transition to turbulence protects the universe from entering the future singularity.

It should be noted that whereas the density is continuous at $t = t_*$ the pressure is not: In the laminar era $p_* = w\rho_* < 0$, while in the turbulent era $p_* = w_{\rm turb}\,\rho_*$ will even be positive, if $w_{\rm turb} > 0$. Thus, we demonstrated the possible role of turbulence to protect the universe from the future singularity. In the same fashion, one can consider its role in protecting the universe from the Rip, i.e. the disintegration of bound structures.

10. On Causal Cosmology

Dissipative dark energy models, in which the negative pressure responsible for the current acceleration is an effective pressure, is a nonequilibrium phenomenon. A viscous pressure can in principle play the role of an agent that drives the present acceleration of the universe; cf. Zimdahl et al. [2001] [65] and Balakin et al. [2003] [66]. Actually, the possibility of a viscosity dominated late epoch of the universe with accelerated expansion was discussed already in 1987 by Padmanabhan and Chitre [23].

Traditionally, the nonequilibrium thermodynamical processes have been formulated in the context of the classic theories of Eckart [1940] [67] and Landau and Lifshitz [1958] [68]. Due to works of Müller [1967] [69], Israel [1976] [70], Israel and Stewart [1979] [71] and others it became clear that Eckart-type formulations suffer from serious drawbacks concerning causality and stability. The reason for this is obviously the restriction of these early theories to first-order deviations from equilibrium. If one includes second order deviations, the problems should most likely disappear. Cosmological implications of second order theories were first considered by Belinskii et al. [1979] [72] and Pavon et al. [1982] [73].

Whereas the influence from a bulk viscosity on the background expansion of the universe is widely investigated in the literature, the perturbative analysis of the viscous cosmological models is not so widely addressed. A common characteristic of investigations of the latter kind is the analysis of a fluid whose equilibrium pressure is small compared with the bulk viscous pressure. Another common feature is that the dissipation is described within Eckart's theory. It is noteworthy that there are choices for the bulk viscosity parameters that give predictions for the matter power spectrum in agreement with observations; cf. Hipolito-Ricaldi et al. [2009] [74], and Hipolito-Ricaldi et al. [2010] [75].

The number of research papers in this area is large, so that we will in the follow focus essentially on the recent extensive paper of Piattella et al. [2011] [76]. That paper relies on cosmological dynamics as realized by one single imperfect fluid with bulk viscosity and vanishing equilibrium pressure. An analysis is carried out showing how this conventional picture is changed when one takes into account the causal Müller-Israel-Stewart theory, instead of Eckart's theory. The gravitational potential is evaluated, and compared with its ΛCDM counterpart. As is usual, it is assumed that the standard ΛCDM model in effect reproduces the observations.

First, let us write the viscous contribution to the total pressure as

$$\Pi = -\theta\xi, \tag{1}$$

where $\theta = 3H$ is the scalar expansion and ξ the bulk viscosity. This is the Eckart theory [67]. By contrast, in the Müller-Israel-Stewart (MIS) theory one has

$$\tau_{\rm rel}\dot{\Pi} + \Pi = -\theta\xi - \frac{1}{2}\Pi\left[\theta - \frac{(\xi/\tau_{\rm rel})^{\cdot}}{(\xi/\tau_{rel})} - \frac{\dot{T}}{T}\right], \tag{2}$$

where $\tau_{\rm rel}$ is a relaxation time and T is the temperature. Overdot means time derivative. This equation is often emplyed in a truncated form, as

$$\tau_{\rm rel}\dot{\Pi} + \Pi = -\theta\xi. \tag{3}$$

Taking Π to be subject to causal thermodynamics implies that the relaxation time $\tau_{\rm rel}$ is introduced as an additional parameter. Also, a propagation velocity for viscous perturbations is introduced that is different from the adiabatic sound velocity.

The variation of the temperature in Eq. (2) is found by means of Gibbs' integrability condition; cf. Maartens [1996] [7]

$$n\frac{\partial T}{\partial n} + (\rho + p)\frac{\partial T}{\partial \rho} = T\frac{\partial p}{\partial \rho}, \tag{4}$$

n being the particle number density. From this, by integration,

$$\frac{\dot{T}}{T} = -\theta\left[\frac{\partial p}{\partial \rho} + \frac{\Pi}{T}\frac{\partial T}{\partial \rho}\right]. \tag{5}$$

Limiting ourselves to barotropic conditions, we have $T = T(\rho)$ and $p = p(\rho)$. With $c_s = \sqrt{\partial p/\partial \rho} = \sqrt{dp/d\rho}$ as the adiabatic sound velocity (in natural units), Eqs. (4) and (5) simplify to

$$\frac{1}{T}\frac{dT}{d\rho} = \frac{c_s^2}{\rho + p}, \tag{6}$$

$$\frac{\dot{T}}{T} = -\theta c_s^2\left(1 + \frac{\Pi}{\rho + p}\right). \tag{7}$$

The MIS transport equation then gets the simplified form

$$\tau_{\rm rel}\dot{\Pi} + \Pi = -\theta\xi - \frac{1}{2}\tau_{\rm rel}\Pi\left[\theta - \frac{(\xi/\tau_{\rm rel})^{\cdot}}{(\xi/\tau_{\rm rel})} + \theta c_s^2\left(1 + \frac{\Pi}{\rho + p}\right)\right]. \tag{8}$$

The coefficients $\tau_{\rm rel}$ and ξ are in general functions of time, but are not completely arbitrary as they are related to the sound velocity; cf. Hiscock and Lindblom [1983] [77] and Maartens [1996] [7],

$$c_b^2 = \frac{\xi}{(\rho + p)\tau_{\rm rel}}. \tag{9}$$

Piattella et al. [2011] [76] now consider the perturbative dynamics of the viscous fluid, making use of the gauge-invariant formalism due to Bardeen. The FRW metric is perturbed in the form

$$ds^2 = a(\eta)^2[-(1 + 2\Phi)d\eta^2 + (1 - 2\Phi)d\mathbf{x}^2], \tag{10}$$

where Φ is the gravitational potential. This potential is calculated in the Eckart, as well as in the MIS, theory. Thus the perturbed bulk viscous contribution $\delta\Pi$ depends on which underlying theory is chosen.

We shall not here go into detail about the perturbation formalism, but mention that the following ansatz is made for the viscous pressure:

$$\frac{\Pi}{\rho} = \rho_0 \left[\mu + (1-\mu)a^{-3n/2}\right]^{2/n}, \tag{11}$$

with μ and n constants. This is the equation of state for a Chaplygin gas. For $n = 2$ the constant pressure case is recovered. From the energy conservation equation, one gets, with $a_0 = 1$ at present time,

$$\rho = \rho_0 \left[\mu + (1-\mu)a^{-3n/2}\right]^{2/n}. \tag{12}$$

We also mention that for the truncated MIS theory, for which one has

$$\dot{\Pi} + \frac{1}{\tau_{\text{rel}}}\Pi = -\frac{3}{\tau_{\text{rel}}}H\xi, \tag{13}$$

the following expression is found for the relaxation time as a function of density:

$$\tau_{\text{rel}}(\rho) = \frac{1}{\sqrt{3\kappa}} \frac{2\mu(\rho/\rho_0)^{-n/2}}{2c_b^2 + \mu(2-n)(\rho/\rho_0)^{-n/2}[1-\mu(\rho/\rho_0)^{-n/2}]}. \tag{14}$$

Piattella et al. [2011] [76] compare their numerical solutions for the gravitational potentials with the results for the standard ΛCDM case. Eckart's and the full causal theory seem to be disfavored, whereas the truncated theory, perhaps surprisingly, turns out to be more promising: it leads to results similar to those of the ΛCDM model for a bulk viscous speed in the interval $10^{-11} \ll c_b^2 \leq 10^{-8}$.

There are, as mentioned, a large number of papers dealing with viscous cosmology. We may mention the review article of Maartens [1995] [6], already referred to earlier, and also papers of Pavon et al. [1991] [78], and Chimento and Jakubi [2012] [79].

11. Summary

Inclusion of viscosity - a most natural way to go from the standpoint of general fluid mechanics - may be of importance both in the initial and in the final stages of the universe. We have surveyed generalized cosmology theories with the inclusion of bulk and shear viscosities, having the property that they tend to smooth out eventual anisotropies in the universe. Turbulence is also a phenomenon which one should expect to be of importance, not least so in the final stages close to the future singularities where the fluid motion is expected to be vigorous. As we have pointed out in Section 9, one possible role of the turbulence in a one-component fluid model is to protect the universe from the future singularity.

Causal cosmology, constructed so as to be better in accordance with basic physics than the 1940 classic theory of Eckart [67], ought to attract interest, although the causal formalism becomes mathematically somewhat more involved than the classic one. As mentioned in Section 10, the truncated version of the causal formalism seems, perhaps surprisingly, to be the variant that most closely gives results in accordance with the standard ΛCDM formalism, and thus with the observations.

References

[1] C. W. Misner: Neutrino Viscosity and the Isotropy of Primordial Black Body Radiation, Phys. Rev. Lett. 19, 533-535 (1967).

[2] Ya. B. Zel'dovich and I. D. Novikov: *Relativistic Astrophysics*, Vol. II (The University of Chicago Press, Chicago 1983).

[3] C. Eckart: The Thermodynamics of irreversible Processes. III. Relativistic Theory of the Simple Fluid, Phys. Rev. 58, 919-924 (1940).

[4] W. Israel and J. M. Stewart: Thermodynamics of non-stationary and transients effects in a relativistic gas, Phys. Lett 58, 213-215 (1976).

[5] Ø. Grøn: Viscous Inflationary Universe models, Astrophysics and Space Science 173, 191-225 (1990).

[6] R. Maartens: Dissipative cosmology, Class. Quant. Grav. 12, 1455-1466 (1995).

[7] R. Maartens: Causal thermodynamics in relativity, in Proceedings of Hanno Rund Workshop on Relativity and Thermodynamics, South Africa June 1996, arXiv: Astro-ph/9609119 (1996).

[8] W. Zimdahl: Bulk viscous cosmology, Phys. Rev. D 53, 5483-5493 (1996).

[9] M. K. Mak: Exact Causal Viscous Cosmologies, Gen. Rel. Grav. 30, 1171-1186 (1998).

[10] B. C. Paul, S. Mukherjee and A. Beesham: Higher Derivative Theory with Viscosity, Int. J. Mod Phys. D7, 499-507 (1998).

[11] A. I. Arbab and A. Beesham: Causal Dissipative Cosmology with Variable G and Λ, Gen. Rel. Grav 32, 615-620 (2000).

[12] S. Lepe, F. Peña and J. Saavedra: Randall-Sundrum model with $\Lambda < 0$ and bulk brane viscosity, Phys. Lett. B 662, 217 - 219 (2008).

[13] Ya. B. Zel'dovich: The cosmological constant and the theory of elementary particles. Sov. Phys. Usp. 11, 209-230 (1968).

[14] Ø. Grøn: Expansion isotropization during the inflationary era, Phys. Rev. D 32, 2522-2527 (1985).

[15] J. A. Belinchón, T. Harko and M. K. Mak: Full causal Bulk-Viscous Cosmological Models with Variable G and Λ. Gravitation & Cosmology 8, 319-326 (2002).

[16] Ø. Grøn: A new standard model of the universe, Eur. J. Phys. 23, 135-144 (2002).

[17] I. S . Kohli: A Bianchi Type IV Viscous Fluid Model of the Early Universe, arXiv: 1206.5438 (2012).

[18] G. F. R. Ellis: On the Raychaudhuri equation, Pramana 69, 15-22 (2007).

[19] M. Jamil and M. U. Farooq: Interacting Holographic Dark energy model. Int. J. Mod. Phys. 49, 42-50 (2010).

[20] D. Pavon and W. Zimdahl: Dark matter and dissipation, Phys. Lett. A 179, 261-265 (1993).

[21] R. Giostri, M. Vargas dos Santos, I. Waga et al.: From cosmic deceleration to acceleration: new constraints from SN Ia and BAO/CM. J. Cosmology and Astropart. Phys., issue 3, article no. 027 (2012).

[22] G. J. Mathews: Late decaying dark matter, bulk viscosity and the cosmic acceleration, Phys. Rev. D 78, 043525 (2008).

[23] T. Padmanabhan and S. M. Chitre: Viscous universes, Phys. Lett. 120, 433-436 (1987).

[24] J. Ren and X. Meng: Cosmological model with viscosity media (dark fluid) described by an effective equation of state, Phys. Lett. B 633, 1-8 (2006).

[25] M. Hu and X. Meng: Bulk viscous cosmology: Statefinder and entropy, Phys. Lett, B635, 186-194 (2006).

[26] N. Mostafapoor and Ø. Grøn: Viscous ΛCDM universe models, Astrophysics and Space Science 333, 357-368 (2011).

[27] I. Brevik and L. T. Heen: Remarks on the viscosity concept in the early universe, Astrophysics and Space Science 219, 99-115 (1994).

[28] I. Brevik and G. Stokkan: Viscosity and matter creation in the early universe. Astrophysics and Space Science 239, 89-96 (1996).

[29] X. Dou and X. H. Meng: Bulk Viscous Cosmology: Unified Dark Matter. Advances in Astronomy 2011 829340 (2011).

[30] A. Avelino and U. Nucamendi: Constraining a matter-dominated cosmological model with bulk viscosity proportional to the Hubble parameter. AIP Conf. Proc. 1083, No 1 (2008).

[31] A. Avelino and U. Nucamendi: Can a matter dominated model with constant bulk viscosity drive the accelerated expansion of the universe? JCAP, 006 (2009).

[32] M Cataldo, N. Cruz and S. Lepe: Viscous dark energy and phantom evolution. Phys. Lett. B 619, 5-10 (2005).

[33] G. M. Kremer and F. P. Devecchi: Viscous cosmological models and accelerated universes. Phys. Rev. D67, 047301 (2003).

[34] J. C. Fabris, S. V. B. Goncalves and R. de Sá Ribeiro: Bulk viscosity driving the acceleration of the universe. Gen. Relativ. Gravit. 38, 495-505 (2006).

[35] I. Brevik and O. Gorbunova: Dark energy and viscous cosmology. Gen. Rel. Grav. 37, 2039 (2005).

[36] Ø. Grøn: Statefinder analysis of universe models with a viscous cosmic fluid and with a fluid having a non-linear equation of state. I: The Casimir Effect and Cosmology. Tomsk State Pedagogical University Press 2009 ISBN 978-5-89428-311-1, p. 75-86.

[37] C. P. Singh: Bulk viscous cosmology in early universe. Pramana 71, 30-48 (2008).

[38] C. P. Singh, A. Beesham and W. S. Mbokazi: Cosmological models with variable G and Λ. Gen. Relativ. Grav. 30, 573-581 (1998).

[39] C. P. Singh, S. Kumar and A. Pradhan: Early viscous universe with variable gravitational and cosmological 'constants', Class. Quantum Grav. 24, 455-474 (2007).

[40] V. A. Belinski and I. M. Khalatnikov: Effect of viscosity on the character of the cosmological singularity, Zh. Eksp. Teor. Fiz. 69, 401-413 (1975).

[41] M. Heller: Singularities in Viscous Universes, Acta Cosmologica 7, 7-15 (1978).

[42] N. Mostafapoor and Ø. Grøn: Bianchi-type I universe models with nonlinear viscosity. Astrophysics and Space Science. DOI 10.1007/s 10509-012-1205-4.

[43] B. Saha: Bianchi type I universe with viscous fluid. Mod. Phys. Lett. A 20, 2127-2143 (2005).

[44] M. K. Mak, T. Harko and P. C. W. Fung: Viscous fluid cosmological models with a Bianchi Type I universe. Astrophys. and Space Science 249, 43-49 (1997).

[45] R. Bali, P. Singh and J. P. Singh: Viscous Bianchi Type I Universe with Stiff Matter and Decaying Vacuum Energy Density, ISRN Mathematical Physics 2012, 704612 (2012).

[46] A. Pradhan and S. K. Srivastava: Tilted Bianchi type V bulk viscous cosmological models in general relativity. Romanian Reports in Physics 59, 749-761 (2007).

[47] J. P. Singh and P. S. Baghel: Bianchi type V bulk viscous cosmological models with time dependent Λ-term. Electronic Journal of Theoretical Physics 22, 85-96 (2009).

[48] S. Ram, M. Singh and M. K. Verma: Totally anisotropic cosmological models with bulk viscosity for variable G and Λ. J. Mod. Phys. 3, 9-15 (2012).

[49] K. Nakamura et al. (Particle Data Group Collaboration): J. Phys. G 37, 075021 (2010).

[50] R. Amanullah et al. (Supernova Cosmology Project Collaboration): Astrophys. J. 76, 712 (2010).

[51] R. R. Caldwell: A phantom menace? Cosmological consequences of a dark energy companion with super-negative equation of state. Phys. Lett. B 545, 23 (2002).

[52] R. R. Caldwell, M. Kamionkowski and N. N. Weinberg: Phantom energy: Dark energy with $w < -1$ causes a cosmic doomsday. Phys. Rev. Lett. 91, 071301 (2003).

[53] J. D. Barrow: Sudden future singularities. Classical Quantum Gravity 21, L79-L82 (2004).

[54] S. Nojiri and S. D. Odintsov: Final state and thermodynamics of a dark energy universe. Phys. Rev. D 70, 103522 (2004).

[55] P. H. Frampton, K. J. Ludwick and R. J. Scherrer: The little rip. Phys. Rev. D 84, 063003 (2011).

[56] P. H. Frampton, K. J. Ludwick and R. J. Scherrer: Pseudo-rip: Cosmological models intermediate between the cosmological constant and the little rip. Phys. Rev. D 85, 083001 (2012).

[57] P. H. Frampton, K. J. Ludwick, S. Nojiri, S. D. Odintsov and R. J. Scherrer: Models for little rip cosmology. Phys. Lett. B 708, 204 (2012).

[58] I. Brevik, E. Elizalde, S. Nojiri and S. D. Odintsov: Viscous little rip cosmology. Phys. Rev. D 84, 103508 (2011).

[59] I. Brevik, R. Myrzakulov, S. Nojiri and S. D. Odintsov: Turbulence and little rip cosmology. Phys. Rev. D 86, 063007 (2012).

[60] S. Panchev: *Random Functions and Turbulence* (Pergamon Press, Oxford, 1971).

[61] I. Brevik: Velocity correlation in isotropic turbulence according to a modified Obukhov theory. ZAMM Z. angew. Math. Mech. 72, 145-148 (1992).

[62] R. A. Carhart and A. B. Kostinski: A generalized von Kármán interpolation formula. Phys. Lett. A 133, 149-153 (1988).

[63] I. Brevik, O. Gorbunova, S. Nojiri and S. D. Odintsov: On isotropic turbulence in the dark energy universe. Europ. Phys. J. C 71, 1629 (2011).

[64] I. Brevik, O. Gorbunova and D. Saez-Gomez: Casimir effects near the Big Bang singularity in viscous cosmology. Gen. Relativ. Gravit. 42, 1513-1522 (2010).

[65] W. Zimdahl, D. J. Schwarz, A. B. Balakin and D. Pavon: Cosmic anti-friction and accelerated expansion. Phys. Rev. D 64, 063501 (2001).

[66] A. B. Balakin, D. Pavon, D. J. Schwarz and W. Zimdahl: Curvature force and and dark energy. New J. Phys. 5, 85 (2003).

[67] C. Eckart: The thermodynamics of irreversible processes. Phys. Rev. 58, 919-924 (1940).

[68] L. D. Landau and E. M. Lifshitz: *Fluid Mechanics* (Reading, MA: Addison-Wesley, 1958).

[69] I. Müller: Zum Paradoxen der Wärmeleitungstheorie. Zeitschrift für Physik A 198, 329-344 (1967).

[70] W. Israel: Nonstationary irreversible thermodynamics: a causal relativistic theory. Annals Phys. (NY) 100, 310-331 (1976).

[71] W. Israel and J. M. Stewart: Transient relativistic thermodynamics and kinetic theory. Annals Phys. (NY) 118, 341-372 (1979).

[72] V. Belinskii, E. Nikomarov and I. Khalatnikov: Investigation of the cosmological evolution of viscoelastic matter with causal thermodynamics. Sov. J. Exp. and Theor. Phys. 50, 213 (1979).

[73] D. Pavon, D. Jou and J. Casas-Vazquez: On a covariant formulation of dissipative phenomena. Ann. Inst. Henri Poincaré (A) Phys. Theor. 36, 79-88 (1982).

[74] W. Hipolito-Ricaldi, H. Velten and W. Zimdahl: Non-adiabatic dark fluid cosmology. JCAP 0906, 016 (2009).

[75] W. Hipolito-Ricaldi, H. Velten and W. Zimdahl: The viscous dark fluid universe. Phys. Rev. D 82, 063507 (2010).

[76] O. F. Piattella, J. C. Fabris and W. Zimdahl: Bulk viscous cosmology with causal transport theory. JCAP 1105, 029 (2011).

[77] W. Hiscock and L. Lindblom: Stability and causality in dissipative relativistic fluids. Annals Phys. (NY) 151, 466-496 (1983).

[78] D. Pavon, J. Bafaluy and D. Jou: Causal Friedmann-Robertson-Walker cosmology. Class. Quantum Grav. 8, 347-360 (1991).

[79] L. P. Chimento and A. S. Jakubi: Evolution of a universe filled with a causal viscous fluid. ArXiv:1208.1548 [gr-qc].

In: Recent Advances in Cosmology
Editors: Anderson Travena and Brady Soren
ISBN: 978-1-62417-943-3

Chapter 5

Cosmology and Science

Jan C. A. Boeyens*
Unit for Advanced Scholarship, University of Pretoria
Pretoria, South Africa

Abstract

The general expectation that the acclaimed observation of a Higgs boson has finally unified the standard theoretical models of cosmology and particle physics is critically examined. On the negative side it is found that the cosmology model consists of several independent, mutually exclusive models, used indiscriminately to address different observations as needed. There is no single model to account for the conservation laws of physics, nature of the gravitational field, space-time symmetry, nucleogenesis, anomalous redshifts, cosmic radiation, microwave background, an expanding universe, dark matter, anti-matter and gamma-ray bursts. A special formulation of the theory is required for each different aspect. Apart from the Gamow-Lemaître hot big bang, there is an inflationary model with a Higgs field and a cold-dark-matter (ΛCDM) alternative, with practically no common ground between them. In particular, the different event chronologies are not synchronized. The resulting mismatch causes a discrepancy in the value of the cosmological constant (Λ) that differs by a factor of 10^{46} as required by the two "symbiotic" standard models.

The cause of all these problems is traced back to two seminal assumptions – the Doppler interpretation of cosmological redshifts and the point model of elementary matter. Without these assumptions there is no need of a big-bang model or a Higgs field. Without the false vacuum of particle physics – an artifact of Euclidean space – there is no basis for inflationary cosmology. An alternative cosmology based on projectively curved space-time is described. It features simple accounts of antimatter, nucleogenesis, the periodicity of atomic matter, black-hole singularities and cosmic self-similarity. The main ideas are conveyed with a minimum of mathematical formalism.

Cosmology in the Western World is increasingly dominated by responses to the most recent deep-space observations made with the Hubble telescope. Most dramatic of these is the measurement of large redshifts associated with distant galaxies and other objects close to the generally accepted limit of 13.7 billion light years that corresponds to the widely

*E-mail address: jan.boeyens@up.ac.za

accepted age of the universe. At the time of writing the recently announced [1] observation of a galaxy at 13.2×10^9 ly recorded in 2003 is only the latest result, most likely to be eclipsed in the near future. The observation of this galaxy just 500 million years from the time of the bigbang, is simply a far too short period for the formation of galaxies for each and every model of star formation. This situation is reminiscent of the decline of phlogiston theory when forced to contemplate negative mass.

The related, and equally recent, ecstatic announcement of a Higgs boson, observed during high-energy hadron collisions, is widely considered sufficient to restore confidence in the wobbly big-bang model. However, commenting on the recognition in this discovery as a rational explanation of the universe, it was stated [2]:

> That such scientific understanding provides a challenge to religion is an idea commonly heard from defenders of science ... Yet scientists who occupy that ground are often too slow to recognize the irrational bases of their own beliefs, and too quick to draw a line between scientific and the irrational.
>
> ... in practical terms, the Higgs is an incomprehensible abstraction, a partial solution to an extraordinarily rarified and perhaps always-incomplete intellectual puzzle.

The inference is clear – widespread scepticism about the big-bang model as a scientific theory abounds. An unbiased assessment of the recent advances in cosmology may therefore benefit by an analysis of the big-bang model as scientific theory.

The analysis is presented in two parts: a critical review of the scientific basis of standard big-bang cosmology and a proposed alternative, which is based on accepted science, and avoids the issue of creation.

Part I
Debunking the Big Bang

1. Big Bang as Theory

By general consensus theories in the physical sciences are subject to several strict criteria:

1. Each theory is based on a number of well-defined, undisputed, experimentally falsifiable assumptions.
2. These assumptions are required to be self-consistent and any conflict is interpreted to invalidate the theory in its entirety.
3. Users that subscribe to the theory do so with respect to all of its assumptions.
4. Should any assumption be falsified in an experimental test, the entire theory must be abandoned.
5. A well-defined theory with agreed mathematical basis may be subject to a degree of conflicting interpretation of its physical meaning, without invalidating the essential theory. The mathematical formulation is of primary importance.

6. Scientific theories do not exist in a vacuum and, unless they are in harmony with other more fundamental seminal theories, have no independent validity.

Being in conflict with each of these criteria, big-bang cosmology cannot qualify as a physically meaningful theory on any single count. Rated on all criteria it performs as follows:

1. The only firm assertion is that the big bang happened. This is a belief rather than an assumption, which, in any case, cannot be falsified by observation.

2. The most disturbing aspect of big-bang cosmology is the lack of agreement on fundamental assumptions. Each expert seems to have a different understanding of the event. Some say that it represents the instantaneous creation of space, time and the sum total of what still exists in the universe. Others maintain that a primeval form of energy and matter appeared within an existing space-time vacuum. The same difference of opinion appears in accounting for the primary assumption of an expanding universe. Whereas the rationalization of some observations seems to require space to expand, in other scenarios galactic objects are assumed to move apart in static space. The consensus, emphasized by all experts, is that the big-bang model is firmly based on the theory of general relativity. However, the field equations, with vanishingly small cosmological constant (Λ), assumed by traditional cosmologists, differ in a fundamental way from the form required by particle physics in support of the big bang. Even the time-table of events, supposed to have shaped future development during the first second, are in spectacular discord. Inflationary expansion, which is essential in some scenarios, is completely absent in others. Two versions of the 'theory' are seen to be in total disagreement about the most essential aspects. By one reconstruction the initial state was characterized by a stupendous density approaching 10^{100} g cm^{-3} at a temperature in excess of 10^{30} K. According to a more recent exposition primordial stuff only acquired mass at a much later stage by interaction with the mysterious Higgs field, to generate 'cold' dark matter.

3. As explained, too many big-bang assertions are mutually contradictory for this proposition to hold. There is no logical way to reconcile the assumptions of big-bang cosmology with those of particle physics.

4. The red-shift identification of galaxies, in existence as early as 4.5×10^8 years after the big bang, must surely represent such an observational test. It calls for the whole 'theory' to be scrapped rather than for the invention of new secondary assumptions to rationalize the impossible.

5. The field equations of general relativity, claimed to be fundamental to big-bang cosmolgy,

$$R_{\mu\nu} + \Lambda g_{\mu\nu} = -\kappa(T_{\mu\nu} - \tfrac{1}{2}Tg_{\mu\nu}) \qquad (1)$$

contain the coefficient Λ as a *constant*. The most bizarre feature of this cosmology is that experts who are in agreement on its validity can assign 'constant' values to Λ that differ by a factor of 10^{46} [3].

6. Big-bang cosmology is in blatant conflict with all theories of physics and chemistry developed over the last three centuries. The most respected of these are the conservation laws of mass and energy, according to which mass and energy can be converted from one form into another, but never destroyed or created from nothing. This means that all matter already existed, in one form or another, at the big-bang moment, $t = 0$, at the beginning of time. Recognition of time as one dimension of space-time implies that the existing matter (bright or dark) was concentrated at a point, with infinite density, which in turn implies infinitely curved space-time. It is hard to imagine a condition more stable than this. Still, it is supposed to have managed to fly apart because of the release of an irresistible amount of latent energy in a phase transformation, triggered by a 'quantum fluctuation' [4].

   This is where the situation becomes complicated. In order to avoid the infinite gravitational field, the primordial stuff is said to have remained massless until the appearance of the Higgs field, which, unlike an infinite gravitational field, cannot exist at a point. To kickstart the expansion away from the original point a phase transition, which implies transition to a lower-energy arrangement, is required. However, a phase transition in a massless system, which is, by definition, in its ground state, has no meaning. There is another complication. The initial events are postulated to antedate the appearance of the electromagnetic field and hence of radiant energy, only supposed to come into being during annihilation of matter and antimatter.

2. Big Bang and Relativity

The one aspect on which big-bang theorists are in firm agreement is that their model is underpinned by the most fundamental theory of physics, known as general relativity. This being the case, the discrepancies pointed out before may all simply arise from different interpretations of the basic mathematical model, without affecting the sound foundation of the theory. Such conflict is not unique and is well known also to occur in rival interpretations of quantum mechanics, the second great theory of physics. The reigning paradigm, known as the Copenhagen interpretation, is by no means generally accepted by all quantum physicists. By this approach, quantum phenomena are interpreted in terms of the probability behaviour of elementary particles, opposed to the wave model invoked in other interpretations.

In order to settle the question it is necessary to examine the mathematical structure of general relativity in relation to the experimental basis of the big-bang model.

2.1. General Relativity

The mathematical statement of the theory of general relativity has already been given in the form of a tensorial set of equations (1). The variables μ and ν, assume integer values of $0 \rightarrow 3$ in all combinations, for the specification of the curvature tensor $R_{\mu\nu}$, the stress tensor $T_{\mu\nu}$ and the metric tensor $g_{\mu\nu}$. Allowing for internal symmetry each tensor contains ten independent components, which together specify the gravitational field of general relativity. The four values of the indices μ and ν refer to the three space coordinates and the time variable, as defined by the Lorentz transformation of special relativity. The treatise by

Bergmann [5] is recommended to the uninitiated as a useful introduction to the theory of relativity.

The famous Lorentz transformation is a set of equations designed to expand the Galilean equations, which describe relative motion, so as to incorporate the motion of electromagnetic signals that propagate through the vacuum at the constant speed demanded by Maxwell's definition of the electromagnetic field. The significance of the Lorentz transformation is that it links space and time coordinates together on an equal footing, conveniently specified in terms of the Minkowski variables: $x_0 = ict$, $x_1 = x$, $x_2 = y$, $x_3 = z$, which define the invariant four-dimensional space-time interval

$$d = \left(\sum_{\mu=0}^{3} x_\mu^2\right)^{\frac{1}{2}} .$$

Minkowski space-time is pseudo-Euclidean, and was demonstrated by Einstein to be restricted to unaccelerated uniform motion. The effect of acceleration was visualized in Einstein's thought experiment [6] of an accelerated disc rotating at constant angular velocity. As the Lorentz transformation predicts contraction of a rigid body in the direction of motion, it follows that, compared to the constant radial distance (r), the ratio of the edge circumference, $C/r < 2\pi$, compared to $C = 2\pi r$ for the static disc in Euclidean space.

The inference is clear — the Euclidean topology of an accelerated disc, rotating with angular velocity approaching the speed of light (c), must be distorted. The demonstrated equivalence between acceleration and gravity was next used to confirm an earlier conjecture by Riemann that a gravitational field is generated in non-Euclidean space-time.

The mathematical manipulation of these ideas, leading to the formulation of the field equations (1), are beyond the scope of this review, but available in any number of reference works, such as [7].

To appreciate the implications of general relativity as the underlying theory of cosmology it is instructive to note the words of the master [6]:

> According to the general theory of relativity, the geometrical properties of space are not independent, but they are determined by matter.
>
> If we are to have in the universe an average density of matter which differs from zero, however small may be that difference, then the universe cannot be quasi-Euclidean. On the contrary, the results of calculation indicate that if matter be distributed uniformly, the universe would necessarily be spherical (or elliptical). Since in reality the detailed distribution of matter is not uniform, the real universe will deviate in individual parts from the spherical, *i.e.* the universe will be quasi-spherical. But it will necessarily be finite.

Although Einstein, in later work, often deviated from these principles, the theory itself has remained in firm support thereof.

2.2. The Experimental Basis

Big-bang cosmology has no experimental support. The conjecture of a miraculous creation event, driven by the "Volition of God", was documented as early as 1848 [8], very likely intended as a hoax. With the observed red shift in the frequency of galactic light and the implied connotation of an expanding universe, the conjecture acquired new meaning. The new interpretation was based on the linear relationship between redshift and galactic distance proposed by Edwin Hubble, who cautioned [9]:

> ... that spatial curvature may have to play a part in the explanation of existing nebular data.

Commentators often try to enlist Arthur Eddington, perhaps the foremost authority of his day, as a big-bang supporter. However, he emphatically distanced himself from Lemaître and his model [10], although he accepted expansion of the universe, with the comment:

> Since I cannot avoid introducing this question of a beginning, it has seemed to me that the most satisfactory theory would be one which made the beginning *not too unæsthetically abrupt*. This condition can only be satisfied by an Einstein universe with all the major forces balanced.

Speaking on the de Sitter-Lemaître theory in September 1931, he said:

> But the theory of the expanding universe is in some respects so preposterous that we naturally hesitate to commit ourselves to it. It contains elements apparently so incredible that I feel almost an indignation that anyone should believe in it–except myself.

He ascribed the reasons for the exception by reference to his almost incomprehensible interprertation of "wave-mechanics". Despite this ambivalence his intervention was apparently enough to turn the hoax into dogma. Still, Hubble's law by itself does not imply expansion of the universe and most decidedly not in three-dimensional Euclidean space. The Doppler interpretation is but one of many possibilities, none of which has been substantiated beyond reasonable doubt. The only reason why it came to be accepted as the undisputed interpretation is because it provided such ready-made support of the big-bang dogma. Even Einstein is purported to have complimented the priest Lemaître on his proposed mechanism of creation, having spotted the consequences of associating Doppler redshifts with an expanding universe. It gained momentum when the practical joker George Gamow persuaded two colleagues, Bethe and Alpher, to publish a spoof by $\alpha\beta\gamma$ on 1 April 1948 in a mainstream physics journal [12], announcing a theory that explains the formation of all the elements and cosmic rays in one spectacular burst at the beginning of time.

The futile protestations of Fred Hoyle, who coined the derogatory term *big bang* and proposed the alternative of a steady-state universe, may have helped to focus attention on the ludicrous notion which was rapidly gaining ground. But the cards were stacked against Hoyle. The mix, consisting of a nuclear physicist, the father of relativity and a man of the cloth, was simply too rich to combat. Gamow's joke was no longer a laughing matter. His April-fool's joke had acquired a life of its own and like the genie refused to return to the bottle.

Subsequent developments are more comic than cosmic. The first evidence to collapse was the $\alpha\beta\gamma$ scheme of nucleogenesis. By elementary arguments it could be shown that the $\alpha\beta\gamma$ scenario cannot produce any nuclides heavier than H, He and Li, nor any cosmic rays.

To salvage the central idea and to add insult to injury, Fred Hoyle's mechanism of stellar synthesis of heavy elements and subsequent distribution by supernovae was unceremoniously incorporated into an updated big-bang theory.

Jumping to the present and to data collected by the Hubble telescope, heavy elements are being detected, associated with galactic redshifts from an epoch, which according to expansion theory, predates the birth of first stars. This is the end of Gamow's prank, but its prodigy survives.

As an interim measure to compensate for the imminent loss of observational support 'the temperature of interstellar space' calculated by Eddington [11] before the big bang, at 3 K, and confirmed by rotational molecular spectroscopy as a black-body spectrum, became sensational news when rediscovered decades later as a microwave background. Suddenly it was hailed as a remnant of the big bang. George Smoot, who won a Nobel prize compared this discovery to "looking at the face of God".

Fast forward to the present where newly revised theories have no further need of a hot big bang for nucleogenesis. The current story is that only cold dark matter emerged from the big bang and part of it was converted into molecular hydrogen which cooled down the universe so fast that within half a million years very massive stars (VMS) with short, but very productive, lives were formed. They produced all the heavy nuclides and, wait for it, all that remains is their starlight which has now cooled down to 3 K. Maybe Smoot was looking at VMS. As a convenient by-product of the new scheme the question of anti-matter is sidestepped without mention. There is, of course, not a shred of experimental evidence to support any of these conjectures.

This is where we fade out on Gamow's joke. No Doppler shift, no expansion, cosmic rays or nucleogenisis – only dogmatic certainty. In place of the "black-body fingerprint of last scattering" there is only the "... void, and darkness was upon the face of the deep."

2.3. The Consilience

The observational basis of the big bang consists of spectroscopically measured galactic redshifts interpreted as Doppler shifts, assumed to occur when galaxies move apart. Many attempts have been made to substantiate the linear relationship inferred by Hubble on the basis of his initial limited data, without any conclusive result.

Statistical analysis of more than 37 000 measured redshifts [13] failed to establish a convincing alternative, but emphatically discounted the linear relationship, which had been challenged before on theoretical grounds by Segal [14]. Such a linear relationship is extremely unlikely in view of the intrinsic nonlinearity of curved space-time [15] that raises a serious objection against the Doppler interpretation.

The expansion conjecture is based on Friedmann's application of the Robertson-Walker line element

$$ds^2 = c^2dt^2 - S^2(t)\left\{\frac{dr^2}{1-Kr^2} + r^2\left(d\theta^2 + \sin^2\theta d\varphi^2\right)\right\}.$$

The parameter K which specifies curvature is seen to operate on space coordinates only, leaving time unaffected. Calling the variable t "cosmic time" does not change the fact that this line element is not defined in the curved space-time of general relativity. The parameter

K in all applications invariably assumes the value $K = 1$, hence the line element is purely Euclidean and reflects simple Newtonian theory – not even Minkowski space-time.

It is not as if the Doppler effect provides the only plausible explanation of galactic redshifts. Emil Wolf [16] demonstrated experimentally that source fluctuations generate redshifts of spectral lines which are indistinguishable from those due to the Doppler effect, even for stationary sources and observers. In fact, galactic redshifts may occur because of several reasons [17], which accounts for the observed scatter in modern Hubble plots [13].

The historical development of the expanding-universe conjecture was critically examined by Capri [18] and shown to be ideologically rather than scientifically based, and demonstrably not supported by the theory of general relativity.

3. Particle Physics

The second leg of standard cosmology, invoked to demonstrate its scientific credibility, comes from particle physics. This branch of physics has developed to rationalize the nature of the elementary building blocks of matter, starting from the unspoken consensus that the only lower level of physical reality is represented by space-time, or the vacuum. From the theory of relativity the relationship between space-time curvature and matter does not directly reveal any property of elementary matter beyond its response to a gravitational field. Weyl's seminal attempt to relate electromagnetic effects to gauged motion of matter through space-time, as modified by Schrödinger as a complex phase factor, provided the link for the development of particle physics.

The fundamental concept is embodied in the equation

$$\Psi' = \Psi e^{i\alpha}$$

that describes the transformation of a state function on transplantation[1] through the vacuum. For constant α the equation defines a *global* gauge transformation, with the special property that the derivative of the field

$$\partial\Psi' = e^{i\alpha}\partial\Psi$$

has the same form as the gauge transformation. If, in the case of a charged particle, the complex field Ψ describes an electron, a charge of $\rho = |\Psi|^2$ is associated with its wave field, which is clearly invariant under the transformation and implies conservation of the total charge.

In the curved space-time of general relativity the gauge factor α is no longer globally fixed, but changes as a function of the local coordinates of the curved space-time manifold,

$$\Psi' = \Psi e^{i\alpha(\mu)} \equiv \Psi e^{i\alpha(x,t)} \tag{2}$$

By forming the derivative

$$\partial_\mu\Psi' = \partial_\mu\Psi \cdot e^{i\alpha} + \partial_\mu\alpha \cdot i\Psi e^{i\alpha}$$

it is seen that the conservation of charge now depends on the appearance of a compensating gauge field.

[1] Comparison with itself at different coordinates.

In the case of an electron the transformed wave function (2), on substitution into Schrödinger's equation

$$i\hbar\frac{\partial\Psi}{\partial t} = -\frac{\hbar^2}{2m}\left(\frac{\partial^2\Psi}{\partial x^2}\right) \tag{3}$$

yields

$$i\hbar\frac{\partial\Psi}{\partial t} = -\frac{\hbar^2}{2m}\left[\frac{\partial^2\Psi}{\partial x^2} + 2i\frac{\partial\alpha}{\partial x}\frac{\partial\Psi}{\partial x} - \Psi\frac{\partial^2\alpha}{\partial x^2} - \Psi\left(\frac{\partial\alpha}{\partial t}\right)^2\right] + \hbar\Psi\frac{\partial\alpha}{\partial t}$$

Abbreviated in vector notation it reads

$$\begin{aligned} i\hbar\frac{\partial\Psi}{\partial t} &= -\frac{\hbar^2}{2m}\left[(\nabla + i\nabla\alpha)^2 - \nabla\cdot\nabla\alpha - \frac{2m}{\hbar}\frac{\partial\alpha}{\partial t}\right]\Psi \\ &\equiv -\frac{\hbar^2}{2m}\left[(\nabla + i\boldsymbol{A})^2 - \boldsymbol{\nabla}\times\boldsymbol{A} - V\right]\Psi \end{aligned} \tag{4}$$

The form of (3) is recovered by defining the vector $\boldsymbol{A}\equiv \partial\alpha/\partial x$ and the scalar $V = (2m/\hbar)(\partial\alpha/\partial t)$.

Equation (4) is recognized as Pauli's equation that describes an electron in an electromagnetic field. $\boldsymbol{A}$ and V are known respectively as the vector and scalar potentials of the electromagnetic field.

The central idea of particle physics is inspired by this successful description of an electron in terms of the electromagnetic gauge field. Having recognized the appearance of strong and weak interactions in atomic nuclei it was argued that these could also serve to characterize the corresponding elementary particles by specifying the appropriate interaction as a gauge field. Whereas Schrödinger's equation correctly describes electronic interaction, the starting point in the analysis of strong and weak interaction is to ensure that the Lagrangian function, which describes the interaction of these particle wave functions, remains invariant under the symmetry transformations that reflect known conservation laws [19].

The next important step is to identify asymmetric solutions of the symmetrical Lagrangian. This procedure corresponds in principle to the modification of global gauge invariance by the recognition of a local gauge field, as before [20]. The process has become known as *spontaneous symmetry breaking* of gauge theories. The most important application in the present context is the broken symmetry that generates the Higgs field, characterized by a massive boson.

This result is important within elementary-particle physics as a mathematical procedure to simulate the effect of space-time curvature, which is known to produce matter. In this sense it is neither unexpected nor mysterious. However, the common interpretation of the effect as a transition between two symmetry states of the vacuum has no observational support. More obviously, the high-symmetry Lagrangian is the correct formulation in hypothetical Euclidean space, whereas the 'broken' symmetry represents physically real curved space-time. Another way of distinguishing between the two symmetry states is in terms of linear and nonlinear formulations.

It is emphasized that the Higgs mechanism does not refer to two physically realizable vacuum states in that pseudo-Euclidean Minkowski space-time is no more than a local tangent approximation to the curved manifold of general relativity. It has no independent existence.

'Symmetry breaking' only has mathematical meaning and it makes no sense to associate it with a phase transition between two possible vacuum states. It is therefore completely inappropriate to invoke the appearance of a Higgs field to kickstart the big bang. A Higgs field is but one mathematical manifestation of space-time curvature.

4. Secondary Assumptions

Having demonstrated that big-bang cosmology is, at best, based on a misreading of both general relativity and particle physics, it remains to examine a number of secondary assumptions, which have been grafted onto the original model in an effort to eliminate some glaring inconsistencies. These are typically the fall-back position of cosmologists in defence of their dogma. It will be shown that for this purpose such assumptions are left (deliberately?) vague. It has been claimed that more than fifty different versions of the inflation assumption can be identified in the cosmology literature. As remarked by Narlikar [21]:

> It is not possible to review them all, especially since they have not yet produced a result that a cosmologist would feel confident about.

Despite these variations, the general idea is clear.

4.1. Inflation

Universal expansion as inferred from galactic redshifts is not always in line with observable features of the cosmos. One aspect that cosmologists find bothersome is the large-scale homogeneity revealed by the microwave background radiation. The rate of expansion is simply too slow to account for this observation. Another problem, faced by all big-bang advocates, is the initial expansion against an infinite gravitational field. What is clearly required to eliminate the problem is antigravity. This may solve the immediate problem, but unless it can be turned off, in the end it creates the bigger problem of blowing the whole thing apart. An attractive alternative is to identify some event that could release sufficient explosive energy for rapid expansion over a limited period.

The postulated device to drive the rapid expansion became known as *inflation* with the most likely source of the inflationary energy, the latent heat released by a cosmic phase transition. Cosmologists explain in detail how undercooled liquid water can freeze instantaneously and deliver the hidden energy that serves as a model of inflation. It is generally agreed to have operated for the period $10^{-35} < t < 10^{-33}$ s during which time the diameter of the universe increased exponentially by a factor of 10^{50}, and through interaction with the Higgs field its mass increased by a comparable factor[2].

According to a leading cosmologist [22] the linear dimension of the universe at the end of the inflationary period amounted to 10 cm. It is not clear which, if any, of the three problems – homogeneity, horizon or flatness – would be solved by this result. However,

[2]This is known as Guth's free lunch.

others estimate a diameter of ~ 1 mm at the onset of inflation, ending up at 10^{47} m $> 10^{30}$ light years, at the end. Given this spread it is evident that inflation can be adapted to fit any taste, and it is.

4.2. The Antimatter Puzzle

One of the most confusing accounts in big-bang cosmology is the 'explanation' of the imbalance between matter and antimatter in the observable universe. By symmetry arguments alone these conjugate forms of matter should be of identical stability, judging by pair-formation and annihilation interactions of the type

$$e^- + e^+ \leftrightarrow 2\gamma\,.$$

Had there been a big-bang creation event the two forms would presumably have occured in equal quantities and, unless smartly separated, would have turned into radiation.

To account for the existence of matter the cosmologists postulate a mysterious imbalance, first between baryons and antibaryons, and later on again between leptons and antileptons. Here is the sequence of events as outlined by one expert [4]:

> During the *Planck epoch*, $t < 5.4 \times 10^{-44}$ s, uncoupling of the gravitational force triggered quantum fluctuation (the formation and annihilation) of particles of matter and antimatter out of vacuum.
>
> Further cooling gave rise to the formation of quarks, antiquarks and gluons ... [which] became glued to form hadrons.
>
> The baryogenesis in this ... *hadron epoch* is believed to have triggered a tiny asymmetry between ... baryons and antibaryons ... responsible for today's predominance of matter over antimatter.
>
> After about 10^{-2} s ... electrons and positrons were created by collision of photons, again with a tiny excess of electrons over positron.

The theories of particle physics [19] explain the imbalance as follows:

> ... if the universe goes through a stage in which reactions occur out of equilibrium, then a non-zero baryon number can be generated. This process is called 'baryogenesis'. Although still very speculative, a convincing chain of reasoning can be advanced[3].

The reasoning goes more or less as follows:

At temperatures of $T > 10^{28}$ K Higgs bosons and their antiparticles are produced in equal proportion during collision of material particles with energies in excess of 10^{15} GeV. They remain in thermal equilibrium, with their rates of production and annihilation,

$$X + X^* \leftrightarrow \text{matter and radiation,}$$

[3]The reported identification of a Higgs boson is now claimed to have turned this speculation into scientific fact.

in balance, until the temperature drops below 10^{28} K when production stops. At about this stage, expansion of the universe also prevents further annihilation, leaving X and X^* in a state of non-equilibrium. From here on the Higgs X and X^* are assumed to decay with $\boldsymbol{CP}$-violation, interpreted to mean that the baryon numbers of X and X^* decay states need not be exact opposites. It is noted in conclusion:

> Although we have examined an appealing simple picture of baryogenesis, it should be realised that the application of GUTs to the early universe is highly speculative.
>
> Clearly GUTs cannot be the final answer as they do not incorporate gravity.

Whatever the agreed explanation, it seems fair to accept that big-bang cosmology denies the appearance of substantial quantities of antimatter anywhere in the universe. It is therefore of interest to note that the only agreed reasonable interpretation [23] of the increasing number of gamma ray bursters, observed in isotropic distribution, appears to be

> ... the annihilation of infalling chunks of antimatter.

4.3. Population III and ΛCDM

A major reformulation of big-bang cosmology has occured in the last decade in an effort to resolve a number of unsolved problems. The increasing Hubble age of recently identified objects has been the major concern and the anomalous mode of galactic rotation, known as Zwicky's paradox could arguably be due to large quantities of dark matter. Both of these issues could be settled tentatively by the postulate of population III stars in the very early pregalactic universe.

This possibility was reviewed in detail by Carr and others [24] on a purely hypothetical basis. Although still without any experimental support, the concept is now in widespread use as the basis of computer simulations to analyze a variety of astrophysical observations and cosmological conjectures. Population III stars correspond to the missing supergiants, high up on the main sequence of stars, which were described by Hoyle [25]p. 192, as follows:

> ... the argument is that stars initially above the upper limit (as at present observed) have by now completed their evolution and disappeared, ...

Carr *et al.* [24] latched onto this statement with the comment:

> The study of "Very Massive Objects" (VMO)... has been somewhat neglected. This is partly because there has been no obvious observational evidence that they exist ... and partly because theoretical reasons have been voiced for why they could never form. However, in the pregalactic context at least, these reasons need no longer apply: the stars would have burnt out long ago (so their present absence need cause no embarrassment) ...

In plain English: As the pregalactic existence of VMOs cannot be disputed by observation we are free to treat it as proven fact without theoretical justification.

This approach underpinned the assumption of Oh *et al.* [26] that an initial population of metal-free very massive stars caused early Fe enrichment and produced the ionizing fluxes that reionized primordial molecular hydrogen, which acted as coolant, needed to promote formation of VMS in the first place. To complete their circular argument it is stated that

> At present, the best evidence for VMS's ... may be from their distictive nucleosynthetic products ...

Suitably emboldened Springel and his team [27] simulated the origin of galaxies from cold dark matter, augmented with a dark-energy field (in the form of a cosmological constant, Λ), together with cosmic inflation, driven by weak density fluctuations in the early expanding universe. The methodology is justified by pointing out that:

> ... critical aspects of galaxy-formation physics, ... such as the ejection and mixing of heavy elements, [are considered] uncertain and beyond the reach of direct simulation; these must be treated by phenomenological models whose form and parameters are adjusted by trial-and-error as part of the overall data-modelling process.

The theory goes that pregalactic VMS were the source of heavy elements in the first galaxies dating from $t \sim 5 \times 10^8$ years. To explain the discrepancy with standard cosmology a new mechanism for the origin of these stars had to be postulated and the convenient answer was found in the role of dark matter.

Having pre-empted any objection against pregalactic population III stars, only the origin of dark matter had to be explained in order to launch ΛCDM cosmology. However, according to the most advanced standard cosmology there is little to explain. Although it does not feature explicitly in the events of the first five minutes dark matter is now assumed to have been around all along, albeit invisibly as its name implies.

The current consensus seems to be [28] that all of this happened

> ... through the growth of structure in the dark matter, by the input of heavy elements from the first stars ...

It is explained that:

> Whereas dark-matter haloes can originate through the action of gravity alone, the formation of luminous objects, such as stars and galaxies, is a more complicated process ...
>
> Trace amounts of molecular hydrogen (H_2) can be produced via a sequence of reactions ... and under the proper conditions this allows the gas to cool and eventually condense into stars.
>
> In the standard CDM model, the minihaloes that were the first sites for star formation are expected to be in place ... when the age of the universe was just a few hundred million years.

These predictions are all based on computer simulation of dark-matter behaviour, without knowing what dark matter is.

5. State of the Art

The hot big bang, on the point of collapse, was revived by Martin Rees with the CDM alternative [29]. This remarkable new theory is based on an entirely hypothetical substance, supported by neither experimental nor theoretical evidence. In a recent review [30] the existence of dark matter is ascribed to the dictates of the hot big-bang model, general relativity and the cosmological principle. General relativity, which never hints at such a substance, is clearly implicated only as a face-saving device. Without the big-bang limit on baryonic matter, the cosmological principle has no need of dark matter.

For good measure the same review also explains dark energy as 75% of the contents of the universe:

> Dark energy is ... simply the negative-pressure fluid postulated to account for cosmic acceleration.

The simplest candidate for dark energy is said to be Einstein's cosmological constant, but now

> ... the observationally required value of the cosmological constant is 10^{120} times smaller than the theoretical expectation.

To clarify matters:

> It seems reasonable to consider the possibility of a 'dark sector', beyond the standard model of particle physics, containing a dark-matter particle and a dark-energy field. Both seem to require unknown sources of gravitational fields, one attractive and the other repulsive.

This type of proposition has an unmistakable ring of desperation. Beyond ΛCDM there is nowhere to hide and the particle-physics angle, like the relativity card, has lost its persuasive edge. As argued before, both of these theories are invoked out of context in support of the big-bang model. The discrepancy between the values of Λ required by the two theories identifies an unbridgeable chasm that should now be recognized as fatally disqualifying the big bang. Although the basic concept is not subject to direct observational scrutiny, the internal contradictions and inconsistencies are too massive to ignore.

These comments are put into perspective by a brief summary of the creation event according to standard cosmology:

> *The entire cosmos came into being at a space-time singularity of infinitely dense dark matter and energy, held together by a single degenerate force. In the first instant, as a gravitational field gained independent existence, it precipitated a phase transition from the initial false vacuum, with the release of sufficient energy to overcome the infinite gravitational field and creation of the Higgs bosons that converted some of the dark matter asymmetrically into baryonic matter, mainly in the form of hydrogen and helium. The exponential expansion smoothed out any incipient turbulence and flattened the universe into*

Euclidean shape. Remnant disturbance of the explosion is still approaching from all directions as microwave background radiation.

At the end of the inflationary epoch the gravitational field managed to slow down the rate of expansion until it was overcome once again by the anti-gravitational effect of dark energy. For the last five billion years expansion has accelerated. All of this is known from spectroscopic measurement of frequency shifts commensurate with the diagnostic Doppler effect due to receding galaxies. Galactic redshifts provide a direct measure, not only of the age of the galaxy, but also of its elemental composition, with two strict provisos: No galaxy can antedate the big bang and no first-generation galaxy may contain any elements other than H *and* He.

All heavy elements, the presence of which is revealed in galactic spectra, were created in a later epoch by nuclear fusion in stars. At the end of its lifetime, when such a star disintegrates, heavy elements are released for incorporation into a next generation of stars and galaxies. Astrophysical analysis of the formation and evolution of stars indicates a few billion years for a typical life cycle and the comparable period for the formation of a galaxy of stars. By this reckoning the appearance of metal-enriched galactic sources could happen no sooner than 5×10^9 *years after the big bang.*

Observations at variance with this conclusion either invalidates the Doppler interpretation of redshifts and hence the big-bang theory, or it calls for a revision of the physics of star formation. Cosmologists favour the second option:

The new physics of star formation in the early universe is not feasible, given the products of the standard hot big bang. The new feedstock consists of **cold** *dark matter, with primordial molecular hydrogen as coolant. Explosion of VMS happens by electron-positron pair instability, which cancels the earlier proposed annihilation of antimatter. The original population III starlight is now the microwave background*

The problems with Λ, required to drive the new physics, speak for themselves. The new ΛCDM is as incompatible with general relativity and particle physics as the hot big bang. Like primitive beliefs, the supporting computer simulations have no rational basis, as pointed out before [2]. Complete collapse of the standard cosmology of the Western World is foreseen and the search has to start all over from what can be observed and understood without dogmatic bias. A brief look at the growth of ancient cosmologies may help to avoid earlier pitfalls.

Part II
A Scientific Alternative

6. The Cosmic Pattern

Looking at the stars has for millenia stirred the imagination of mankind and fueled a desire for knowledge and understanding. Their hopeless inaccessibility has turned the stars into symbols of heaven, inspiration, gods and destiny. The mystical link between macro and microcosm has been used since time immemorial to interpret human affairs in harmony with the changing pattern of heavenly panoramas. The affairs of man and beast were clearly perceived to be controlled by diurnal cycles, lunar variability and seasonal changes, superimposed on the wandering of the planets through the zodiac of fixed stars and precession of the equinoxes.

Countless volumes, exploring the details of astronomical phenomena, continue to appear and will continue to appear while the stars are there. Despite phenomenal improvement of observational means and theoretical understanding, it is fair to say that the stars remain as elusive as ever and the interpretation of astronomical data a game of faith and conjecture. The mathematical models whereby the mechanics of the solar system can be mapped with astonishing accuracy remain almost powerless beyond the orbit of Pluto, where conjecture takes over.

General relativity, the most elegant theory of modern physics, dictates that the impression of a three-dimensional world in which we live, is only valid in a local environment that does not extend to the galactic scale. The picture we get by looking at the stars is not what it seems. The further we look, the larger the distortion. In simple terms, the theory tells us that space-time is curved, without specifying the topology of the curvature. To make sense of what we see, any evidence that points at the topology of space-time therefore becomes of crucial importance. To make matters worse there are no theoretical criteria for identification of the most likely topology and direct observation at the cosmic level is not technologically feasible. The only hope is that some fingerprints of the cosmical topology would be detectable in our three-dimensional world. To understand this it is only necessary to appreciate that there is nothing special about our solar system, which, like the rest, carries an imprint of the large-scale topology, albeit difficult to recognize during casual observation. The problem is knowing where to look and how to interpret what we see..

The theory of relativity provides some clues. In essence, it shows that matter cannot exist in what is traditionally assumed to be flat space of infinite extent. This is the space familiar to all and described by the axioms of Euclidean geometry, universally known from high-school mathematics. For centuries the world remained ignorant of other possibilities, on the strength of Euclid's authority. The cosmical implications of non-Euclidean geometry was first revealed by the genius of Riemann, but, as pointed out by Jonathan Swift[4]:

> When a true genius appears in the world, you may know him by this sign, that the dunces are all in confederacy against him.

[4]Thoughts on various subjects (1706).

The dunces are still arrayed against Einstein, who gave physical content to Riemann's geometry with the theory of general relativity, almost a century ago.

Following Riemann and Einstein, the presence of matter, wherever we look, therefore implies that the geometry of space is non-Euclidean. In the same way that curvature of space generates matter, matter causes space to curve – in a typical chicken-and-egg situation. The theory of relativity, if correctly interpreted, shows that the curvature must be positive, which hints at a closed topology. We note in passing that the popular theory of an expanding universe, with its Euclidean basis, can be disregarded directly for our purpose. The implied closed topology is consistent with either of two non-Euclidean geometries, known as elliptic and spherical geometry, respectively [6]. We need physical evidence to distinguish between these possibilities.

Riemannian geometry is a differential geometry and defines a manifold that embraces both of the possible constant-curvature geometries. Therein lies the promise of describing gravity as a geometrical property of space-time. Even Zwicky's paradox could arguably be traced back to the analysis of large-scale rotation in Euclidean, rather than curved space, without postulating dark matter. As a working model of cosmic space we accept a closed manifold, with a mean global curvature that ensures closure, and local irregularities in the vicinity of large accumulations of matter – not unlike the surface of planet Earth. The local excursions constitute the gravitational field and the global curvature ensures the stability of matter. This reasoning implies that the intrinsic nature of all matter, on large or small scale, should be the same throughout the entire cosmos. Gravitational fields are responsible for no more than minor variation in the structure of matter. We look for similarities between material structures at the atomic, solar and galactic levels that could be seen as a common effect of global curvature.

Precisely this type of self-similarity has been observed and documented. The most striking example is found on comparison of the structures of nautilus shells, hurricanes and spiral galaxies, all of which are correctly described by the same mathematical formula. On closer examination it turns out that the same pattern is manifest in the arrangement of nucleons in an atomic nucleus, the extranuclear arrangement of electrons, seeds in sunflower heads, the petals of botanical structures, the pattern on pineapple skins, the spirals of kudu horns, the curvature of elephant tusks, the arrangement of natural satellites in the solar system, and many more. The design feature, which is common to all these structures, is known as the *golden ratio*. If we can find a functional relationship between this golden ratio and some aspect of space curvature it would be possible, in principle, to identify the topological structure of the cosmos.

6.1. Seen from Inside

A burden of consciousness is the constant hunger to understand how we fit into the grand scheme of things. For a start we could try to imagine what the world looks like through the eyes of an outsider, but the picture we conjure up can never be more detailed than the observations we make from inside. The quality of the final image depends exclusively on the available data that we can feed in, and for which we rely on our senses. What we see must be reflected in the final picture.

We see a solid earth bathed in an aqueous ocean, with birds flying in the sky, and circled on a daily basis, by the sun, moon and stars, high overhead. Such observations are embodied in an early Babylonian model of the cosmos and also reflected in the Hebrew Bible. In this model the solid earth and the surrounding ocean are strictly two-dimensional, and flat, stretching indefinitely into the unknown.

The problem of up and down was handled in the same way, as reaching into the unknown: heaven above and the underworld below. This is as far as logic can go. More detailed description is hampered by limited observational data. The unanswered questions of where the sun goes at night and on what the solid earth is supported, remained conjectural for millenia. The most convincing of such conjectures gradually gained general acceptance in the form of enduring creation myths and a pantheon of gods in control of unexplained phenomena. There is always a sungod, a moongod(dess) and a god of the underworld. With growing conviction there appeared gods and goddesses in charge of human practices and emotions: a god of war, a goddess of love, a god of thunder and an original mother of all things. These beliefs are perpetuated in the western world in the form of patron saints.

As conjecture evolved into religion the emergence of dogmatic truth impeded further enquiry and prohibited dissenting views. This was the challenge faced by scholars of the Renaissance and, again, by open-minded scientists of the 21st century. There is little difference between the methods of the Inquisition and the ostracism of modern-day science establishments. It has been said that Galileo was denounced by fellow scientists rather than the catholic church.

An intrinsic danger of dogmatic certainty on conjectural issues is the selective handling of available data. A striking example is the way in which theological ideas, superimposed on Ptolemy's Almagest and Aristotelian physics, developed into a geocentric medieval cosmology, clearly at variance even with elementary naked-eye astronomy. At the same time (1600CE) that European philosophers, such as Giordano Bruno were incinerated in Rome as punishment for their heretical ideas, William Gilbert, not under the whip of the Roman Inquisition, published a stinging refutation of medieval cosmology in Britain [31][5]. He accepted rotation of the earth and was particularly severe in his denunciation of the *primum mobile* and the rotation of celestial spheres around the earth. *Primum mobile*, the supposed outermost sphere, was added in the Middle Ages to the Ptolemaic system of astronomy, and supposed to revolve around the earth from east to west in 24 hours, carrying with it the (eight or nine) embedded spheres.

It is important to realize that Gilbert wrote at a time before the invention of telescopes. His conclusions were based on naked-eye astronomy, which by itself, is sufficient to refute the fanciful peripatetic[6] notions of the Christian philosopher saints. Gilbert is seen here as a role model for some modern cosmologist.

7. Alternative Cosmology

In order to develop an alternative modern cosmology it will be necessary to denounce all prevailing irrational beliefs and reconstruct a falsifiable model of the cosmos based on sound

[5]Not that the Protestants were more tolerant against progressive ideas. The protestant leader Calvin is on record of ordering the physician Michael Servetus to be burned at the stake.

[6]Following Aristotle

scientific principles, rather than agonise over unanalysables such as the beginning and the end of time. It would be necessary to consider all data made available by astronomy, physics and chemistry. Above all the new cosmology should reflect the relevant concepts from advanced mathematics. Preliminary ideas are outline in the following.

7.1. A Mathematical Model

In historical perspective there is a clear parallel between the cosmology of any civilization and the sophistication of their number system. Current consensus among mathematicians [32] is that normed division algebras, which could be the basis of high-dimensional physical vector spaces, are restricted to 1, 2, 4 and 8 dimensions. Sensible physics or cosmology is therefore possible only in the 1, 2, 4 or 8 dimensional spaces: $\mathbb{R}^1$, $\mathbb{C}^2$, $\mathbb{H}^4$ and $\mathbb{O}^8$, known as the real line, complex, quaternion and octonion space, respectively.

7.1.1. The Real Line

Real numbers can only describe interactions in one dimension, *i.e.* interactions that depend on the linear separation of two points. In the simplest case the interaction is independent of direction, as exemplified by the gravitational interaction between two mass points. It is noted that interaction at a point is undefined and the singularity of an infinite gravitational field is naturally avoided by the impossibility of two objects at the same point. Singular creation of the universe at a point is undefined. Gravitational interaction in the real world is measured along the four-dimensional geodesic of curved space-time.

Electrostatic interaction depends on distance and direction, which introduces the concept of polarity and the distinction between attractive and repulsive interaction, still in one dimension. Like gravity, electrostatic interaction is described by a scalar potential that varies along the geodesic between the charges. The gradient of this potential defines an electrostatic field, $\boldsymbol{E}= -\partial\varphi/\partial x$.

7.1.2. Complex Numbers

It is well known that the complex number $x + iy$ is given in polar form by Euler's equation:

$$r(\cos\theta + i\sin\theta) = re^{i\theta}\,.$$

The product

$$z_1 z_2 = r_1 e^{i\theta_1}\cdot r_2 e^{i\theta_2} = r_1 r_2 e^{i(\theta_1+\theta_2)}$$

is recognized immediately as the rotation of z_1 through the angle θ_2 and increase of its length by a factor r_2.

Electromagnetic interaction between moving charges depends on the distance between the charges and the variation of this distance with time, known as Faraday's law. A moving charge is said to generate a magnetic field that depends on the variation of a vector potential, $\boldsymbol{B}= \partial\boldsymbol{A}/\partial t$.

The total interaction depends on both potentials, which together define a complex field vector [33]:

$$\boldsymbol{Q} = \boldsymbol{B} + i\sqrt{\epsilon\mu}\boldsymbol{E} = \boldsymbol{B} + \frac{i}{c}\boldsymbol{E}.$$

The scalar and vector potentials are linked together inseparably in the complex plane and their total effect is like a single force, rather than a force with electric and magnetic components.

7.1.3. Quaternions

Extension of the complex formalism to more dimensions suggests the definition of related *hypercomplex* numbers. On multiplication of two three-dimensional vectors, without defining the mathematical properties of unit vectors $\boldsymbol{i}$, $\boldsymbol{j}$, $\boldsymbol{k}$, the formal result is:

$$\begin{aligned} q &= (\boldsymbol{i}x_1 + \boldsymbol{j}y_1 + \boldsymbol{k}z_1)(\boldsymbol{i}x_2 + \boldsymbol{j}y_2 + \boldsymbol{k}z_2) \\ &= \boldsymbol{i}^2 x_1x_2 + \boldsymbol{j}^2 y_1y_2 + \boldsymbol{k}^2 z_1z_2 \\ &+ \boldsymbol{ij}x_1y_2 + \boldsymbol{ji}y_1x_2 + \boldsymbol{ik}x_1z_2 + \boldsymbol{ki}z_1x_2 + \boldsymbol{jk}y_1z_2 + \boldsymbol{kj}z_1y_2 \end{aligned}$$

This expression is rearranged into the same form as a complex product by defining

$$\begin{aligned} &\boldsymbol{i}^2 = \boldsymbol{j}^2 = \boldsymbol{k}^2 = -1 \\ &\boldsymbol{ij} = \boldsymbol{k}\,,\ \boldsymbol{jk} = \boldsymbol{i}\,,\ \boldsymbol{ki} = \boldsymbol{j} \\ &\boldsymbol{ji} = -\boldsymbol{k}\,,\ \boldsymbol{kj} = -\boldsymbol{i}\,,\ \boldsymbol{ik} = -\boldsymbol{j}\,, \end{aligned}$$

the result first obtained by William Hamilton who defined

$$q = -(x_1x_2 + y_1y_2 + z_1z_2) + i(y_1z_2 - y_2z_1) + j(z_1x_2 - z_2x_1) + k(x_1y_2 - x_2y_1)$$

with the rule of composition: $i^2 = j^2 = k^2 = ijk = -1$.
A hypercomplex number of unit norm can now be defined in the form $z = a_0 + a_ie_i$, where the e_i are generalizations of $\sqrt{-1}$. For $i = 1, 3$ it is called a quaternion. The unit number $\mathbf{1}$ represents the unit vector in the fourth dimension.

To represent a quaternion in matrix form it is written as a pair of complex numbers, $q = (u, v)$, in the same way that the complex number $a + ib$ is written as $(a, b) = \begin{pmatrix} a & -b \\ b & a \end{pmatrix}$. As for complex numbers the product of (u, v) and its conjugate should be $(|u|^2 + |v|^2, 0)$. This result is obtained by defining the conjugate $q^* = (u^*, -v)$ and the matrix equivalent of (u, v) as $\begin{pmatrix} u & v^* \\ -v & u^* \end{pmatrix}$. This way

$$\begin{aligned} qq^* = \begin{pmatrix} u & v^* \\ -v & u^* \end{pmatrix}\begin{pmatrix} u^* & -v \\ v & u \end{pmatrix} &= \begin{pmatrix} uu^* + v^*v & -uv^* + v^*u \\ -vu^* + u^*v & vv^* + u^*u \end{pmatrix} \\ &\equiv (|u|^2 + |v|^2, 0) \end{aligned}$$

The modulus of q, $\sqrt{q^*q} = \sqrt{u^2 + v^2}$. Notice that the determinant of the matrix of $q = (u, v)$, gives

$$\begin{vmatrix} u & v^* \\ -v & u^* \end{vmatrix} = |u|^2 + |v|^2\,.$$

By expanding the quaternion in terms of the complex pair, $u = a + ib$ and $v = c + id$, it follows directly that $q^2 = a^2 + b^2 + c^2 + d^2$. In terms of four quaternion components:

$$q \cdot q = q_0^2 + q_1^2 + q_2^2 + q_3^2\,.$$

For $|u|^2 + |v|^2 = 1$, the product of the quaternion and its conjugate (or adjoint) is the unitary matrix $\begin{pmatrix} 1 & 0 \\ 0 & 1 \end{pmatrix}$. Unit quaternions (of modulus 1) are therefore seen to define rotations of $\mathbb{C}^2$, just as the unit complex numbers correspond to rotations of $\mathbb{R}^2$.

In Hamilton's notation the conjugate of the general quaternion $q = a + ib + jc + kd$ is $q^* = a - ib - jc - kd$, so that the square of the magnitude, $|q|^2 = q^*q = a^2 + b^2 + c^2 + d^2$ is a simple extension of the rule for complex numbers.

As with complex numbers the product of two quaternions

$$\begin{aligned}(p = \alpha + i\beta + j\gamma + k\delta)q &= (\alpha a - \beta b - \gamma c - \delta d) + i(\alpha b + \beta a + \gamma d - \delta c) \\ &+ j(\alpha c - \beta d + \gamma a + \delta b) + k(\alpha d + \beta c - \gamma b + \delta a)\end{aligned}$$

is another quaternion. Quaternion multiplication is clearly not commutative, as $pq \neq qp$.

By expanding the exponential and using the rules for multiplication of i, j, k, it can be shown that, for $\alpha^2 + \beta^2 + \gamma^2 = 1$, a quaternion equivalent of Euler's equation for the imaginary exponent is obtained as:

$$e^{\theta(i\alpha + j\beta + k\gamma)} = \cos\theta + \sin\theta(i\alpha + j\beta + k\gamma)\,.$$

This exponential is of unit length; the exponent represents a space vector of length θ, with direction as given by the direction cosines α, β, γ.

Any quaternion can now be represented in the form $Qe^{\theta(i\alpha+j\beta+k\gamma)}$, where Q is the length of a four vector with direction as fixed by θ, α, β, γ. By analogy with the complex exponential it is anticipated that this operator (with $Q = 1$) describes a rotation. The argument on which it operates is also a quaternion. In three dimensions the vector $f = ix + jk + kz$, in quaternion notation, is rotated by an angle θ about an axis of direction cosines α, β, γ into f' according to:

$$f' = ix' + jy' + kz' = e^{(\theta/2)(i\alpha+j\beta+k\gamma)} f e^{-(\theta/2)(i\alpha+j\beta+k\gamma)} = q \cdot f \cdot q^{-1}\,,$$

where $\alpha^2 + \beta^2 + \gamma^2 = 1$.

The equation

$$\Box^2\Phi = 0 \tag{5}$$

has unitary quaternion solutions which represent the $SU(2)$ rotational Lie group. The four-dimensional eigenfunction is known as the spin function.

The most general rotation of a four-vector represented by $q = w + ix + jy + kz$ is therefore described by the equation

$$q' = e^{\theta(i\alpha+j\beta+k\gamma)} q e^{-\varphi(i\lambda+j\mu+k\nu)}$$

where both $\alpha^2 + \beta^2 + \gamma^2 = \lambda^2 + \mu^2 + \nu^2 = 1$. When both θ and φ are imaginary angles, the transformation corresponds to the Lorentz transformation. This is a mathematical proof that quantum theory and special relativity originate from the same basis as the most general solution of (5) must obviously be a quaternion function.

Four-dimensional interactions are described by quaternions. The best-known example is the electromagnetic field, which derives directly from (5). In practice, equation (5) is routinely solved by separating space and time variables, whereby the equation is reduced to a three-dimensional wave equation. Rather than quaternion solutions that describe the spin function, the wave equation has complex matter-wave solutions based on Schrödinger's equation (3). Because of this defect wave-mechanical models of elementary matter are inadequate [34]. A spin operator, superimposed on the wave functions effects some improvement, but cannot restore the Lorentzian entanglement between space and time coordinates. The mysterious properties of quantum systems, such as non-commuting variables, all stem from this defect, which is shared by the broken-symmetry solutions of elementary-particle physics. It is noted that the simulation of electroweak unification by this procedure depends on the empirical assumption of non-Abelian $SU(2)$ symmetry [35]. This property appears naturally in quaternion algebra. With the addition of extra terms the three-dimensional Lagrangian remains a partial solution, in the same way that *ad hoc* correction factors to Schrödinger wave functions can never restore the four-dimensional nature that was lost on separation of the variables.

It is important to note that quaternions describe rotations in all dimensions less than five. This means that the same mathematics, which describes four-dimensional interaction, also describe interactions in the complex plane and along the real line. The grand unification strategy of combining symmetry groups, $U(1) \times U(2) \times U(3)$, to provide a unified description of strong and electroweak fields is recognized as another measure to approximate a higher-dimensional field in terms of complex and linear fields. Until the theory can be reformulated in four (or eight) dimensions it cannot stand in support of any cosmology.

7.1.4. Octonions

Grassmann's original idea of hypercomplex numbers in n dimensions has been extended beyond 4, to 8 dimensions. The rationale is to define octonions as an ordered pair of quaternions (q, r) with a rule of multiplication without zero divisors; in the same way that quaternions are defined as an ordered pair of complex numbers. The formula

$$(q, r)(s, t) = (qs - tr^*, q^*t + sr)$$

by which

$$(q, r)(q^*, r) = (|q|^2 + |r|^2, 0)\,,$$

does produce a useful number system, denoted $\mathbb{O}^8$. Whereas quaternion multiplication is not commutative, octonion multiplication is not even associative.

The quark-gluon theory of strong nuclear interaction is simulated by the broken symmetry of a Lagrangian with so-called GUT[7] symmetry. Apart from spin and isospin, a range of new flavour and colour parameters, with special rules of combination, emerge in the formalism. It is rather obvious that these features cannot be accommodated in four-dimensional hypercomplex algebra, but highly suggestive that their proper description could be achieved with octonions.

An octonion, like quaternions and complex numbers, also describes symmetry operations in fewer dimensions as special cases. Octonion algebra provides an overarching description of all the known interactions in Nature, without the assumption that all of them

[7] Grand unified theory assumes the supergroup $SU(1) \times SU(2) \times SU(3)$.

operated as a single interaction, associated with a mysterious singularity, at the beginning of time.

The one feature that all of these interactions have in common is that they only appear in curved space-time. What particle physicists refer to as symmetry breaking simply amounts to the reformulation of global gauge invariance in local curved space-time.

As a preliminary to any cosmological speculation it could be helpful to examine the evidence for curved space-time and to what extent it accounts for astrophysical observations.

7.2. Topology of Space-Time

The only observations that reveal something about the topology of space-time are those that support the theory of general relativity. The connection with topology was pointed out precisely by Einstein [6], before he introduced simplifying assumptions in the formulation of cosmological models. The original conclusion that the space-time of general relativity must be positively curved, still stands. That such curvature inevitably leads to a closed system with either spherical or elliptic topology remains equally valid. The arguments that produced the notion of an expanding universe distorted Einstein's sound conclusions to the point where he supported an Euclidean cosmology, despite Hubble's admonition [9] that spectroscopic redshifts could be due to space-time curvature. This advice has been ignored for too long.

The fundamental chronometric analysis of Segal [14], to demonstrate this effect, is based on the distinction between a globally curved space-time manifold and a local Euclidean tangent space with universal time [8]. An elementary demonstration [36] of how curved space causes redshift has also been ignored. Without the Doppler interpretation Einstein's original two alternatives are back on the table.

7.2.1. Projective Geometry

The four-dimensionally closed spherical and elliptical manifolds are equally difficult to visualize. A compelling argument that favours elliptical topology is the appearance of antimatter and other evidence of space-time chirality, such as the electromagnetic right-hand rule. Only this alternative, also known as projective geometry, is therefore considered in the following.

The Euclidean postulate of parallelism prevents consistent definition of the concepts point and line. Any two lines intersect in a point, except when the lines are parallel, and any two points are connected by a single line, except when the points are infinitely far apart. Two parallel lines go through the same points at $\pm\infty$ and intersect in two points.

All of these anomalies are eliminated by adding a special point at infinity to each line, assuming plus and minus infinity to coincide. The point at infinity cannot be described correctly by cartesian coordinates and to ensure that all points on the line are described in equivalent fashion, points in n–dimensional projective space are described in terms of $n+1$ homogeneous coordinates. The meaning of homogeneous coordinates is illustrated well in one dimension by the barycentric coordinates, first introduced by Möbius to describe static equilibrium by the well–known lever rule.

[8]Note the subtle difference in the use of the words *global* and *local* in gauge theories.

The centre of gravity is fixed by the equation $b \cdot PB = a \cdot AP$ and the barycentric coordinates of $P = [a/(a+b), b/(a+b)]$ are normalized to the homogeneous coordinates $x_1 = a$, $x_2 = b$, provided $a + b \neq 0$. Two homogeneous coordinates are required to identify any point (including the point at infinity) on a one–dimensional line. The balance is not disturbed when both weights are increased by the same factor $\mu \neq 0$, such that

$$\frac{AP}{PB} = \frac{b}{a} = \frac{b\mu}{a\mu}\,.$$

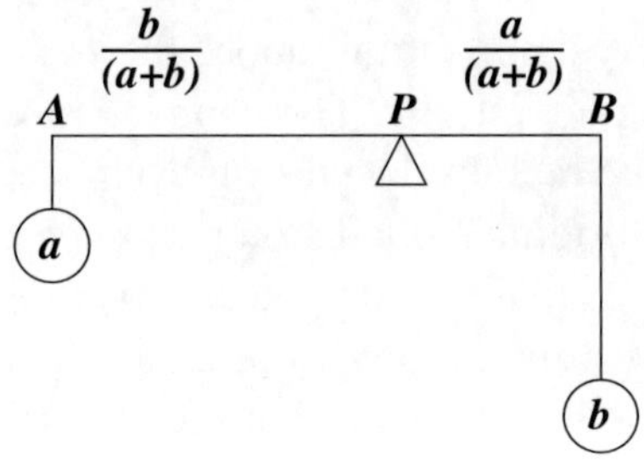

The homogeneous coordinates x^i of points in projective space remain valid when multiplied by an arbitrary factor, k, *i.e.*

$$x^i = kx^i\,.$$

This property is akin to Weyl's idea of a gauge factor that affects vector displacement and it is useful to define the proportionality factor as $k = e^{x^0}$. The parameter x^0 is known as a gauge variable. It is of fundamental importance in electromagnetic theory where it functions as a phase factor and it is important to note that it only emerges naturally in projective space. This phase factor provides the missing link between wave mechanics, electromagnetism and the gravitational field.

The simplest example of projective space is the projective plane P^2, familiar to most as a Möbius strip, which is a section through P^2. It has the celebrated properties of one-sidedness and non-orientability. Using a paper model it is readily demonstrated that the two–dimensional Möbius surface cannot be embedded in two–dimensional space, with the inference that P^2 cannot be embedded in $\mathbb{R}^3$. It is the three-dimensional equivalent of this surface, known as the double cover, that will be considered as the proposed model of space-time.

7.2.2. Projective Cosmology

Minkowski space in the real world cannot be Euclidean on a large scale. In order to generate a connected surface it is necessary that all of space should be curved appropriately. The curvature should be such that the isotropic lines that appear to diverge in two–dimensional projection should close onto themselves in more dimensions. A useful clue is provided [37] by the pseudocircles in Minkowski space, $\mathbb{M}^2$, $x^2 - c^2t^2 = \pm r^2$, which appear as a pair of hyperbolas that may join up into two Möbius bands as shown.

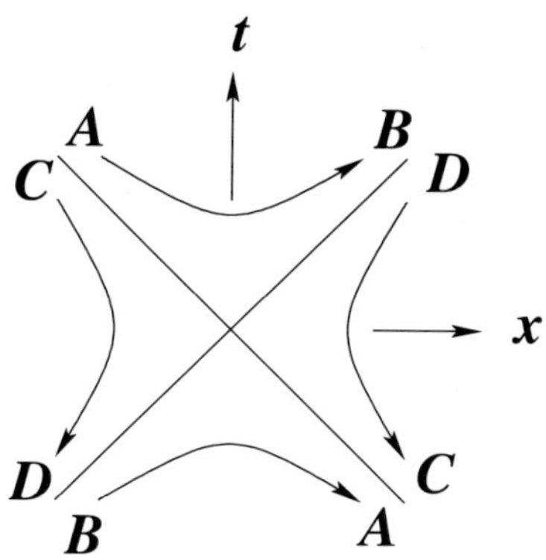

The four-dimensional equivalent defines a projective plane, an elliptic space, which, like a Möbius band is one-sided and non-orientable. Motion along the double cover of the projective plane (or the Möbius band) goes through an involution which entails an inversion of chirality. It thereby defines an isotropic interface that separates regions of matter and antimatter.

In the non-Euclidean space of general relativity the Minkowskian interface closes on itself and transforms into a projective plane with involution, to distinguish between generated matter and antimatter, and introduces the local gauge factor that defines the electromagnetic field. The realization that neither matter (gravity) nor the electromagnetic field can exist in Euclidean space should be the final theoretical vindication of general relativity, which strangely enough, is still disputed in the 21st century.

Nowhere is the equivalence of relativity and quantum theories demonstrated more convincingly than in the transformation of the projective relativistic field equations of Veblen [38] to affine tangent space [39]. In gravitation-free Minkowski space of signature $\{\ 1\ 1\ 1\ -c^2\}$ the field equations reduce to a form that closely resembles the relativistic Klein-Gordon wave equation in the electromagnetic field. Conversion of the four-potential to tangent space takes the form

$$(\tau + \tfrac{1}{2})\phi_j = \left(\frac{e}{mc^2}\right) V_j\,,$$

in which the golden ratio, τ appears as a proportionality constant [17]. The quantum condition is introduced by scaling the arbitrary gauge index, N, of the projective curvature tensor as in

$$(\tau + \tfrac{1}{2})N = -\frac{3}{4}\left(\frac{mc}{\hbar i}\right)\,,$$

which is equivalent to the stipulation of gauge invariance according to Schrö-dinger [40]. Noting that $\tau^2 + \tau = 1$, the product of these two expressions reduces to

$$\frac{5}{3}\phi_j N = \left(\frac{ie}{\hbar c}\right) V_j.$$

The quantity in brackets is the same as the phase factor associated with a five–dimensional electronic wave function [41], or equivalently, the four–dimensional function in homogeneous projective coordinates [38].

As a model of the universe the projective topology provides a simple explanation of the observed preponderance of matter over antimatter. On interpretation of the interface as an element of $\boldsymbol{CPT}$ symmetry, it follows that transplantation of 2π along the double cover, through the involution, gradually converts matter into antimatter by an unobservable phase (gauge) transformation. Equal amounts of matter and antimatter therefore occur in the universe without fixed boundaries that separate antipodal conjugate regions.

Not only the big-bang singularity, but also the problem with black-hole singularities are naturally avoided in projective space-time. It is foreseen that antipodal regions of opposite chirality could be connected by the equivalent of an Einstein–Rosen bridge [42] across the vacuum interface, at an outlet for excess matter that accumulated in the black hole. In the (anti)world the discharge would appear as a white hole and the source of new (anti)matter, which develops into cosmic activity as observed in quasars, globular clusters or Seyfert galaxies. This idea has been discussed in more detail elsewhere [17].

It is an attractive possibility that a black-white-hole system could be responsible for the synthesis of heavy elements. It is convincingly demonstrated that the periodicity of atomic matter points at nucleogenesis from α-particles in an equilibrium process that produces atomic nuclides in massratios that match the observed cosmic abundances [43], [17] p.179. This proposal has much in common with the $\alpha\beta\gamma$ proposal [12] for which there is not sufficient time in a hot big bang. In a black hole, where temporal and space coordinates become interchanged [44], there is no such time constraint.

The question of cosmic self-similarity is an aspect which is not adequately explored in any other known cosmology. In projective cosmology everything depends on the curvature of space-time, and there are strong indications that this is closely related to the golden ratio. The results of Veblen's projective relativity that points at such a relationship has already been mentioned. Viewed in conjunction with the structure of spiral galaxies and the simulation of the Bode–Titius law, which describes the structure of the solar system [45], both of which depend on golden-spiral optimization, self-similarity becomes obvious. Any growing structure, big or small, develops on a template conditioned by general space-time curvature and the golden ratio.

Acknowledgments

I sincerely thank Casper Schutte and Demi Levendis for their critical comments on the first draft and Faan Naude for his friendly information retrieval service.

References

[1] SPIEGELONLINE *Der tiefste Blick im All.* 26 September 2012. http://www.spiegel.de/wissenschaft/weltall/nasa-weltraumteleskop-ins-all-a-858002.html.

[2] D. Sarewitz, *Sometimes science must give way to religion*, Nature 488 (2012) 431.

[3] L. Abbott, *The mystery of the cosmological constant*, Sci. Amer. 256 (1988) 106–113.

[4] D. Rehder, *Chemistry in space*, Wiley-VCH, Weinheim, 2010.

[5] P.G. Bergmann, *Introduction to the theory of relativity*, Dover, NY, 1976.

[6] A Einstein, *Relativity. The special and general theory*, (Translated by R.W. Lawson), Methuen, London, 1920.

[7] R. Adler, M. Bazin and M. Schiffer, *Introduction to general relativity*, McGraw-Hill, NY, 1965.

[8] E.A. Poe, *Eureka*, Pullman, NY, 1848; Reprinted in *The science Fiction of Edgar Allan Poe*, Penguin Books, Harmondsworth, England, 1976.

[9] E. Hubble and R.C. Tolman, *Two methods of investigating the nature of nebular red-shift*, Astrophys. J. 82 (1935) 302–337.

[10] A.S. Eddington, *The Expanding Universe*, Penguin Books, Harmondsworth, England, 1940.

[11] A.S. Eddington, *The internal constitution of the stars*, Univ. Press, Cambridge, 1930.

[12] R.A. Alpher, H. Bethe and G, Gamow, *The origin of chemical elements*, Phys. Rev. 73 (1948) 80–82.

[13] V.S. Troitskij, *Observational test of the cosmological theory testifies to the static universe and a new redshift-distance relation*, Astrophys. Space Sci. 240 (1996) 89–121.

[14] I.E. Segal, *Mathematical Cosmology and Extragalactic Astronomy*, Academic Press, NY, 1976.

[15] D. Finkelstein and C.W. Misner, *Some new conservation laws*, Ann. Phys. 6 (1959) 230–242.

[16] D.F.V. James and E. Wolf, *A class of scattering media which generate Doppler-like frequency shifts of spectral lines*, Phys. Lett. 188 (1994) 239–244.

[17] J.C.A. Boeyens, *Chemical Cosmology*, Springer.com, 2010.

[18] M.M. Capria, *The rebirth of cosmology from the static to the expanding universe*, in M.M. Capria (ed.) *Physics berfore and after Einstein*, IOS Press, Amsterdam, 2005.

[19] G.D. Coughlan and J.E. Dodd, *The ideas of particle physics*, 2nd ed., Univ. Press, Cambridge, 1991.

[20] P.W. Higgs, *Spontaneous symmetry breakdown without massless bosons*, Phys. Rev. 145 (1966) 1156–1163.

[21] J.V. Narlikar, *An Introduction to Cosmolgy*, 3rd ed., Univ. Press, Cambridge, 2002.

[22] H. Genz, *Die Entdeckung des Nichts*, Carl Hauser, München, 1994.

[23] B. Schwarzschild, *Compton observatory data deepen the gamma ray burster mystery*, Physics Today, 45 (1992) 21–24.

[24] B.J. Carr, J.R. Bond and W.D. Arnett, *Cosmological consequences of population III stars*, Astrophys. J., 277 (1984) 445–469.

[25] F. Hoyle, *Frontiers of Astronomy*, Heinemann, London, 1955.

[26] S.P. Oh, K.M. Nollett, P. Madau and G.T. Wasserburg, *Did very massive stars preenrich and reionize the universe?* Astrophys. J. 562 (2001) L1–4.

[27] V. Springel and 16 co-authors, *Simulations of the formation, evolution and clustering of galaxies and quasars*, Nature 435 (2005) 629–636.

[28] V. Bromm, N. Yoshida, L. Hernquist and C.F. McKee, *The formation of the first stars and galaxies*, Nature 459 (2009) 49–54.

[29] G.R. Blumenthal, S.M. Faber, J.R. Primack and M.J. Rees, *Formation of galaxies and large-scale struture with cold dark matter*, Nature 311 1984 517–525.

[30] R. Caldwell and M. Kamionkowski, *Dark matter and dark energy*, Nature 458 (2009) 587–589.

[31] G.Gilbert, *On the Loadstone and Magnetic Bodies*, (1600). Translated by P.F. Motteley, Encyclopædia Btitannica, 1952, Chicago.

[32] T. Gowers (ed.), *The Princeton Companion to Mathematics*, University Press, Princeton (2008).

[33] J.A. Stratton, *Electromagnetic Theory*, McGraw-Hill, NY, 1941.

[34] J.C.A. Boeyens, *Chemistry in Four Dimensions*, Struct. Bond. 148 (2013) 25–47.

[35] F. Halzen and A.D. Martin, *Quarks and Leptons: An introductory course in modern particle physics*, Wiley, NY, 1984.

[36] J.C.A. Boeyens, *Red shifts in a curved space*, S.A. Jour. Sci. 91 (1995) 220.

[37] G.A. Jennings, *Modern Geometry with Applications*, Springer-Verlag, NY, 1994.

[38] O. Veblen, *Projektive Relativitätstheorie*, Julius Springer, Berlin (1933). (This seminal monograph appears in English translation in [17])

[39] O. Veblen and B. Hoffmann, *Projective Relativity*, Phys. Rev., 36 (1930) 810–822.

[40] E. Schrödinger, *Über eine bemerkenswerte Eigenschaft der Quantenbahnen eines einzelnen Elektrons*, Z. für Physik, 12 (1922) 13–23.

[41] F. London, *Quantenmechanische Deutung der Theorie von Weyl*, Z. für Phys., 42 (1927) 275–389.

[42] A. Einstein and N. Rosen, *The particle problem in the general theory of relativity*, Phys. Rev. 48 (1935) 73–77.

[43] J.C.A. Boeyens and D.C. Levendis, *Number theory and the periodicity of matter*, Springer.com, 2008.

[44] M.D. Kruskal, *Maximal extension of Schwarzschild metric*, Phys. Rev. 119 (1960) 1743–1745.

[45] J.C.A. Boeyens, *Commensurability in the solar system*, Phys. Essays 22 (2009) 493–499.

In: Recent Advances in Cosmology
Editors: Anderson Travena and Brady Soren
ISBN: 978-1-62417-943-3

Chapter 6

IMPLEMENTING HILLTOP F-TERM HYBRID INFLATION IN SUPERGRAVITY

***R. Armillis*[1]* *and C. Pallis*[2]†**
[1] Institut de Théorie des Phénomenès Physiques,
École Polytechnique Fédérale de Lausanne,Lausanne, Switzerland
[2] Department of Physics,University of Cyprus,
Nicosia, Cyprus

Abstract

F-term hybrid inflation (FHI) of the hilltop type can generate a scalar spectral index, $n_{\rm s}$, in agreement with the fitting of the seven-year Wilkinson microwave anisotropy probe data by the standard power-law cosmological model with cold dark matter and a cosmological constant, ΛCDM. We investigate the realization of this type of FHI by using quasi-canonical Kähler potentials with or without the inclusion of extra hidden-sector fields. In the first case, acceptable results can be obtained by constraining the coefficients of the quadratic and/or quartic supergravity correction to the inflationary potential and therefore a mild tuning of the relevant term of the Kähler potential is unavoidable. Possible reduction of $n_{\rm s}$ without generating maxima and minima of the potential on the inflationary path is also possible in a limited region of the available parameter space. The tuning of the terms of the Kähler potential can be avoided with the adoption of a simple class of string-inspired Kähler potentials for the hidden-sector fields which ensures a resolution to the η problem of FHI and allows acceptable values for the spectral index, constraining the coefficient of the quartic supergravity correction to the inflationary potential. Performing a four-point test of the analyzed models, we single out the most promising of these.

PACS 98.80.Cq, 11.30.Pb

Keywords: Cosmology, Inflation

*E-mail address: roberta.armillis@epfl.ch
†E-mail address: cpallis@ucy.ac.cy

1 Prologue

Inflation [1] has been incredibly successful in providing solutions to the problems of the *Standard Big Bang cosmology* (SBB). It can set the initial conditions, which give rise to the high degree of flatness and homogeneity that we observe in the universe today. From particle physics motivated models, it not only yields a mechanism for accelerated expansion but also explains, through quantum fluctuations, the origin of the temperature anisotropies in the *Cosmic Microwave Background* (CMB) and the seeds for the observed Large Scale Structure – for reviews see e.g. Refs. [2, 3].

We focus on a set of well-motivated, popular and quite natural models of *supersymmetric* (SUSY) *F-term hybrid inflation* (FHI) [4]. Namely, we consider the standard [5] FHI and some of its specific versions: the shifted [6] and smooth [7] FHI. They are realized [5] at (or close to) the SUSY *Grand Unified Theory* (GUT) scale $M_{\rm GUT} \simeq 2.86 \cdot 10^{16}$ GeV and can be easily linked to several extensions [3] of the *Minimal Supersymmetric Standard Model* (MSSM) which have a rich structure. Namely, the μ-problem of MSSM is solved via a direct coupling of the inflaton to Higgs superfields [8] or via a Peccei-Quinn symmetry [9], baryon number conservation is an automatic consequence [8] of an R symmetry and the baryon asymmetry of the universe is generated via leptogenesis which takes place [10] through the out-of-equilibrium decays of the inflaton's decay products.

Although quite successful, these models have at least two shortcomings:

(i) The problem of the enhanced (scalar) spectral index, $n_{\rm s}$. It is well-known that under the assumption that the problems of SBB are resolved exclusively by FHI, these models predict $n_{\rm s}$ just marginally consistent with the fitting of the seven-year results [11] from the *Wilkinson Microwave Anisotropy Probe Satellite* (WMAP7) data with the standard power-law cosmological model with *cold dark matter and a cosmological constant* (ΛCDM).

(ii) The so-called η problem. This problem is tied [2, 4, 12] on the expectation that *supergravity* (SUGRA) corrections generate a mass squared for the inflaton of the order of the Hubble parameter during FHI and so, the η criterion is generically violated, ruining thereby FHI. Inclusion of SUGRA corrections with canonical Kähler potential prevents [4, 13, 14] the generation of such a mass term due to a mutual cancellation. However, despite its simplicity, the canonical Kähler potential can be regarded [4] as fine tuning to some extent and, in all cases, increases $n_{\rm s}$ even more.

In this topical review we reconsider one set of possible resolutions (for other proposals, see Ref. [15–18]) of the tension between FHI and the data. This is relied on the utilization of three types of quasi-canonical [19] Kähler potential with or without the inclusion of extra fields, h_m. The term "extra fields" refers to hidden-sector [20] fields or fields which do not participate [21] in the inflationary superpotential but may only affect the Kähler potential. The consideration of extra fields assists us in solving the η problem of FHI as well. In particular, we review the following embeddings of FHI in SUGRA:

(i) FHI in *next-to-minimal SUGRA* (nmSUGRA) – see Sec. 5. A convenient choice of the next-to-minimal term [22–24] of the Kähler potential leads to a negative mass (quadratic) term for the inflaton and therefore $n_{\rm s}$ can be diminished sizeably.

(ii) FHI in *next-to-next-to-minimal SUGRA* (nnmSUGRA) – see Sec. 6. A convenient choice of the next-to-minimal and the next-to-next-to-minimal term generates [25,26] a positive mass (quadratic) term for the inflaton and a sizeable negative quartic term which yield acceptable n_s enhancing somehow the running of n_s, α_s.

(iii) FHI with extra fields, h_m, obeying a string-inspired Kähler potential (hSUGRA) – see Sec. 7. In the presence of h_m's, we can establish [27] a type of FHI which avoids the tuning – required in the cases (i) and (ii) above – of the quadratic SUGRA correction and is largely dominated by the quartic SUGRA correction. Namely, the coefficients of the Kähler potential are constrained to natural values (of order unity) so as the mass term of the inflaton field is identically zero.

In all the cases above and in the largest part of the parameter space the inflationary potential acquires a local maximum and minimum. Then, FHI of the *hilltop* type [28] can occur as the inflaton rolls from this maximum down to smaller values. However, the value of the inflaton field at the maximum is to be sufficiently close to the value that this field acquires when the pivot scale crosses outside the inflationary horizon. Therefore, n_s can become consistent with data, but only at the cost of an extra indispensable mild tuning [22] of the initial conditions. Another possible complication is that the system may get trapped near the minimum of the inflationary potential, thereby jeopardizing the attainment of FHI. On the other hand, we can show [23] that acceptable n_s's can be obtained even maintaining the monotonicity of the inflationary potential, i.e. without this minimum-maximum problem in the case of nmSUGRA.

In this presentation we reexamine the above ideas for the reduction of n_s within FHI, updating our results in Ref. [24,27] and incorporating recent related developments in Ref. [25]. In particular, the text is organized as follows: In Sec. 2, we review the basic FHI models and in Sec. 4 we recall the results holding for FHI in *minimal SUGRA* (mSUGRA). In the following we demonstrate how we can obtain hilltop FHI using various types of Kähler potentials – see Secs 5, 6 and 7. Our conclusions are summarized in Sec. 8. Throughout the text, charge conjugation is denoted by a star and brackets are, also, used by applying disjunctive correspondence.

2 FHI within SUGRA

We outline the salient features of the basic types of FHI. Namely we present the relevant superpotentials in Sec 2.1 and the SUSY potentials in Sec. 2.2. We then (in Sec. 2.3) describe the embedding of these models in SUGRA and extract the relevant inflationary potential in Sec. 2.4.

2.1 The Relevant Superpotential

The F-term hybrid inflation can be realized adopting one of the superpotentials below:

$$W=\widehat{W}+W_{\rm FHI}\ \text{with}\ W_{\rm FHI}=\begin{cases}\widehat{\kappa}S\left(\bar\Phi\Phi-M^2\right) & \text{for standard FHI,}\\ \widehat{\kappa}S\left(\bar\Phi\Phi-M^2\right)-S\frac{(\bar\Phi\Phi)^2}{M_{\rm S}^2} & \text{for shifted FHI,}\\ S\left(\frac{(\bar\Phi\Phi)^2}{M_{\rm S}^2}-\widehat{\mu}_{\rm S}^2\right) & \text{for smooth FHI,}\end{cases}\tag{1}$$

where we allow for the presence of a part, $\widehat{W}$, which depends exclusively on the hidden sector superfields, h_m. Here (and hereafter) we use the hat to denote such quantities. To keep our analysis as general as possible, we do not adopt any particular form for $\widehat{W}$ – for some proposals see Ref. [29, 30]. Note that our construction remains intact even if we set $\widehat{W} = 0$ as it was supposed in Ref. [21]. This is due to the fact that $\widehat{W}$ is expected to be much smaller than the inflationary energy density – see Sec. 2.3. For $\widehat{W} \neq 0$, though, we need to assume that h_m's are stabilized before the onset of FHI by some mechanism not consistently taken into account here. As a consequence, we neglect the dependence of $\widehat{W}$, $\widehat{\kappa}$ and $\widehat{\mu}_S$ on h_m and so, these quantities are treated as constants. We further assume that the D-terms due to h_m's vanish – contrary to the strategy adopted in Ref. [21].

The remaining symbols in the *right hand side* (r.h.s) of Eq. (1) are identified as follows:

- S is a left handed superfield, singlet under a GUT gauge group G;
- $\bar{\Phi}$, Φ is a pair of left handed superfields belonging to non-trivial conjugate representations of G, and reducing its rank by their *vacuum expectation values* (v.e.vs);
- $\widehat{M}_S \sim 5\times 10^{17}$ GeV is an effective cutoff scale comparable with the string scale;
- $\widehat{\kappa}$ and $\widehat{M}$, $\widehat{\mu}_S$ $(\sim M_{\rm GUT})$ are parameters which can be made positive by field redefinitions.

The superpotential in Eq. (1) for standard FHI is the most general renormalizable superpotential consistent with a continuous R-symmetry [5] under which

$$S \to e^{ir}\,S,\ \bar{\Phi}\Phi \to \bar{\Phi}\Phi,\ W \to e^{ir}\,W. \tag{2}$$

After including in this superpotential the leading non-renormalizable term, one obtains the superpotential of shifted [6] FHI in Eq. (1). Finally, the superpotential of smooth [7] FHI can be obtained if we impose an extra Z_2 symmetry under which the combination $\bar{\Phi}\Phi$ has unit charge.

2.2 The SUSY Potential

The SUSY potential, $V_{\rm SUSY}$, extracted (see e.g. ref. [2]) from $W_{\rm FHI}$ in Eq. (1) includes F and D-term contributions. Namely,

$$V_{\rm SUSY} = V_{\rm F} + V_{\rm D}, \tag{3}$$

where

(i) The F-term contribution can be written as:

$$V_{\rm F} = \begin{cases} \kappa^2 M^4\left((\Phi^2-1)^2 + 2S^2\Phi^2\right) & \text{for standard FHI,}\\ \kappa^2 M^4\left((\Phi^2-1-\xi\Phi^4)^2 + 2S^2\Phi^2(1-2\xi\Phi^2)^2\right) & \text{for shifted FHI,}\\ \mu_S^4\left((1-\Phi^4)^2+8S^2\Phi^6\right) & \text{for smooth FHI,}\end{cases} \tag{4}$$

where the scalar components of the superfields are denoted by the same symbols as the corresponding superfields and $\xi = M^2/\kappa M_S^2$ with $4 < 1/\xi < 7.2$ [6]. In order to recover the

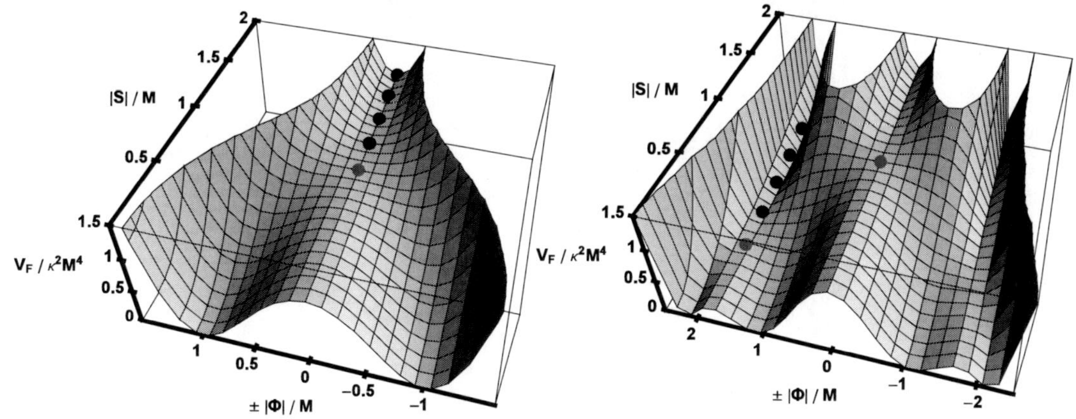

Figure 1: *The three dimensional plot of the (dimensionless) F-term potential $V_{\rm F}/\kappa^2M^4$ versus $\mathsf{S}=|S|/M$ and $\pm\Phi=\pm|\Phi|/M$ for standard [shifted with $\xi=1/6$] FHI (left [right]). The inflationary trajectories are also depicted by black points whereas the critical points (of the shifted and standard trajectories) are depicted by red/light points.*

properly normalized energy density during FHI, we absorb in the constants of Eq. (4) some normalization pre-factors emerging from the SUGRA potential $V_{\rm SUGRA}$ – see below – so that their definition is

$$\kappa=e^{\widehat{K}/2m_{\rm P}^2}\widehat{Z}^{-1/2}\widehat{\kappa},\;\;\mu_{\rm S}=e^{\widehat{K}/4m_{\rm P}^2}\widehat{Z}^{-1/4}\widehat{\mu}_{\rm S}\;\;\text{and}\;\;M_{\rm S}=e^{-\widehat{K}/4m_{\rm P}^2}\widehat{Z}^{1/4}\widehat{M}_{\rm S} \tag{5}$$

where $\widehat{K}$ and $\widehat{Z}$ are the h_m-dependent parts of the Kähler potential, K, considered in Sec. 2.3. The last relation is introduced so as $\kappa M_{\rm S}^2=\widehat{\kappa}\widehat{M}_{\rm S}^2$ and $\mu_{\rm S}M_{\rm S}=\widehat{\mu}_{\rm S}\widehat{M}_{\rm S}$. Also, we use [6,7] the following dimensionless quantities

$$\begin{cases}\Phi=|\Phi|/M\;\text{ and }\;\mathsf{S}=\widehat{Z}^{1/2}|S|/M & \text{for standard or shifted FHI,}\\ \Phi=|\Phi|/\sqrt{\mu_{\rm S}M_{\rm S}}\;\text{ and }\;\mathsf{S}=\widehat{Z}^{1/2}|S|/\sqrt{\mu_{\rm S}M_{\rm S}} & \text{for smooth FHI.}\end{cases} \tag{6}$$

In Fig. 1-(a), Fig. 1-(b) and 2 we present the three dimensional plot of $V_{\rm F}$ versus $\pm\Phi$ and S for standard, shifted and smooth FHI, respectively.

(ii) The D-term contribution $V_{\rm D}$ vanishes for $|\bar{\Phi}|=|\Phi|$ since $V_{\rm D}$ has the form:

$$V_{\rm D}=\frac{1}{2}g^2\sum_a D_aD_a\;\;\text{with}\;\;D_a=\phi_M\left(T_a\right)^M_{\bar N}K^{\bar N}\simeq|\Phi|^2-|\bar{\Phi}|^2, \tag{7}$$

where g is the (unified) gauge coupling constant, T_a are the generators of G and the notation used is explained below Eq. (11) – recall that Φ and $\bar{\Phi}$ belong to the conjugate representation of G.

From the form of $V_{\rm SUSY}$ in Eq. (4), we can understand that W in Eq. (1) plays a twofold crucial role:

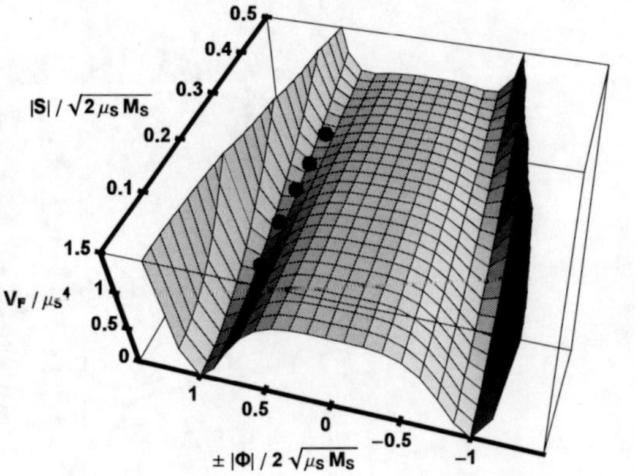

Figure 2: *The three dimensional plot of the (dimensionless) F-term potential $V_{\rm F}/\mu_{\rm S}^4$ for smooth FHI versus $\mathsf{S} = |S|/\sqrt{\mu_{\rm S}M_{\rm S}}$ and $\pm\Phi = \pm|\Phi|/\sqrt{\mu_{\rm S}M_{\rm S}}$. The inflationary trajectory is also depicted by black points.*

(i) It leads to the spontaneous breaking of G. Indeed, the vanishing of $V_{\rm F}$ gives the v.e.vs of the fields in the SUSY vacuum. Namely,

$$\langle S\rangle = 0 \;\text{ and }\; |\langle\bar\Phi\rangle| = |\langle\Phi\rangle| = v_{_G} = \begin{cases} M & \text{for standard FHI,}\\ \frac{M\sqrt{1-\sqrt{1-4\xi}}}{\sqrt{2\xi}} & \text{for shifted FHI,}\\ \sqrt{\mu_{\rm S}M_{\rm S}} & \text{for smooth FHI}\end{cases} \tag{8}$$

(in the case where $\bar\Phi$, Φ are not *Standard Model* (SM) singlets, $\langle\bar\Phi\rangle$, $\langle\Phi\rangle$ stand for the v.e.vs of their SM singlet directions). The non-zero value of the v.e.v $v_{_G}$ signalizes the spontaneous breaking of G.

(ii) It gives rise to FHI. This is due to the fact that, for large enough values of $|S|$, there exist valleys of local minima of the classical potential with constant (or almost constant in the case of smooth FHI) values of $V_{\rm F}$. In particular, we can observe that $V_{\rm F}$ takes the following constant value

$$V_{\rm HI0} = \begin{cases}\kappa^2M^4\\ \kappa^2M_\xi^4\\ \mu_{\rm S}^4\end{cases} \text{along the direction(s): } \Phi = \begin{cases} 0 & \text{for standard FHI,}\\ 0 \text{ or } 1/\sqrt{2\xi} & \text{for shifted FHI,}\\ 0 \text{ or } 1/2\sqrt{3}\mathsf{S} & \text{for smooth FHI,}\end{cases} \tag{9}$$

with $M_\xi = M\sqrt{1/4\xi-1}$. From Figs. 1 and 2 we deduce that the flat direction $\Phi = 0$ corresponds to a minimum of $V_{\rm F}$, for $|S| \gg M$, in the cases of standard and shifted FHI and to a maximum of $V_{\rm F}$ in the case of smooth FHI. The inflationary trajectories are depicted by bold points, whereas the critical points by red/light points. Note that critical points exist only in the case of standard – for $\mathsf{S} = 1$ – and shifted – for $\mathsf{S} = (1/4\xi-1)/2$ – FHI but not for smooth FHI. In the case of Fig. 1-(b), the implementation of shifted FHI is ensured by restricting $1/\xi$ in the range $(4-7.2)$ [6]. Under this assumption, the shifted track lies lower that the trivial one and so, it is energetically more favorable to drive FHI.

Since FHI can be attained along a minimum of $V_{\rm F}$, we infer that, during standard FHI, the GUT gauge group G is necessarily restored. As a consequence, topological defects such as strings [22,31], monopoles, or domain walls may be produced [7] via the Kibble mechanism [32] during the spontaneous breaking of G at the end of FHI. This can be avoided in the other two cases, since the form of V_F allows for non-trivial inflationary valleys along which G is spontaneously broken due to non-zero values that $\bar{\Phi}$ and Φ acquire during FHI. Therefore, no topological defects are produced in these cases.

2.3 SUGRA Corrections

The consequences that SUGRA has on the models of FHI can be investigated by restricting ourselves to the inflationary trajectory $\Phi = \bar{\Phi} \simeq 0$ (possible corrections due to the non-vanishing Φ and $\bar{\Phi}$ in the cases of shifted and smooth FHI are expected to be negligible). Therefore, W in Eq. (1) takes the form

$$W = \widehat{W} + I\,, \text{ where } I = -\widehat{V}_{\rm HI0}^{1/2} S \text{ with } \widehat{V}_{\rm HI0} = e^{-\widehat{K}/m_{\rm P}^2}\widehat{Z}V_{\rm HI0}. \tag{10}$$

The SUGRA scalar potential (without the D-terms) is given (see, e.g., Ref. [2]) by

$$V_{\rm SUGRA} = e^{\frac{K}{m_{\rm P}^2}}\left(K^{M\bar{N}}F_M F_{\bar{N}}^* - 3\frac{|W|^2}{m_{\rm P}^2}\right) \text{ where } F_N = W_N + K_N\frac{W}{m_{\rm P}^2} \tag{11}$$

are the SUGRA-generalized F-terms, the subscript M $[\bar{M}]$ denotes derivation *with respect to* (w.r.t) the complex scalar field ϕ_M $[\phi_M^*]$ which corresponds to the chiral superfield ϕ_M with $\phi_M = h_m, S, \Phi, \bar{\Phi}$ and the matrix $K^{M\bar{N}}$ is the inverse of the Kähler metric $K_{M\bar{N}}$. In this paper we consider a quite generic form of Kähler potentials, which do not deviate much from the canonical one and respect the R symmetry of Eq. (2). Namely we take

$$\begin{aligned} K \;=\; & \widehat{K} + \widehat{Z}|S|^2 + \frac{1}{4}k_{4S}\widehat{Z}^2\frac{|S|^4}{m_{\rm P}^2} + \frac{1}{6}k_{6S}\widehat{Z}^3\frac{|S|^6}{m_{\rm P}^4} + \frac{1}{8}k_{8S}\widehat{Z}^4\frac{|S|^8}{m_{\rm P}^6} \\ & + \frac{1}{10}k_{10S}\widehat{Z}^5\frac{|S|^{10}}{m_{\rm P}^8} + \frac{1}{12}k_{12S}\widehat{Z}^6\frac{|S|^{12}}{m_{\rm P}^{10}} + |\Phi|^2 + |\bar{\Phi}|^2 + \cdots\,, \end{aligned} \tag{12}$$

where $k_{4S}, k_{6S}, k_{8S}, k_{10S}$ and k_{12S} are positive or negative constants of order unity and the ellipsis represents higher order terms involving the waterfall fields (Φ and $\bar{\Phi}$) and S. We can neglect these terms since they are irrelevant along the inflationary path.

Substituting Eq. (12) into Eq. (11) and expanding $V_{\rm SUGRA}$ in powers of $|S|$, we end up with an expansion of the form:

$$V_{\rm SUGRA} \simeq V_{\rm HI0}\left(1 - 2c_{1K}e^{-\frac{\widehat{K}}{2m_{\rm P}^2}}\frac{|\widehat{W}|\widehat{Z}^{1/2}}{\sqrt{V_{\rm HI0}}}|S|\cos\theta + \sum_{\nu=1}^{5}(-1)^\nu c_{2\nu K}\widehat{Z}^\nu\left(\frac{|S|}{m_{\rm P}}\right)^{2\nu}\right) \tag{13}$$

where the phase θ reads $\theta = \arg\widehat{W} + \arg S + \arg\widehat{V}_{\rm HI0}^{1/2}$. In the r.h.s of the expression above, we neglect terms proportional to $|\widehat{W}|^2$ which are certainly subdominant compared with those which are proportional to $V_{\rm HI}$. From the terms proportional to $|\widehat{W}|V_{\rm HI0}^{1/2}$ we present the second term of the r.h.s of Eq. (13) which expresses the most important contribution [14,29] to the inflationary potential from the soft SUSY breaking terms. For natural values of $\widehat{W}$ and $e^{\widehat{K}/2m_{\rm P}^2}$ this term starts [14] playing an important role in the case of standard FHI in mSUGRA – see Sec. 4.2 – for $\kappa \lesssim 5\cdot 10^{-4}$ whereas it has [14] no significant effect in the cases of shifted and smooth FHI.

Taking in Eq. (12) $\widehat{K}=0$ and $\widehat{Z}=1$ (as in Secs. 4 – 6) the coefficients $c_{\nu K}=c_{\nu K}^{(0)}$ are found to be

$$c_{1K}^{(0)} = -2, \tag{14a}$$

$$c_{2K}^{(0)} = k_{4S}, \tag{14b}$$

$$c_{4K}^{(0)} = \frac{1}{2}-\frac{7k_{4S}}{4}+k_{4S}^2-\frac{3k_{6S}}{2}, \tag{14c}$$

$$c_{6K}^{(0)} = -\frac{2}{3}+\frac{3k_{4S}}{2}-\frac{7k_{4S}^2}{4}+k_{4S}^3+\frac{10k_{6S}}{3}-3k_{4S}k_{6S}+2k_{8S}, \tag{14d}$$

$$\begin{aligned} c_{8K}^{(0)} &= \frac{3}{8}-\frac{5k_{10S}}{2}-\frac{13k_{4S}}{24}+\frac{41k_{4S}^2}{32}-\frac{7k_{4S}^3}{4}+k_{4S}^4-\frac{13k_{6S}}{4} \\ &+\frac{143k_{4S}k_{6S}}{24}-\frac{9k_{4S}^2k_{6S}}{2}+\frac{9k_{6S}^2}{4}-\frac{39k_{8S}}{8}+4k_{4S}k_{8S}, \end{aligned} \tag{14e}$$

$$\begin{aligned} c_{10K}^{(0)} &= -\frac{2}{15}+\frac{32k_{10S}}{5}+3k_{12S}+\frac{k_{4S}}{24}-5k_{10S}k_{4S}-\frac{13k_{4S}^2}{24}+\frac{41k_{4S}^3}{32} \\ &-\frac{7k_{4S}^4}{4}+k_{4S}^5+\frac{5k_{6S}}{3}-\frac{29k_{4S}k_{6S}}{6}+\frac{103k_{4S}^2k_{6S}}{12}-6k_{4S}^3k_{6S}-5k_{6S}^2 \\ &+\frac{27k_{4S}k_{6S}^2}{4}+5k_{8S}-\frac{67k_{4S}k_{8S}}{8}+6k_{4S}^2k_{8S}-6k_{6S}k_{8S}. \end{aligned} \tag{14f}$$

We observe that terms of order ν in the expansion of Eq. (12) give rise to contributions of order equal or greater than $(\nu-2)$ in the expansion of Eq. (13).

2.4 The Inflationary Potential

The general form of the potential which can drive the various versions of FHI reads

$$V_{\rm HI} \simeq V_{\rm HI0}\left(1+c_{\rm HI}-{\rm a}_S\frac{\sigma}{\sqrt{2V_{\rm HI0}}}+\sum_{\nu=1}^{5}(-1)^{\nu}c_{2\nu K}\left(\frac{\sigma}{\sqrt{2}m_{\rm P}}\right)^{2\nu}\right), \tag{15}$$

where $\sigma=\sqrt{2}\widehat{Z}^{1/2}|S|$ is the canonically (up to the order $|S|^2$) normalized inflaton field and we take $\theta=\pi$ which minimizes $V_{\rm SUGRA}$ for given σ. To facilitate our numerical analysis, we introduce the real tadpole parameter ${\rm a}_S$ defined in terms of the $V_{\rm SUGRA}$ parameter, by the relation

$${\rm a}_S = 2c_{1K}e^{-\widehat{K}/2m_{\rm P}^2}|\widehat{W}|. \tag{16}$$

In Eq. (15), besides the contributions originating from $V_{\rm SUGRA}$ in Eq. (13), we include the term $c_{\rm HI}V_{\rm HI0}$ which represents a correction to $V_{\rm HI}$ resulting from the SUSY breaking on the inflationary valley, in the cases of standard [5] and shifted [6] FHI, or from the structure of the classical potential in the case of smooth [7] FHI. Indeed, $V_{\rm HI0}>0$ breaks SUSY and gives rise to logarithmic radiative corrections to the potential originating from a mass splitting in the $\Phi-\bar{\Phi}$ supermultiplets. On the other hand, in the case of smooth [7] FHI, the inflationary valleys are not classically flat and, thus, the radiative corrections are expected to be subdominant. The term $c_{\rm HI}$ can be written as follows:

$$c_{\rm HI} = \begin{cases} \kappa^2 {\sf N}\left[2\ln\left(\kappa^2 x M^2/Q^2\right) + f_{\rm rc}(x)\right]/32\pi^2 & \text{for standard FHI,} \\ \kappa^2 \left[2\ln\left(\kappa^2 x_\xi M_\xi^2/Q^2\right) + f_{\rm rc}(x_\xi)\right]/16\pi^2 & \text{for shifted FHI,} \\ -2\mu_{\rm S}^2 M_{\rm S}^2/27\sigma^4 & \text{for smooth FHI,} \end{cases} \qquad (17)$$

with $x = \sigma^2/2M^2$, $x_\xi = \sigma^2/M_\xi^2$ and

$$f_{\rm rc}(x) = (x+1)^2\ln(1+1/x) + (x-1)^2\ln(1-1/x) \Rightarrow f_{\rm rc}(x) \simeq 3 \text{ for } x \gg 1. \qquad (18)$$

Also ${\sf N}$ is the dimensionality of the representations to which $\bar{\Phi}$ and Φ belong and Q is a renormalization scale. For the values of κ encountered in our work renormalization group effects [33] remain negligible.

In our applications in Secs. 4.2, 5.3, 6.3 and 7.3 we take ${\sf N} = 2$. This choice corresponds to the left-right symmetric GUT gauge group $SU(3)_c \times SU(2)_L \times SU(2)_R \times U(1)_{B-L}$ with $\bar{\Phi}$ and Φ belonging to $SU(2)_R$ doublets with $B-L = -1$ and 1 respectively. No cosmic strings are produced during the GUT phase transition and, consequently, no extra restrictions on the parameters (as e.g. in Refs. [31]) have to be imposed. As regards the case of shifted [6] FHI we identify G with the Pati-Salam gauge group $SU(4)_c \times SU(2)_L \times SU(2)_R$. Needless to say that the case of smooth FHI is independent on the adopted GUT since the inclination of the inflationary path is generated at the classical level and the addition of any radiative correction is expected to be subdominant. Negligible is also the third term in the r.h.s of Eq. (15) for ${\rm a}_S \sim 1$ TeV, besides the case of standard FHI in mSUGRA – see Sec. 4 – where it may be important for $\kappa \leq 5\cdot 10^{-4}$. For simplicity, we neglect it, in the analysis of the remaining cases – see Secs. 5, 6 and 7.

3 Constraining FHI

The parameters of FHI models can be restricted imposing a number of observational constraints described in Secs. 3.1 and 3.2. Additional theoretical considerations presented in Sec. 3.3 can impose further limitations.

3.1 Inflationary Observables

Applying standard formulae – see e.g. Refs. [2, 3] – we can estimate the inflationary observables of FHI. Namely, we can find:

(i) The number of e-foldings $N_{\rm HI*}$ that the scale $k_* = 0.002/{\rm Mpc}$ suffers during FHI,

$$N_{\rm HI*} = \frac{1}{m_{\rm P}^2}\int_{\sigma_{\rm f}}^{\sigma_*} d\sigma\, \frac{V_{\rm HI}}{V'_{\rm HI}}, \qquad (19)$$

where the prime denotes derivation w.r.t σ, σ_* is the value of σ when the scale k_* crosses outside the horizon of FHI, and $\sigma_{\rm f}$ is the value of σ at the end of FHI, which can be found, in the slow roll approximation, from the condition

$$\max\{\epsilon(\sigma_{\rm f}), |\eta(\sigma_{\rm f})|\} = 1, \text{ where } \epsilon \simeq \frac{m_{\rm P}^2}{2}\left(\frac{V'_{\rm HI}}{V_{\rm HI}}\right)^2 \text{ and } \eta \simeq m_{\rm P}^2\,\frac{V''_{\rm HI}}{V_{\rm HI}}. \qquad (20)$$

In the cases of standard [5] and shifted [6] FHI and in the parameter space where the terms in Eq. (11) do not play an important role, the end of inflation coincides with the onset of the GUT phase transition, i.e. the slow roll conditions are violated close to the critical point $\sigma_c = \sqrt{2}M$ [$\sigma_c = M_\xi$] for standard [shifted] FHI, where the waterfall regime commences. On the contrary, the end of smooth [7] FHI is not abrupt since the inflationary path is stable w.r.t $\Phi - \bar{\Phi}$ for all σ's and σ_f is found from Eq. (20). An accurate enough estimation of σ_f's – suitable for our analytical expressions presented below – is

$$\sigma_f \simeq \begin{cases} \sqrt{2}M & \text{for standard FHI,} \\ M_\xi & \text{for shifted FHI,} \\ \sqrt{2\sqrt[3]{5}/3\sqrt[3]{\mu_S M_S m_P}} & \text{for smooth FHI.} \end{cases} \tag{21}$$

(ii) The power spectrum $\Delta^2_{\mathcal{R}}$ of the curvature perturbations generated by σ at the pivot scale k_*

$$\Delta_{\mathcal{R}} = \frac{1}{2\sqrt{3}\,\pi m_P^3} \left.\frac{V_{HI}^{3/2}}{|V'_{HI}|}\right|_{\sigma=\sigma_*}. \tag{22}$$

(iii) The spectral index

$$n_s = 1 - 6\epsilon_* + 2\eta_*, \tag{23a}$$

its running

$$\alpha_s = \frac{2}{3}\left(4\eta_*^2 - (n_s - 1)^2\right) - 2\xi_*, \tag{23b}$$

with $\xi \simeq m_P^4\, V'_{HI}V'''_{HI}/V_{HI}^2$ and the scalar-to-tensor ratio

$$r = 16\epsilon_* \tag{23c}$$

where all the variables with the subscript $*$ are evaluated at $\sigma = \sigma_*$.

3.2 Observational Constraints

Under the assumption that the contribution in Eq. (22) is solely responsible for the observed curvature perturbation – i.e. there are no contributions to $\Delta_{\mathcal{R}}$ from curvatons [2] or topological defects [31] – and (ii) there is a conventional cosmological evolution after FHI – see point (i) below –, the parameters of the FHI models can be restricted imposing the following requirements:

(i) The number of e-foldings N_{HI*} computed by means of Eq. (19) has to be set equal to the number of e-foldings N_* elapsed between the horizon crossing of the observationally relevant mode k_* and the end of FHI. N_* can be found as follows [2]:

$$\frac{k_*}{H_0R_0} = \frac{H_*R_*}{H_0R_0} = \frac{H_*}{H_0}\frac{R_*}{R_{\rm Hf}}\frac{R_{\rm Hf}}{R_{\rm rh}}\frac{R_{\rm rh}}{R_{\rm eq}}\frac{R_{\rm eq}}{R_0}$$
$$= \sqrt{\frac{V_{\rm HI0}}{\rho_{\rm c0}}}e^{-N_*}\left(\frac{V_{\rm HI0}}{\rho_{\rm rh}}\right)^{-1/3}\left(\frac{\rho_{\rm rh}}{\rho_{\rm eq}}\right)^{-1/4}\left(\frac{\rho_{\rm eq}}{\rho_{\rm m0}}\right)^{-1/3}$$
$$\Rightarrow\quad N_* \simeq \ln\frac{H_0R_0}{k_*} + 24.72 + \frac{2}{3}\ln\frac{V_{\rm HI0}^{1/4}}{1\ {\rm GeV}} + \frac{1}{3}\ln\frac{T_{\rm rh}}{1\ {\rm GeV}}, \tag{24}$$

where $T_{\rm rh}$ is the reheating temperature after the completion of the FHI. Moreover, R is the scale factor, $H = \dot{R}/R$ is the Hubble rate, ρ is the energy density and the subscripts 0, k, Hf, rh, eq and m denote respectively values at the present (except for the symbol $V_{\rm HI0}$), at the horizon crossing ($k = R_kH_k$) of the mode k, at the end of FHI, at the end of the reheating period, at the radiation-matter equidensity point and in the matter dominated era. In our calculation we take into account that $R \propto \rho^{-1/3}$ for decaying-particle domination or matter dominated era and $R \propto \rho^{-1/4}$ for radiation dominated era. We use the following numerical values:

$$\rho_{\rm c0} = 8.099\times 10^{-47}h_0^2\ {\rm GeV}^4 \ \text{ with } h_0 = 0.71, \tag{25a}$$
$$\rho_{\rm rh} = \frac{\pi^2}{30}g_{\rho*}T_{\rm rh}^4 \ \text{ with } g_{\rho*} = 228.75, \tag{25b}$$
$$\rho_{\rm eq} = 2\Omega_{\rm m0}(1-z_{\rm eq})^3\rho_{\rm c0} \ \text{ with } \Omega_{\rm m0} = 0.26 \ \text{ and } z_{\rm eq} = 3135. \tag{25c}$$

Setting $H_0 = 2.37\cdot 10^{-4}/{\rm Mpc}$ and $k/R_0 = 0.002/{\rm Mpc}$ in Eq. (24) we arrive at

$$N_{\rm HI*} \simeq 22.6 + \frac{2}{3}\ln\frac{V_{\rm HI0}^{1/4}}{1\ {\rm GeV}} + \frac{1}{3}\ln\frac{T_{\rm rh}}{1\ {\rm GeV}}. \tag{26}$$

Throughout our investigation we take $T_{\rm rh} \simeq 10^9$ GeV as in the majority of these models [3, 10, 14] saturating conservatively the gravitino constraint [34]. This choice for $T_{\rm rh}$ does not affect crucially our results, since $T_{\rm rh}$ appears in Eq. (26) under its logarithm raised to the one third power and therefore, its variation over two or three orders of magnitude has a minor influence on the final value of $N_{\rm HI*}$.

(ii) The power spectrum of the curvature perturbations given by Eq. (22) is to be confronted with the WMAP7 data [11]:

$$\Delta_{\mathcal{R}} \simeq 4.93\cdot 10^{-5} \ \text{ at } k_* = 0.002/{\rm Mpc}. \tag{27}$$

(iii) According to the fitting of the WMAP7 results by the cosmological model ΛCDM, $n_{\rm s}$ at the pivot scale $k_* = 0.002/{\rm Mpc}$ has to fall within the following range of values [11]:

$$n_{\rm s} = 0.968 \pm 0.024 \Rightarrow 0.944 \lesssim n_{\rm s} \lesssim 0.992 \ \text{ at } 95\%\ \text{c.l.} \tag{28}$$

(iv) Limiting ourselves to a_s's consistent with the assumptions of the power-law ΛCDM cosmological model, we have to ensure that $|a_s|$ remains negligible. Since, within the cosmological models with running spectral index, $|a_s|$'s of order 0.01 are encountered [11], we impose the following upper bound:

$$|a_s| \ll 0.01. \tag{29}$$

3.3 Theoretical Considerations

From a more theoretical point of view, the models of (hilltop) FHI can be better refined using the following criteria:

(i) Gauge coupling unification. When G contains non-abelian factors (beyond the SM one), the mass, gv_G, of the lightest gauge boson at the SUSY vacuum, Eq. (8) is to take the value dictated by the unification of the gauge coupling constants within MSSM, i.e.,

$$gv_G \simeq 2 \cdot 10^{16}\ \text{GeV} \Rightarrow v_G \simeq 2.86 \cdot 10^{16}\ \text{GeV} \ \ \text{with} \ \ g \simeq 0.7, \tag{30}$$

being the value of the unified gauge coupling constant. However, we display in the following results for standard FHI which do not fulfill Eq. (30). This is allowed since the relevant restriction can be evaded if G includes only abelian factors (beyond the SM one) which do not disturb the gauge coupling unification. Otherwise, threshold corrections may be taken into account in order to restore the unification.

(ii) Boundness of $V_{\rm HI}$. The inflationary potential is expected to be bounded from below. This requirement lets open the possibility that the inflaton may give rise to an inflationary expansion under generic initial conditions set at $\sigma \simeq m_{\rm P}$.

(iii) Convergence of $V_{\rm HI}$. The expression of $V_{\rm HI}$ in Eq. (15) is expected to converge at least for $\sigma \sim \sigma_*$. This fact can be ensured if, for $\sigma \sim \sigma_*$, each successive term in the expansion of $V_{\rm SUGRA}$ (and K) Eq. (13) (and Eq. (12)) is smaller than the previous one. In practice, this objective can be easily accomplished if the k's in Eq. (12) – or Eq. (15) – are sufficiently low.

(iv) Monotonicity of $V_{\rm HI}$. Depending on the values of the coefficients k's in Eq. (15), $V_{\rm HI}$ is a monotonic function of σ or develops a local minimum and maximum. The latter case leads to the possible complication in which the system gets trapped near the minimum of the inflationary potential and, consequently, no FHI takes place. It is, therefore, crucial to check if we can avoid the minimum-maximum structure of $V_{\rm HI}$. In such a case the system can start its slow rolling from any point on the inflationary path without the danger of getting trapped. This can be achieved, if we require that $V_{\rm HI}$ is a monotonically increasing function of σ, i.e. $V'_{\rm HI} > 0$ for any σ or, equivalently,

$$V'_{\rm HI}(\bar\sigma_{\rm min}) > 0 \ \ \text{with} \ \ V''_{\rm HI}(\bar\sigma_{\rm min}) = 0 \ \ \text{and} \ \ V'''_{\rm HI}(\bar\sigma_{\rm min}) > 0 \tag{31}$$

where $\bar\sigma_{\rm min}$ is the value of σ at which the minimum of $V'_{\rm HI}$ lies.

(v) Tuning of the initial conditions. When hilltop FHI occurs with σ rolling from the region of the maximum down to smaller values, a mild tuning of the initial conditions is required [22] in order to obtain acceptable n_s's. In particular, the lower n_s we want to obtain, the closer we must set σ_* to σ_{max}, where σ_{max} is the value of σ at which the maximum of V_{HI} lies. To quantify somehow the amount of this tuning in the initial conditions, we define [22] the quantity:

$$\Delta_{m*} = (\sigma_{max} - \sigma_*)/\sigma_{max}. \tag{32}$$

The naturalness of the attainment of FHI increases with Δ_{m*}.

4 FHI in mSUGRA

The simplest choice of Kähler potential emerging from the expression of Eq. (12) is the one which assures canonical kinetic terms for the inflaton field, S, with the minimal number of terms. This choice is specified in Sec. 4.1,and our results are discussed in Sec. 4.2.

4.1 The Relevant Set-up

The used Kähler potential in this case can be derived from Eq. (12) by setting:

$$\widehat{K} = 0,\ \widehat{Z} = 1 \ \text{ and } \ k_{4S} = k_{6S} = k_{8S} = k_{10S} = k_{12S} = 0. \tag{33}$$

Upon substituting Eqs. (14a) – (14f) into Eq. (15) we infer that the resulting V_{HI} takes the form

$$V_{HI} \simeq V_{HI0}\left(1 + c_{HI} - a_S\frac{\sigma}{\sqrt{2V_{HI0}}} + \frac{\sigma^4}{8m_P}\right), \tag{34}$$

since in this case $c_{2K}^{(0)} = 0$ and $c_{4K}^{(0)} = 1/2$. It is worth mentioning that mSUGRA is, in principle, beneficial for the implementation of FHI, since it does not generate any new contribution in the η parameter, Eq. (20), due to a miraculous cancellation emerging in the computation of $c_{2K}^{(0)}$. Despite this fact, fixing all the remaining terms in Eq. (12) beyond the quadratic term equal to zero can be regarded as an ugly tuning.

4.2 Results

The investigation of this model of FHI depends on the parameters:

$$\sigma_*, v_G, a_S \ \text{ and } \begin{cases} \kappa & \text{for standard and shifted FHI,} \\ M_S & \text{for smooth FHI,} \end{cases}$$

where we fix $M_S = 5 \cdot 10^{17}$ GeV in the case of shifted FHI. In our computation, we use as input parameters a_S, and κ or M_S. We then restrict v_G and σ_* so as Eqs. (26) and (27) are fulfilled. Using Eqs. (23a) and (23b) we can extract n_s and α_s respectively. Our findings for standard [shifted and smooth] FHI are displayed in Sec. 4.2.1 [Sec. 4.2.2].

We can obtain a rather accurate estimation of the expected n_s's if we omit the third term in the r.h.s of Eq. (34), calculate analytically the integral in Eq. (19), replace the σ_f's by

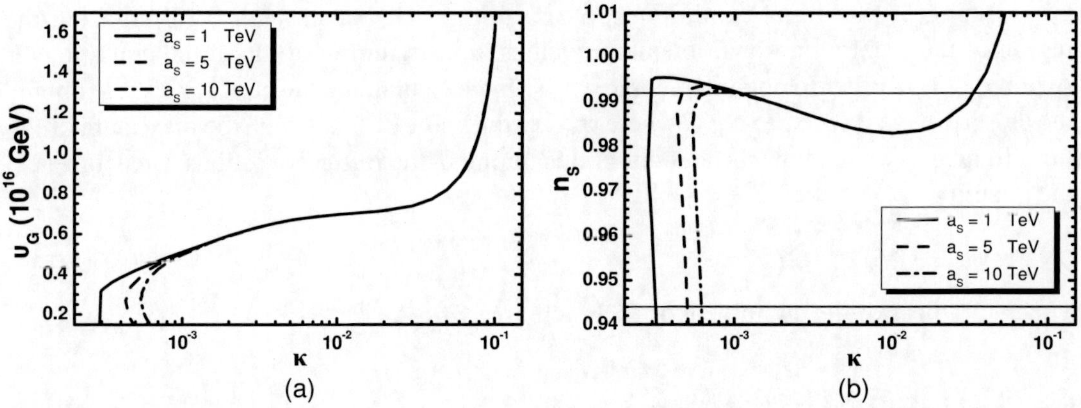

Figure 3: *The allowed by Eqs. (26) and (27) values of v_G [n_s] versus κ for standard FHI with* $\mathsf{N}=2$ *and* $a_S = 1$ TeV *(solid lines),* $a_S = 5$ TeV *(dashed lines) and* $a_S = 10$ TeV *(dot-dashed lines). The region of Eq. (28) is also limited by thin lines.*

their values in Eq. (21) and solve the resulting equation w.r.t σ_*. Taking into account that $\varepsilon < \eta$ we can extract n_s from Eq. (23a) and find

$$n_s \simeq \begin{cases} 1-1/N_{\rm HI*}+3\kappa^2 \mathsf{N} N_{\rm HI*}/4\pi^2 & \text{for standard FHI,} \\ 1-1/N_{\rm HI*}+3\kappa^2 N_{\rm HI*}/2\pi^2 & \text{for shifted FHI,} \\ 1-5/3N_{\rm HI*}+2\left(6\mu_S^2 M_S^2 N_{\rm HI*}/m_{\rm P}^4\right)^{1/3} & \text{for smooth FHI.} \end{cases} \quad (35)$$

Observing that the last term in the r.h.s of the expressions above arise from the last term in the r.h.s of Eq. (34), we can easily infer that mSUGRA increases significantly n_s for relatively large κ's or M_S's.

On the other hand, the third term in the r.h.s of Eq. (34) can be important for $\kappa \leq 6\cdot 10^{-4}$ and $M \leq 10^{15}$ GeV – cf. Ref. [14]– since it becomes comparable with the second term there. In this regime, the required by Eq. (26) σ_* becomes comparable to σ_c and the approximation of $f_{\rm rc}$ in Eq. (18) is no longer valid. Instead, we here have

$$f_{\rm rc}(x) = 3-\frac{x^{-2}}{6}-\frac{x^{-4}}{30}-\frac{x^{-6}}{84}-\frac{x^{-8}}{180}-\frac{x^{-10}}{330}-\frac{x^{-12}}{546}-\frac{x^{-14}}{840}-\frac{x^{-16}}{1224}-\frac{x^{-18}}{1710}-\frac{x^{-20}}{2310}-\cdots($$

As a consequence, $V'_{\rm HI}$ decreases sharply (enhancing $N_{\rm HI*}$) whereas $|V''_{\rm HI}|$ (or η) increases adequately, lowering thereby n_s to an acceptable level.

4.2.1 Standard FHI

In the case of standard FHI (with $\mathsf{N}=2$), we display in Fig. 3-(a) the allowed by Eqs. (26) and (27) values of v_G versus κ. The corresponding variation of n_s versus κ is depicted in Fig. 3-(b) where the observationally compatible region of Eq. (28) is also delimited by thin lines. Solid, dashed and dot-dashed lines stand for the results obtained for $a_S = 1, 5$ and 10 TeV respectively. We observe that the various lines coincide for $\kappa \gtrsim 6\cdot 10^{-4}$. For the sake of clarity we do not show in Fig. 3 solutions with $v_G > 2\cdot 10^{16}$ GeV or $\kappa < 2\cdot 10^{-4}$ – cf. Ref. [14] – which are totally excluded by Eq. (28). The third [last] term in the r.h.s of

Table 1: Input and output parameters consistent with Eqs. (26), (27) and (30) for shifted (with $M_S = 5\cdot 10^{17}$ GeV) or smooth FHI within mSUGRA

Shifted FHI		Smooth FHI	
$\kappa/10^{-3}$	9.2	$M_S/5\cdot 10^{17}$ GeV	0.79
$\sigma_*/10^{16}$ GeV	5.37	$\sigma_*/10^{16}$ GeV	32.9
$M/10^{16}$ GeV	2.3	$\mu_S/10^{16}$ GeV	0.21
$1/\xi$	4.36	$\sigma_f/10^{16}$ GeV	13.4
N_{HI*}	52.2	N_{HI*}	53
n_s	0.982	n_s	1.04
$-\alpha_s/10^{-4}$	3.4	$-\alpha_s/10^{-4}$	16.6

Eq. (34) become important for $\kappa \gtrsim 0.01$ [$\kappa \lesssim 6\cdot 10^{-4}$] whereas for $6\cdot 10^{-4} \lesssim \kappa \lesssim 0.01$, the second term in the r.h.s of Eq. (34) becomes prominent. As a consequence, the last term in the r.h.s of Eq. (34) drives n_s to values close to or larger than unity whereas the third one succeed in reconciling it with Eq. (28) for discriminated κ's related to the chosen a_S's. Namely, from Fig. 3 we deduce that there is a marginally allowed area for

$$0.0015 \lesssim \kappa \lesssim 0.032,\ \ 5.5 \lesssim v_G/(10^{15}\ \mathrm{GeV}) \lesssim 7.5, \tag{37a}$$
$$0.983 \lesssim n_s \lesssim 0.99 \ \text{ and } \ 1.2 \lesssim |\alpha_s|/10^{-4} \lesssim 3.5. \tag{37b}$$

In addition, from Fig. 3 we find isolated corridors consistent with Eq. (28), e.g.

$$\kappa \simeq 3\cdot 10^{-4},\ 5\cdot 10^{-4},\ 6\cdot 10^{-4} \text{ with } v_G \lesssim 3\cdot 10^{15}\,\mathrm{GeV} \text{ for } a_S = 1, 5 \text{ and } 10\,\mathrm{TeV}, \tag{37c}$$

respectively. We remark that the v_G's allowed here lie well below the ones required by Eq. (30). In conclusion, although standard FHI in mSUGRA can not be excluded, it can be considered as rather disfavored since the allowed region is extremely limited.

4.2.2 Shifted and Smooth FHI

In the cases of shifted and smooth FHI we confine ourselves to the values of the parameters consistent with Eq. (30) and display the solutions fulfilling Eqs. (26) and (27) in Table 1. Given that $a_S \sim 1$ TeV plays no role in the determination of the inflationary observables, we conclude that the resulting n_s's and α_s's are obviously predictions of these FHI models – without the possibility of altering them by some adjustment. We observe that the required κ, in the case of shifted FHI, is rather low and so, the last terms in the r.h.s of Eq. (34) is more or less negligible. As a result, η is exclusively determined by $c_{HI}V_{HI0}$ and n_s remains within the range of Eq. (28) although outside the 68% c.l. region. On the contrary, in the case of smooth FHI, η strongly depends on the last term in the r.h.s of Eq. (34) enhancing thereby n_s beyond the range of Eq. (28). In the latter case, $|\alpha_s|$ is also considerably enhanced.

5 Hilltop FHI in nmSUGRA

The fitting of the WMAP7 data with the ΛCDM model enforces [22,23] us to consider more complicated (and possibly more general) forms of Kälher potentials. The simplest choice is to consider of a moderate deviation from mSUGRA, named [22] nmSUGRA, according to which the next-to-minimal term is tuned so as an adequately small, negative mass squared for the inflaton is generated. This setting is outlined in Sec. 5.1 and the structure of the resulting $V_{\rm HI}$ is analyzed in Sec. 5.2. Our results are exhibited in Sec. 5.3.

5.1 The Relevant Set-up

In this scenario, the form of the relevant Kähler potential is given by Eq. (12), setting

$$\widehat{K} = k_{6S} = k_{8S} = k_{10S} = k_{12S} = 0 \;\text{ and }\; \widehat{Z} = 1. \tag{38a}$$

Therefore, $V_{\rm HI}$ takes the form of Eq. (15) with

$$c_{2K} = c_{2K}^{(0)} = k_{4S},\; c_{4K} = c_{4K}^{(0)}(k_{6S} = 0) \;\text{ and }\; c_{2\nu K} = 0 \;\text{ for }\; \nu \geq 3. \tag{38b}$$

Note that the expansion of $V_{\rm SUGRA}$ in Eq. (13) terminates at the terms of order $|S|^4$ consistently with the fact that the expansion of K terminates at the terms of order $|S|^6$.

5.2 Structure of the Inflationary Potential

For σ close to σ_*, $V_{\rm HI}$ given by Eq. (15) can be approximated as

$$V_{\rm HI} \simeq V_{\rm HI0}\left(1 + c_{\rm HI} - k_{4S}\frac{\sigma^2}{2m_{\rm P}^2} + c_{4K}\frac{\sigma^4}{4m_{\rm P}^4}\right). \tag{39}$$

Given that $c_{4K} = c_{4K}^{(0)} \simeq 1/2$ is much larger than k_{4S}, the boundedness of $V_{\rm HI}$ is ensured in this scenario, by construction.

The monotonicity of $V_{\rm HI}$ in Eq. (39) can be investigated applying Eq. (31). In particular, we can find approximately – note that in the formulas below, we have $c_{4K} \simeq 1/2$:

$$\bar{\sigma}_{\rm min} \simeq \begin{cases} \sqrt{k_{4S}/3c_{4K}}\, m_{\rm P} & \text{for standard and shifted FHI,} \\ \sqrt{2m_{\rm P}/3}\left(\sqrt{5/2c_{4K}}\,\mu_{\rm S} M_{\rm S}\right)^{1/4} & \text{for smooth FHI.} \end{cases} \tag{40}$$

Inserting Eq. (40) into Eq. (31), we find that $V_{\rm HI}$ remains monotonic for

$$k_{4S} < k_{4S}^{\max} \;\text{ with }\; k_{4S}^{\max} = \begin{cases} 3\kappa\sqrt{c_{4K}\mathsf{N}}/4\pi & \text{for standard FHI,} \\ 3\kappa\sqrt{c_{4K}}/2\sqrt{2}\pi & \text{for shifted FHI,} \\ (8/3)(2c_{4K}/5)^{3/4}\sqrt{\mu_{\rm S} M_{\rm S}}/m_{\rm P} & \text{for smooth FHI.} \end{cases} \tag{41}$$

For $k_{4S} > k_{4S}^{\max}$, $V_{\rm HI}$ reaches a local minimum [maximum] at the inflaton-field value $\sigma_{\rm min}$ [$\sigma_{\rm max}$] which can be estimated as follows:

$$\sigma_{\rm min} \simeq \sqrt{\frac{k_{4S}}{c_{4K}}}\, m_{\rm P} \;\text{ and }\; \sigma_{\rm max} \simeq \begin{cases} \kappa m_{\rm P}\sqrt{\mathsf{N}}/2\sqrt{2k_{4S}}\pi & \text{for standard FHI,} \\ \kappa m_{\rm P}/2\sqrt{k_{4S}}\pi & \text{for shifted FHI,} \\ \sqrt{2/3\sqrt[3]{k_{4S}}}(\mu_{\rm S} M_{\rm S} m_{\rm P})^{1/3} & \text{for smooth FHI.} \end{cases} \tag{42}$$

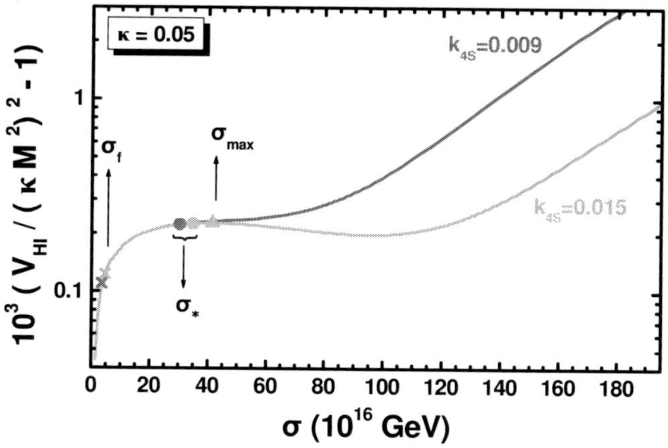

Figure 4: *The variation of* $V_{\rm HI}$ *in Eq. (39) as a function of* σ *for standard FHI in nmSUGRA,* $\kappa = 0.05$ *and* $k_{4S} = 0.009$ *(*$n_{\rm s} = 0.968$*) [*$k_{4S} = 0.015$ *(*$n_{\rm s} = 0.944$*)](gray [light gray] line). The values of* $\sigma_*, \sigma_{\rm f}$ *and* $\sigma_{\rm max}$ *are also depicted.*

The structure of $V_{\rm HI}$ is depicted in Fig. 4 where we display $V_{\rm HI}$ as a function of σ for standard FHI in nmSUGRA, $\kappa = 0.05$ and $k_{4S} = 0.009$ (gray line) or $k_{4S} = 0.015$ (light gray line). In the first case (gray line) we obtain $n_{\rm s} = 0.968$ with $k_{4S} < k_{4S}^{\rm max} \simeq 0.011$ and therefore $V_{\rm HI}$ remains monotonic. On the contrary, for $k_{4S} = 0.015$ we get $n_{\rm s} = 0.944$ with $k_{4S} > k_{4S}^{\rm max}$ and therefore $V_{\rm HI}$ develops the minimum-maximum structure with the maximum being located at $\sigma_{\rm max} = 4.15 \cdot 10^{17}$ GeV. The resulting $\Delta_{\rm m*}$ in the latter case is $\Delta_{\rm m*} = 0.16$. The values of σ_* and $\sigma_{\rm f}$ are also depicted.

5.3 Results

Our strategy in the numerical investigation of the nmSUGRA scenario is the one described in Sec. 4.2 – recall that we fix $a_S = 1$ TeV henceforth and no impact from that term on our results is detected. In addition to the parameters manipulated there, we have here the parameter k_{4S} which can be adjusted so as to achieve $n_{\rm s}$ in the range of Eq. (28). We check also the fulfillment of Eq. (31). Our findings for standard [shifted and smooth] FHI are accommodated in Sec. 5.3.1 [Sec. 5.3.2].

Employing the procedure outlined in Sec. 4.2 above Eq. (35) we can take a flavor for the expected $n_{\rm s}$'s in the nmSUGRA scenario, for any k_{4S}:

$$n_{\rm s} \simeq \begin{cases} 1 - 2k_{4S}\,(1-1/c_N) - 6c_{4K}\kappa^2 {\rm N} c_N/4k_{4S}\pi^2 & \text{for standard FHI,} \\ 1 - 2k_{4S}\,(1-1/c_N) - 6c_{4K}\kappa^2 c_N/2k_{4S}\pi^2 & \text{for shifted FHI,} \\ 1 - 5/3N_{\rm HI*} + 2\tilde{c}_N - (2\tilde{c}_N N_{\rm HI*} + 7)\,k_{4S} & \text{for smooth FHI,} \end{cases} \tag{43}$$

$$\text{with } c_N = 1 - \sqrt{1 + 4k_{4S}N_{\rm HI*}} \text{ and } \tilde{c}_N = 2c_{4K}\left(6\mu_{\rm S}^2 M_{\rm S}^2 N_{\rm HI*}/m_{\rm P}^4\right)^{1/3}.$$

We can clearly appreciate the contribution of a positive k_{4S} in lowering of $n_{\rm s}$.

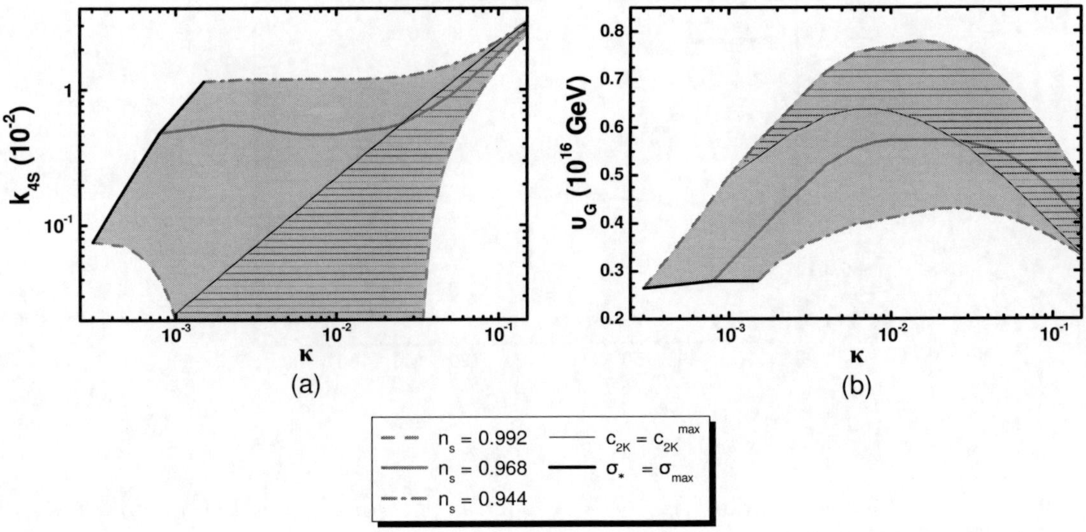

Figure 5: *Allowed (lightly gray shaded) region as determined by Eqs. (26)-(29) in the* $\kappa - c_{2K}$ *[*$\kappa - v_G$*] plane (a) [(b)] for standard FHI within the nmSUGRA. Ruled are the regions where the inflationary potential remains monotonic. The conventions adopted for the various lines are also shown.*

5.3.1 Standard FHI

In the case of standard FHI (with $\mathsf{N} = 2$), we delineate the (lightly gray shaded) region allowed by Eqs. (26)-(29) in the $\kappa - k_{4S}$ [$\kappa - v_G$] plane – see Fig. 5-(a) [Fig. 5-(b)]. The conventions adopted for the various lines are also shown. In particular, the dashed [dot-dashed] lines correspond to $n_s = 0.992$ [$n_s = 0.944$], whereas the gray solid lines are obtained by fixing $n_s = 0.968$ – see Eq. (28). Below the black solid lines, our initial assumption $\sigma_* < \sigma_{\max}$ is violated. In the hatched regions, Eq. (31) is also satisfied and along their boundaries designed by thin, black, solid lines Eq. (41) is saturated. We observe that the optimistic constraint of Eq. (31) can be met in a rather wide fraction of the allowed area. In particular, for $n_s = 0.968$ we find

$$\{0.8\}\, 2.5 \lesssim \frac{\kappa}{10^{-2}} \lesssim 15, \;\; \{11\}\, 5.7 \gtrsim \frac{v_G}{10^{15}\ \mathrm{GeV}} \gtrsim 4.1, \tag{44a}$$

$$\{0.48\}\, 0.56 \lesssim \frac{k_{4S}}{10^{-2}} \lesssim 2.9 \;\text{ and }\; \{0.2\}\, 0.39 \lesssim -\frac{\alpha_s}{10^{-3}} \lesssim 1.1, \tag{44b}$$

where the limiting values obtained without imposing Eq. (31) are indicated in curly brackets. In the corresponding region, Δ_{m*} ranges between 0 and 50%. Note that the v_G's encountered here are lower that those required by Eq. (30).

5.3.2 Shifted and Smooth FHI

In the cases of shifted and smooth FHI we confine ourselves to the values of the parameters which satisfy Eq. (30) and display in Table 2 their values which are consistent with

Table 2: Input and output parameters consistent with Eqs. (26) - (30) for shifted (with $M_S = 5 \cdot 10^{17}$ GeV) or smooth FHI in nmSUGRA

SHIFTED FHI				SMOOTH FHI			
n_s	0.944	0.968	0.992	n_s	0.944	0.968	0.992
$k_{4S}/10^{-3}$	11.7	4.5	-3.2	$k_{4S}/10^{-3}$	9.65	7.6	5.4
$k_{4S}^{\max}/10^{-3}$	1.8	1.9	2.2	$k_{4S}^{\max}/10^{-3}$	9.1	9.2	9.2
$\Delta_{m*}/10^{-2}$	16	39	–	$\Delta_{m*}/10^{-2}$	7	–	–
$\sigma_*/10^{16}$ GeV	2.45	3.1	4.31	$\sigma_*/10^{16}$ GeV	23.8	24.9	26.6
$\kappa/10^{-3}$	8.1	8.7	9.7	$M_S/5\cdot 10^{17}$ GeV	2.4	1.83	1.4
$M/10^{16}$ GeV	2.2	2.26	2.3	$\mu_S/10^{16}$ GeV	0.07	0.09	0.12
$1/\xi$	4.1	4.3	4.5	$\sigma_f/10^{16}$ GeV	13.4	13.4	13.4
N_{HI*}	51.9	52.1	52.3	N_{HI*}	52.3	52.4	52.6
$-\alpha_s/10^{-4}$	3.1	3.5	3.5	$-\alpha_s/10^{-3}$	0.7	0.9	1.1

Eqs. (26)-(29) as well. In the case of shifted FHI, we observe that (i) we need positive k_{4S} to obtain $n_s = 0.992$ since the mSUGRA result is lower – see Table 1; (ii) the lowest possible n_s compatible with the conditions of Eq. (31) is 0.976 and so, $n_s = 0.968$ is not consistent with Eq. (31). In the case of smooth FHI, we see that a reduction of n_s consistently with Eq. (31) can be achieved for $n_s \gtrsim 0.951$ and so $n_s = 0.968$ can be obtained without complications.

6 Hilltop FHI in nnmSUGRA

Another possible SUGRA set-up which can accommodate an observationally viable FHI is the one, first proposed in Ref. [25], which we here name nnmSUGRA. In this case, a convenient choice, specified in Sec. 6.1, of the next-to-minimal and the next-to-next-to-minimal terms in the Kähler potential is employed. The structure of the resulting V_{HI} is studied in Sec. 6.2 and our related results are exhibited in Sec. 6.2.

6.1 The Relevant Set-up

In this scenario, the form of the relevant Kähler potential is given by Eq. (12) after setting

$$\widehat{K} = 0 \text{ and } \widehat{Z} = 1 \tag{45a}$$

and so V_{HI} takes the form of Eq. (15) with

$$c_{2K} = c_{2K}^{(0)} = k_{4S},\ c_{4K} = c_{4K}^{(0)},\ c_{6K} = c_{6K}^{(0)},\ c_{8K} = c_{8K}^{(0)} \text{ and } c_{10K} = c_{10K}^{(0)}. \tag{45b}$$

In other words, this scenario is the most general one which arises from the Kähler potential of Eq. (12) in the absence of h_m's. The crucial difference between nnmSUGRA and nmSUGRA, however is the sign of $c_{2K} = k_{4S}$ which is here negative. As a consequence, fulfilling of Eq. (28) requires negative c_{4K} or positive k_{6S} – see Eq. (14d). The inclusion of higher order terms in the expansion of Eq. (13) prevents the runaway behavior of the resulting $V_{\rm HI}$ – see Eq. (15).

6.2 Structure of the Inflationary Potential

For σ close to σ_*, $V_{\rm HI}$ given by Eq. (15) can be approximated as

$$V_{\rm HI} \simeq V_{\rm HI0}\left(1 + c_{\rm HI} - k_{4S}\frac{\sigma^2}{2m_{\rm P}^2} + c_{4K}\frac{\sigma^4}{4m_{\rm P}^4} - c_{6K}\frac{\sigma^6}{8m_{\rm P}^6}\right). \tag{46}$$

The monotonicity of $V_{\rm HI}$ here can be checked only numerically – due to the numerous terms involved in $V'_{\rm HI}$ and $V''_{\rm HI}$ – by applying the criterion in Eq. (31). In the case of a non-monotonic $V_{\rm HI}$, we can show that it reaches a local maximum at the inflaton-field value:

$$\sigma_{\rm max} \simeq \begin{cases} \frac{m_{\rm P}\sqrt{2\pi|k_{4S}|+\sqrt{2}\sqrt{2k_{4S}^2\pi^2+{\rm N}k^2|c_{4K}|}}}{2\sqrt{\pi|c_{4K}|}} & \text{for standard FHI,} \\ \frac{m_{\rm P}\sqrt{\pi|k_{4S}|+\sqrt{k_{4S}^2\pi^2+k^2|c_{4K}|}}}{\sqrt{2\pi|c_{4K}|}} & \text{for shifted FHI,} \\ (2/3)^{3/8}\sqrt{m_{\rm P}\sqrt{\mu_{\rm S}M_{\rm S}}}/|c_{8K}|^{1/8} & \text{for smooth FHI.} \end{cases} \tag{47}$$

and a local minimum at the inflaton-field value:

$$\sigma_{\rm min} \simeq m_{\rm P}\frac{\sqrt{-3|c_{6K}|+\sqrt{9c_{6K}^2+32|c_{4K}c_{6K}|}}}{2\sqrt{|c_{8K}|}}. \tag{48}$$

The last result holds for all the types of FHI, since for σ's close $\sigma_{\rm min}$, $V_{\rm HI}$ is dominated by the last terms of the expansion in the r.h.s of Eq. (15) and so, any depedence on $c_{\rm HI}$, which essentially indentifies the type of FHI, is switched off.

The structure of $V_{\rm HI}$ is depicted in Fig. 6 where we display $V_{\rm HI}$ as a function of σ for standard FHI in nnmSUGRA, $\kappa = 0.021, k_{8S} = -1.5, k_{10S} = -1, k_{12S} = 0.5$ and $k_{4S} = -0.048, k_{6S} = 0.755$ (gray line) or $k_{4S} = -0.052, k_{6S} = 0.784$ (light gray line). In the first case (gray line) we obtain $n_{\rm s} = 0.968$ and $V_{\rm HI}$ remains monotonic. On the contrary, in the second case (light gray line) $V_{\rm HI}$ develops the minimum-maximum structure and we get $n_{\rm s} = 0.944$. The maximum of $V_{\rm HI}$ is located at $\sigma_{\rm max} = 2.8\cdot 10^{17}$ GeV and we get $\Delta_{\rm m*} = 0.29$. The values of σ_* and $\sigma_{\rm f}$ are also depicted.

6.3 Results

Our strategy in the numerical investigation of the nnmSUGRA scenario is the one described in Sec. 4.2. In addition to the parameters manipulated there, we have here the parameters k_{4S} and k_{6S} which can be adjusted in order to fulfill Eq. (28) whereas the boundedness of $V_{\rm HI}$ is controlled by the $k_{2\nu S}$'s with $4 \leq \nu \leq 6$. We check also the validity of Eq. (31). Our findings for standard [shifted and smooth] FHI are arranged in Sec. 6.3.1 [Sec. 6.3.2].

Preliminarily results on the $n_{\rm s}$'s expected, however, can be extracted by applying the procedure highlighted above Eq. (35). Namely, for standard FHI we find

$$n_{\rm s} \simeq 1 - \frac{1}{2}\left(2|k_{4S}| + \frac{1}{\pi}\left(3\sqrt{4\Delta_{1K}}\tanh\frac{N_{\rm HI*}\sqrt{\Delta_{1K}}}{\pi} + {\rm arctanh}\frac{\Delta_{2K}\pi}{\sqrt{\Delta_{1K}}}\right)\right) \tag{49a}$$

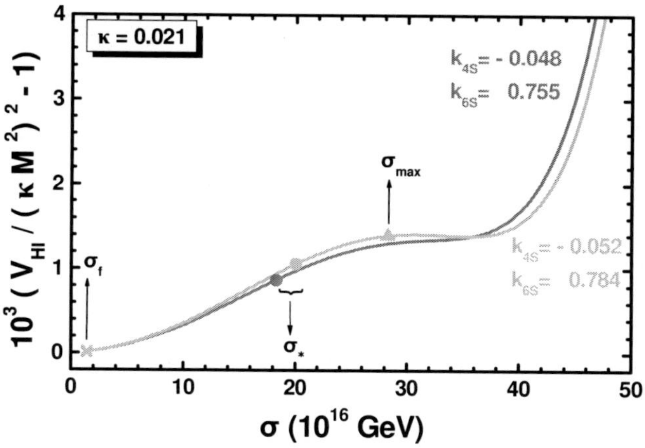

Figure 6: *The variation of $V_{\rm HI}$ in Eq. (46) as a function of σ for standard FHI in nnmSUGRA, $\kappa = 0.021$, $k_{8S} = -1.5, k_{10S} = -1, k_{12S} = 0.5$ and $k_{4S} = -0.048, k_{6S} = 0.755$ ($n_s = 0.968$) [$k_{4S} = -0.052, k_{6S} = 0.784$ ($n_s = 0.944$)] (gray [light gray] line). The values of σ_*, σ_f and σ_{max} are also depicted.*

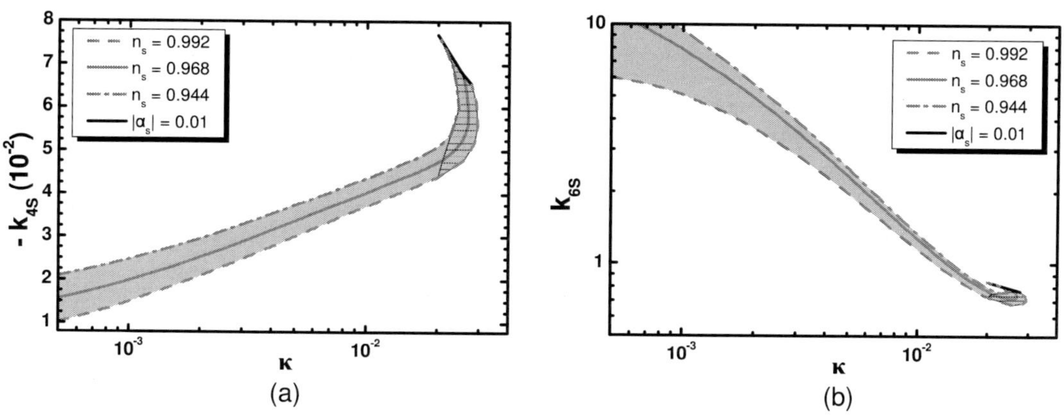

Figure 7: *Allowed (lightly gray shaded) region, as determined by Eqs. (26)-(30), in the $\kappa - (-k_{4S})$ [$\kappa - k_{6S}$] plane (a) [(b)] for standard FHI in nnmSUGRA. We take $k_{8S} = -1.5, k_{10S} = -1$ and $k_{12S} = 0.5$. Hatched are the regions where $V_{\rm HI}$ remains monotonic. The conventions adopted for the various lines are also shown.*

where

$$\Delta_{1K} = (|c_{4K}|k^2\mathsf{N} + 2k_{4S}^2\pi^2)/2 \ \text{ and } \ \Delta_{2K} = -4|c_{4K}|M^2/m_{\rm P}^2 + |k_{4S}|. \tag{49b}$$

The result for shifted FHI can be obtained from the expression above setting $\mathsf{N} = 2$ and replacing 4 with 2 in the formula of Δ_{2K} in Eq. (49b). We did not succeed to obtain similar formulas for smooth FHI due to complications related to the numerus significant terms in $V_{\rm HI}$, Eq. (46).

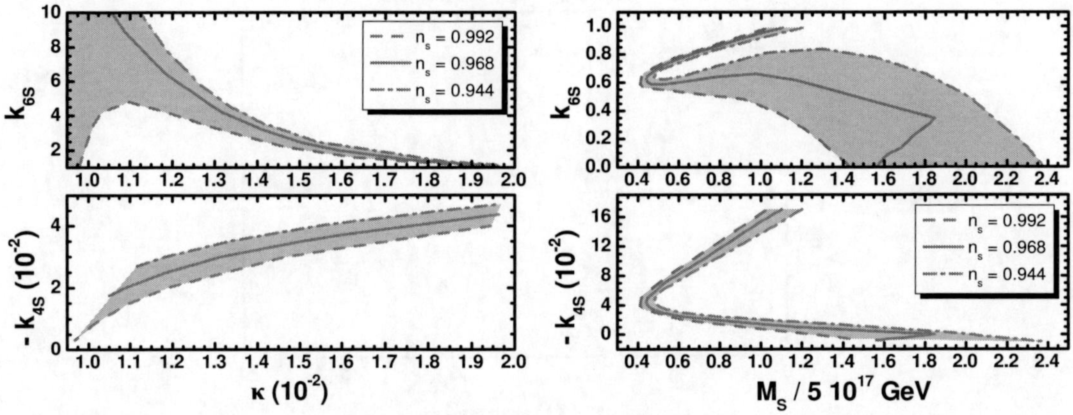

Figure 8: *Allowed (lightly gray shaded) regions, as determined by Eqs. (26)-(30), in the* $\kappa-(-k_{4S})$ *and* $\kappa-k_{6S}$ *planes for shifted FHI (left graphs) or in the* $M_S-(-k_{4S})$ *and* M_S-k_{6S} *planes for smooth FHI (right graphs) in nnmSUGRA. We take* $k_{8S}=-1.5, k_{10S}=-1$ *and* $k_{12S}=0.5$. *The conventions adopted for the various lines are also shown.*

6.3.1 Standard FHI

One of the outstanding advantages of the realization of standard FHI within nnmSUGRA is that Eq. (30) can be attained – cf. Figs. 3-(a), 5-(b) and 11-(b), below. Therefore, in this scenario, we are able to display regions (lightly gray shaded) allowed by Eqs. (26) – (30) in the $\kappa-(-k_{4S})$ [$\kappa-k_{6S}$] plane – see Fig. 7-(a) [Fig. 7-(b)] – for $k_{8S}=-1.5, k_{10S}=-1, k_{12S}=0.5$. The conventions adopted for the various lines are also shown. In particular, the gray dashed [dot-dashed] lines correspond to $n_s=0.992$ [$n_s=0.944$], whereas the gray solid lines have been obtained by fixing $n_s=0.968$ – see Eq. (28). We remark that increasing $|k_{4S}|$'s the required k_{6S}'s drop. We observe that the optimistic constraint of Eq. (31) can be met in a very limited slice of the allowed area. In this region also σ_* turns out to be rather large ($\sim 10^{17}$ GeV) and therefore we observe a mild dependence of our results on c_{6K} (or k_{8S}) too. Also there is a remarkable augmentation of α_s which saturates the bound of Eq. (29) along the thick black solid line. Namely, for $n_s=0.968$ we find

$$\{0.05\}\, 2.1 \lesssim \frac{\kappa}{10^{-2}} \lesssim 2.5, \;\; \{1.5\}\, 4.8 \lesssim \frac{-k_{4S}}{10^{-2}} \lesssim 6.8 \tag{50a}$$

$$7.2 \lesssim \frac{k_{6S}}{10^{-1}} \lesssim 7.55\{11.2\} \;\text{and}\; \{1.5\cdot 10^{-3}\}\, 0.48 \lesssim -\frac{\alpha_s}{10^{-2}} \lesssim 1, \tag{50b}$$

where the limiting values obtained without imposing Eq. (31) are indicated in curly brackets. In the corresponding region, Δ_{m*} ranges between 18 and 38%. As can be deduced from Fig. 7-(a), Δ_{m*} increases with $|k_{4S}|$ or as k_{6S} drops.

6.3.2 Shifted and Smooth FHI

As in the other scenaria, shifted and smooth FHI can become consistent with Eq. (30). Contrary to the the other scenaria, though, the consideration of two parameters (k_{4S} and k_{6S}) allows here to find wider regions of parameters compatible with Eqs. (26)-(30). These are shown in Fig. 8 for shifted [smooth] FHI, $k_{8S} = -1.5, k_{10S} = -1$ and $k_{12S} = 0.5$ (left [right] panel). The solid, dashed and dot-dashed gray lines correspond, as usual, to $n_s = 0.968, 0.992$ and 0.944 respectively. More specifically,

- In the case of shifted FHI, we display the allowed regions in the $\kappa - (-k_{4S})$ and $\kappa - k_{6S}$ planes. The left and right boundaries of the allowed regions come from the bounds on ξ – see Sec. 2.1. The results are pretty stable against variations of k_{8S} since the used κ's are rather low whereas Eq. (31) is violated throughout. For $n_s = 0.968$ we obtain $\Delta_{m*} \simeq 0.29 - 0.33$ and $\alpha_s \simeq (1.2 - 5.7) \cdot 10^{-3}$ with

$$1.05 \lesssim \kappa/10^{-2} \lesssim 1.96, \;\; 1.7 \lesssim -k_{4S}/10^{-2} \lesssim 4 \;\text{ and }\; 1.1 \gtrsim k_{6S} \gtrsim 10. \qquad (51a)$$

- In the case of smooth FHI, we display the allowed areas in the $M_S - (-k_{4S})$ and $M_S - k_{6S}$ planes. It is worth mentioning that V_{HI} remains monotonic almost in the whole allowed region depicted in the graphs. Only minor portions close the dot-dashed line violate Eq. (31). For $n_s = 0.968$ we obtain $0.6 \lesssim |\alpha_s|/10^{-3} \lesssim 8$ with

$$0.4 \lesssim M_S/5 \cdot 10^{17}\ \text{GeV} \lesssim 1.82, \;\; 0 \lesssim -k_{4S} \lesssim 0.17 \;\text{ and }\; 0.3 \lesssim k_{6S} \lesssim 0.99. \qquad (51b)$$

In both cases above we observe that k_{4S} is tuned to rather low values whereas k_{6S} can be adjusted to rather natural values of order one. This naturalness is removed only in a very minor slice of the allowed region where positive k_{4S}'s can be considered and the required k_{6S}'s are tuned to values of order 0.01.

7 Hilltop FHI in hSUGRA

Another, more drastic (and perhaps more radical) way to circumvent the n_s problem of FHI is the inclusion of extra fields in K. This proposition [16] is based on the observation that these fields provide extra terms in the expressions of $c_{\nu K}$ in Eq. (13). As a bonus, this construction gives us the opportunity to elude the notorious η problem of FHI. The Kähler potential of these extra field is specified in Eq. (7.1) and the structure of the resulting V_{HI} is studied in Sec. 7.2. Our results are presented in Sec. 7.3.

7.1 The Relevant Set-up

As we mention in Sec. 2.3, the dependence of K on the h_m's is encoded in the elements $\widehat{K}$ and $\widehat{Z}$. Motivated by several superstring and D-brane models [35], we seek the following ansatz for them:

$$\widehat{K} = m_P^2 \sum_{m=1}^{\mathsf{M}} \beta_m \ln \frac{h_m + h_m^*}{m_P} \;\text{ and }\; \widehat{Z} = k_Z \prod_{m=1}^{\mathsf{M}} \left(\frac{h_m + h_m^*}{m_P}\right)^{\alpha_m} \;\text{ with }\; \beta_m < 0. \qquad (52)$$

The last restriction in Eq. (52) is demanded so as to obtain positivity of the various kinetic energies. We further assume that β_m's have to be integers and α_m's have to be rational numbers. Also M measures the number of hidden sector fields. We here restrict ourselves to its lowest possible value, $\mathrm{M}=1$, defining $\alpha := \alpha_1$ and $\beta := \beta_1$. Note, in passing, that the form used here for $\widehat{K}$ and $\widehat{Z}$ has been initially proposed in Ref. [36] in order to justify the saddle point condition needed for the attainment of A-term or MSSM inflation [37].

In the presence of $\widehat{K}$ and $\widehat{Z}$ in Eq. (52), the coefficients c_{1K} and $c_{2\nu K}$ in the series of Eqs. (13) and (15) receive extra contributions beyond those exposed in Eqs. (14b)–(14f). The total expressions for c_{1K} and $c_{2\nu K}$ are found to be

$$c_{1K} = c_{1K}^{(0)} + \alpha + \beta, \tag{53a}$$

$$c_{2K} = c_{2K}^{(0)} + \frac{(\alpha-\beta)^2}{\beta}, \tag{53b}$$

$$c_{4K} = c_{4K}^{(0)} + \frac{(\alpha-\beta)^3}{\beta^2}, \tag{53c}$$

$$c_{6K} = c_{6K}^{(0)} + \frac{(\alpha-\beta)^2(2\alpha^2-2\alpha\beta+\beta^2)}{2\beta^3} - \left(\alpha + \frac{\alpha^3}{2\beta^2} - \frac{5\alpha^2}{4\beta} - \frac{\beta}{4}\right) k_{4S}, \tag{53d}$$

$$\begin{aligned} c_{8K} = {} & c_{8K}^{(0)} + \frac{(\alpha-\beta)^2(6\alpha^3-6\alpha^2\beta+3\alpha\beta^2-\beta^3)}{6\beta^4} \\ & + \left(\frac{5\alpha}{4} - \frac{\alpha^4}{\beta^3} + \frac{11\alpha^3}{4\beta^2} - \frac{11\alpha^2}{4\beta} - \frac{\beta}{4}\right) k_{4S} \\ & + \left(\frac{5\alpha}{6} + \frac{\alpha^3}{2\beta^2} - \frac{7\alpha^2}{6\beta} - \frac{\beta}{6}\right) k_{6S}, \end{aligned} \tag{53e}$$

$$\begin{aligned} c_{10K} = {} & c_{10K}^{(0)} + \frac{(\alpha-\beta)^2(24\alpha^4-24\alpha^3\beta+12\alpha^2\beta^2-4\alpha\beta^3+\beta^4)}{24\beta^5} \\ & - \left(\frac{3\alpha}{4} + \frac{3\alpha^5}{2\beta^4} - \frac{17\alpha^4}{4\beta^3} + \frac{9\alpha^3}{2\beta^2} - \frac{19\alpha^2}{8\beta} - \frac{\beta}{8}\right) k_{4S} \\ & - \left(\frac{3\alpha}{16} - \frac{\alpha^4}{4\beta^3} + \frac{5\alpha^3}{8\beta^2} - \frac{17\alpha^2}{32\beta} - \frac{\beta}{32}\right) k_{4S}^2 \\ & - \left(\alpha + \frac{\alpha^4}{\beta^3} - \frac{8\alpha^3}{3\beta^2} + \frac{5\alpha^2}{2\beta} + \frac{\beta}{6}\right) k_{6S} \\ & - \left(\frac{3\alpha}{4} + \frac{\alpha^3}{2\beta^2} - \frac{9\alpha^2}{8\beta} - \frac{\beta}{8}\right) k_{8S}. \end{aligned} \tag{53f}$$

From Eqs. (13) and (20) we infer that a resolution to the η problem of FHI requires $V''_{\rm HI}=0$ – needless to say that there is no contribution to η from the term including the c_{1K} coefficient in Eq. (13). Consequently, the η problem of FHI can be alleviated, if we demand:

$$c_{2K} = 0 \tag{54}$$

Moreover, the favored by the data on $n_{\rm s}$ hilltop FHI can be attained for $c_{4K}<0$. Solutions of Eq. (54) satisfying the latter restriction are listed in Table 3-(a). We observe that $k_{6S}>0$ is beneficial for the latter result, since it decreases c_{4K}, without disturbing the fulfilment of

Table 3: Solutions to Eq. (54) for $k_{6S} < 0$ and $c_{2K} \geq 0$ (a) or $k_{6S} > 0$ and $c_{2K} < 0$ (b)

$-\alpha$	$-\beta$	k_{6S}	k_{4S}	$-c_{4K}$
4	3	1/6	1/3	1/3
9/2	3	1/4	3/4	1
4	2	1	2	2.5
6/5	2	1/4	8/25	0.205

(a)

$-\alpha$	$-\beta$	$-k_{6S}$	k_{4S}	c_{4K}
3	2	1/4	1/2	0
6	4	1/2	1	0
3	2	1/3	1/2	1/8
1/2	1	1/4	9/4	5

(b)

Eq. (54). Another set of solutions can be taken for $\alpha = 0$. In this case – which resembles the cases studied in Ref. [21] –, we get

$$c_{4K} = 3/4,\ 0,\ -3,\ -6, -9 \ \text{ for } k_{4S} = -\beta = 1 \ \text{ and } k_{6S} = 0, 1/2, 5/2, 9/2, 13/2. \tag{55}$$

On the other hand, $c_{4K} \geq 0$ is still marginally allowed. Solutions to Eq. (54) with the latter resulting c_{4K}'s are arranged in Table 3-(b). In both cases, the extracted k_{4S} are confined in the range $0.1 - 10$, which we consider as being natural. Note that the realization of FHI in nmSUGRA – see Sec. 5.3 – or nnmSUGRA – see Sec. 6.3 – requires a significantly lower $|k_{4S}|$, i.e., $10^{-3} \lesssim |k_{4S}| \lesssim 0.1$.

More generically, taking α, β and k_{6S} as input parameters we can assure the fulfillment of Eq. (54) constraining k_{4S} via Eq. (53b) and find c_{4K} through Eq. (53c). Working this way, we plot in Fig. 9 (left [right] graph) $-c_{4K}$ [k_{4S}] versus $-\alpha$ for $\beta = -2$ and various k_{6S}'s (black points) or $k_{6S} = 0.5$ and various β's (gray points). The adopted values for all the free parameters employed are also shown. From the right plot of Fig. 9 we remark that the derived k_{4S}'s can be characterized as natural since the majority of them are of order 1. These are independent from the used k_{6S}'s since Eq. (14b) does not contain k_{6S}. From the left plot of Fig. 9, we notice that a wide range of negative c_{4K}'s can be produced which, however, can be bounded from above by the result $|c_{4K}| \leq 16$. As we show in Sec. 7.3, these c_{4K}'s can assist us to achieve hilltop FHI consistently with Eq. (28) for a broad range of κ's or M_S's.

7.2 Structure of the Inflationary Potential

For σ close to σ_*, $V_{\rm HI}$ given in Eq. (15) can be approximated as

$$V_{\rm HI} \simeq V_{\rm HI0}\left(1 + c_{\rm HI} + c_{4K}\frac{\sigma^4}{4m_{\rm P}^4} - c_{6K}\frac{\sigma^6}{8m_{\rm P}^6}\right), \tag{56}$$

where Eq. (54) is taken into account. As in the case of nnmSUGRA, a possible ugly runaway behavior of the resulting $V_{\rm HI}$ can be evaded by the inclusion of higher order terms in the expansion of Eq. (13) – see Eq. (15).

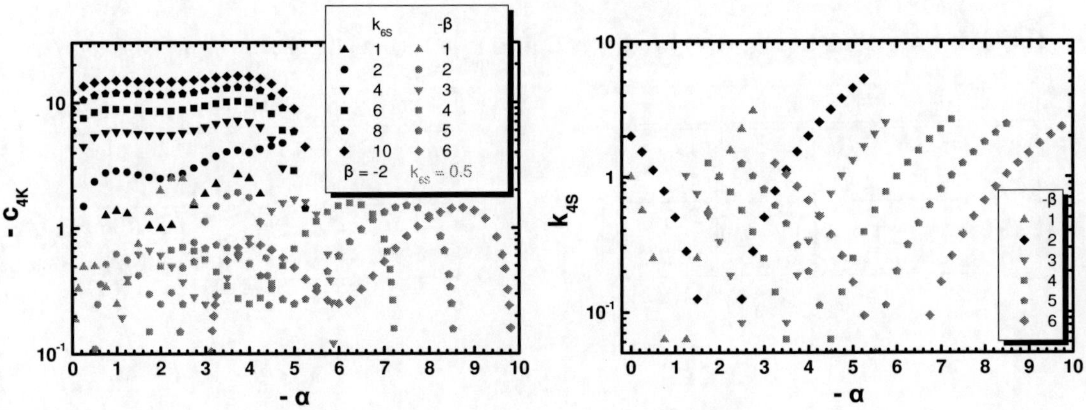

Figure 9: *Values of $-c_{4K}$ obtained from Eqs. (53c) and (54) (left plot) and the resulting k_{4S}'s from Eqs. (53b) and (54) (right plot) versus $-\alpha$ for $\beta = -2$ and various k_{6S}'s (black points) or $k_{6S} = 0.5$ and various β's (gray points). The adopted values for the remaining parameters – k_{6S} and β – are also shown.*

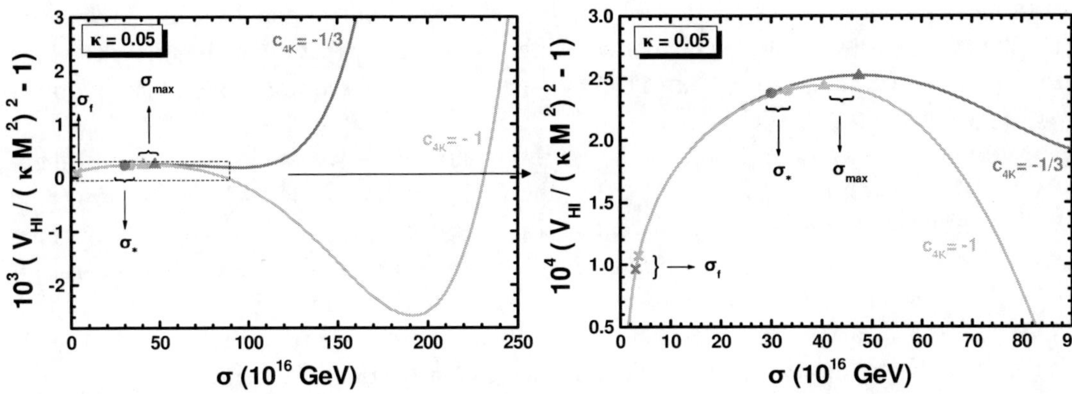

Figure 10: *The variation of $V_{\rm HI}$ in Eq. (56) as a function of σ for standard FHI in hSUGRA, $\kappa = 0.05, k_{8S} = -3, k_{10S} = -k_{12S} = -0.5$ and $c_{4K} = -1/3$ ($n_{\rm s} = 0.969$) [$c_{4K} = -1$ ($n_{\rm s} = 0.945$)](gray [light gray] line). The values of $\sigma_{\rm max}$, σ_* and $\sigma_{\rm f}$ are also depicted.*

For $c_{4K} < 0$, $V_{\rm HI}$ reaches a maximum at $\sigma = \sigma_{\rm max}$ which can be estimated as follows:

$$V'_{\rm HI}(\sigma_{\rm max}) = 0 \;\Rightarrow\; \sigma_{\rm max} \simeq \begin{cases} \left(\kappa^2 {\sf N}/8\pi^2\,|c_{4K}|\right)^{1/4} & \text{for standard FHI,} \\ \left(\kappa^2/4\pi^2\,|c_{4K}|\right)^{1/4} & \text{for shifted FHI,} \\ \left(8\mu_{\rm S}^2 M_{\rm S}^2/27\,|c_{4K}|\right)^{1/8} & \text{for smooth FHI,} \end{cases} \tag{57}$$

with $V''_{\rm HI}(\sigma_{\rm max}) < 0$. Since the behavior of $V_{\rm HI}$ at large σ's is dominated by the higher powers of σ in Eq. (15), as in the case of Sec. 6, $V_{\rm HI}$ can develop a minimum which is located at $\sigma = \sigma_{\rm min}$ given by Eq. (48).

This structure of $V_{\rm HI}$ is visualized in Fig. 10 where we display its variation as a function of σ for $\sigma_{\rm c} \le \sigma \le m_{\rm P}$ [$\sigma_{\rm c} \le \sigma \le 2.5\sigma_*$] (left [right] plot) for standard FHI in hSUGRA,

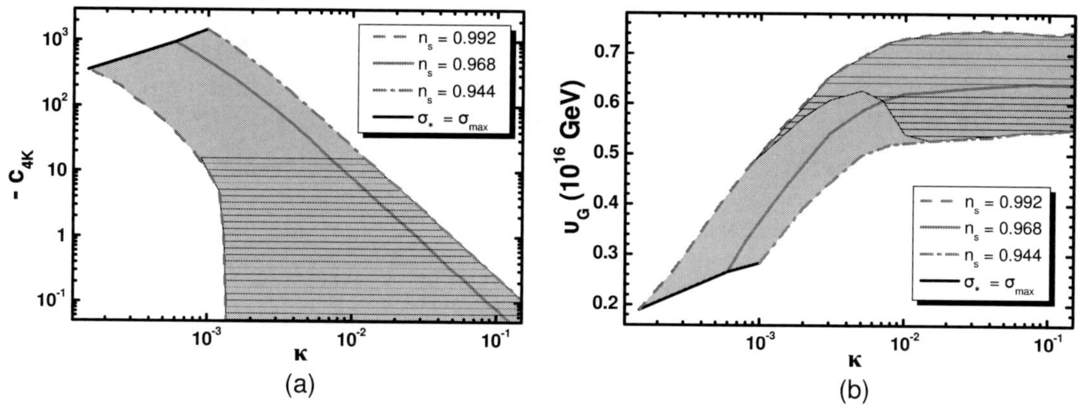

Figure 11: *Allowed (lightly gray shaded) regions, as determined by Eqs. (26)-(29), in the* $\kappa - c_{4K}$ *[*$\kappa - v_G$*] plane (a) [(b)] for standard FHI in hSUGRA with* $c_{2K} = 0$*. Ruled is the region which can be covered by the values of* α, β *and* k_{4S} *depicted in Fig. 9. The conventions adopted for the various lines are also shown.*

$\kappa = 0.05$, $k_{8S} = -3$, $k_{10S} = -k_{12S} = -0.5$ and $c_{4K} = -1/3$ (gray line) or $c_{4K} = -1$ (light gray line). The parameters α, β, k_{4S} and k_{6S} which lead to the selected c_{4K}'s are shown in Table 3-(a). In the first case (gray line) we obtain $n_s = 0.969$ with $\Delta_{m*} = 0.32$ whereas in the second one, we get $n_s = 0.945$ with $\Delta_{m*} = 0.17$. This result signalizes the presence of a rather severe tuning needed in order to implement hilltop FHI as anticipated in Sec. 3.3. It is also clear that the minimum-maximum structure of $V_{\rm HI}$ remains in both cases with the second case being much more evident. The values of σ_* and $\sigma_{\rm f}$ are also depicted.

7.3 Results

Our strategy in the numerical investigation of the hSUGRA scenario is the one described in Sec. 4.2. In addition to the parameters manipulated there, we have here the parameter c_{4K} which can be adjusted in order to fulfill Eq. (28) whereas the boundedness of $V_{\rm HI}$ is controlled by the $k_{2\nu S}$'s with $3 \leq \nu \leq 6$. We finally check if the required c_{4K}'s can be derived from Eqs. (53b), (53c) and (54) and the validity of Eq. (31). Our numerical results are presented in Sec. 7.3.1 for standard FHI and in Sec. 7.3.2 for shifted and smooth FHI.

We can, however, do some preliminary estimations for the expected n_s's, following the steps described above Eq. (35). In particular we find:

$$n_{\rm s} \simeq \begin{cases} 1 - 1/N_{\rm HI*} + 3\kappa^2 {\rm N} N_{\rm HI*} c_{4K}/4\pi^2 & \text{for standard FHI,} \\ 1 - 1/N_{\rm HI*} + 3\kappa^2 N_{\rm HI*} c_{4K}/2\pi^2 & \text{for shifted FHI,} \\ 1 - 5/3N_{\rm HI*} + 4c_{4K}\left(6\mu_{\rm S}^2 M_{\rm S}^2 N_{\rm HI*}\right)^{1/3} & \text{for smooth FHI.} \end{cases} \tag{58}$$

From the expressions above, we can easily infer that $c_{4K} < 0$ can diminish significantly n_s. To this end, in the cases of standard and shifted FHI, $|c_{4K}|$ has to be of order unity for relatively large κ's and much larger for lower κ's whereas, for smooth FHI, a rather low $|c_{4K}|$ is enough.

Table 4: Input and output parameters consistent with Eqs. (26)-(30) for shifted (with $M_S = 5\cdot 10^{17}$ GeV) or smooth FHI in hSUGRA and selected c_{4K}'s indicated in Table 3. To ensure the boundedness of V_{HI} in the case of shifted [smooth] FHI we take $k_{8S} = -15, k_{10S} = -8$ and $k_{12S} = -0.5$ [$k_{8S} = -1$ and $k_{10S} = -k_{12S} = -0.5$]

SHIFTED FHI				SMOOTH FHI			
c_{4K}	-6	-2.5	3	c_{4K}	-0.205	0	$1/8$
$\Delta_{m*}/10^{-1}$	3.9	5	–	$\Delta_{m*}/10^{-2}$	8.3	–	–
$\sigma_*/10^{16}$ GeV	3.8	3.7	3.8	$\sigma_*/10^{16}$ GeV	25.6	27	28.1
$\kappa/10^{-3}$	9	9.1	9.3	$M_S/5\cdot 10^{17}$ GeV	1.97	1.5	1.3
$M/10^{16}$ GeV	2.28	2.29	2.3	$\mu_S/10^{16}$ GeV	0.083	0.11	0.13
$1/\xi$	4.32	4.35	4.4	$\sigma_f/10^{16}$ GeV	13.4	13.4	13.4
N_{HI*}	52.2	52.2	52.3	N_{HI*}	52.4	52.6	52.7
n_s	0.973	0.978	0.986	n_s	0.946	0.974	0.991
$-\alpha_s/10^{-4}$	1.9	2.6	4.3	$-\alpha_s/10^{-4}$	4.7	6.5	8.1

7.3.1 Standard FHI

In Fig. 11-(a) [Fig. 11-(b)] we delineate the (lightly gray shaded) regions allowed by Eqs. (26) – (29), in the $\kappa - c_{4K}$ [$\kappa - v_G$] plane for standard FHI. The conventions adopted for the various lines are also shown in the r.h.s of each graph. In particular, the black solid [dashed] lines correspond to $n_s = 0.992$ [$n_s = 0.944$], whereas the gray solid lines have been obtained by fixing $n_s = 0.968$ – see Eq. (28). Below the black solid line, our initial assumption $\sigma_* < \sigma_{max}$ is violated. The various lines terminate at $\kappa = 0.15$, since for larger κ's the two restrictions in Eqs. (26) and (27) cannot be simultaneously met. Note that for $n_s = 0.992$ and $1.3\cdot 10^{-3} \lesssim \kappa \lesssim 0.15$ the curve is obtained for positive $0 \lesssim c_{4K} \lesssim 0.025$, not displayed in Fig. 11-(a).

From our data, we can deduce that (i) v_G, c_{4K} and Δ_{m*} increase with n_s, for fixed κ and (ii) c_{4K} and Δ_{m*} increase with κ, for fixed n_s. Comparing Fig. 11-(a) and Fig. 9-(a), we observe that the required c_{4K}'s, in order to achieve n_s's within the range of Eq. (28), can be derived from the fundamental parameters of the proposed Kähler potentials – see Eqs. (12) and (52) – in a wide range of parameters which is depicted as hatched portions of the light gray areas in Fig. 11. In particular, for $n_s = 0.968$ we obtain

$$0.7 \lesssim \frac{\kappa}{10^{-2}} \lesssim 15,\ 6.1 \lesssim \frac{v_G}{10^{15}\ \mathrm{GeV}} \lesssim 6.4, \tag{59a}$$

$$16 \gtrsim -c_{4K} \gtrsim 0.035 \text{ and } 0.30 \lesssim \Delta_{m*} \lesssim 0.33. \tag{59b}$$

Note that the v_G's encountered here are lower than those required by Eq. (30). In this case also, as for nmSUGRA and nnmSUGRA – see Secs 5.3.1 and 6.3.1 –, a certain degree of tuning is required as can been seen from the values of Δ_{m*} above.

7.3.2 Shifted and Smooth FHI

In the cases of shifted and smooth FHI, the achievement of Eq. (30) is possible and so, we can confine ourselves to solutions consistent with Eqs. (26) – (30) in Table 4. The selected c_{4K}'s here can be generated by the initial parameters $(\alpha, \beta, k_{4S}$ and $k_{6S})$ of our model as shown in Table 3-(a) and Eq. (55) for $c_{4K} < 0$ and Table 3-(b) for $c_{4K} \geq 0$. The entries without a value assigned for $\Delta_{\rm m*}$ refer to cases in which $V_{\rm HI}$ has no distinguishable maximum. From the data collected in Table 4 we observe the following:

- In the case of shifted FHI, the required κ's for fulfilling Eq. (30) come out to be rather low and so, the reduction of $n_{\rm s}$ to the level dictated by Eq. (28) requires rather high c_{4K}'s which in turn ask for large k_{6S}'s too. As a consequence, the boundedness of $V_{\rm HI}$ is affected since $-c_{6K}$ in Eq. (15) becomes negative and rather large k_{8S}'s, k_{10S}'s and k_{12S}'s (we here pose $k_{8S} = -15, k_{10S} = -8$ and $k_{12S} = -0.5$). The lowest possible $n_{\rm s}$ achieved with bounded $V_{\rm HI}$ from below is 0.973 which lies within the 68% c.l. observationally allowed margin – see Eq. (28).
- In the case of smooth FHI, $n_{\rm s}$ turns out to be quite close to its central value in Eq. (28) even with $c_{4K} = 0$. Therefore, in order to reach the central and the lowest value of $n_{\rm s}$ in Eq. (28), one needs rather small c_{4K}'s, which may be obtained from our initial parameters – see Fig. 9-(a). However, the resulting $\Delta_{\rm m*}$'s are lower than those of shifted FHI. On the other hand, the boundedness of $V_{\rm HI}$ is not disturbed here and can be assured for natural values of k_{8S}, k_{10S} and k_{12S} (we use $k_{8S} = -1$ and $k_{10S} = -k_{12S} = -0.5$).

Conclusion

We reviewed the basic types (standard, shifted and smooth) of FHI employing four possible embeddings in SUGRA. Each of these can be characterized by the adopted Kähler potential. In our work, we considered a quite generic Kähler potential in Eq. (12), from which the various SUGRA scenaria can be deduced. In particular, the Kähler potential of mSUGRA, nmSUGRA and nnmSUGRA can be determined by Eq. (12) substituting Eqs. (33), (38a) and (45a) respectively, whereas the one of hSUGRA can be derived by Eqs. (12) and (52). A crucial difference between hSUGRA and the other scenaria is that in hSUGRA we have taken into account contributions to inflationary potential, $V_{\rm HI}$ in Eq. (15), originating by extra (hidden-sector) fields obeying a string-inspired Kähler potential. The resulting forms of $V_{\rm HI}$ implementing FHI in mSUGRA, nmSUGRA, nnmSUGRA and hSUGRA are written in Eqs. (34), (39), (46) and (56) respectively.

We confronted the considered models of FHI with a number of observational data and theoretical requirements which are (i) the need for a solution to the horizon and flatness problems of the SSB cosmology – Eq. (26) (ii) the constraints on $\Delta_{\mathcal{R}}$, $n_{\rm s}$ and $\alpha_{\rm s}$ as result fitting the WMAP7 data by the ΛCDM model – see Eqs. (27), (28) and (29); (iii) the grand unification of the gauge coupling constants – Eq. (30); (iv) the boundedness, the convergence and the monotonicity of $V_{\rm HI}$ – see Eq. (31). Our findings can be summarized as follows:

- FHI in mSUGRA: The predicted n_s is just marginally consistent with the observational data, since Eq. (28) is fulfilled either beyond its 68% c.l. or for rather tuned values of κ close to 10^{-5} due to the presence of the tadpole term in V_{HI} with $a_S \sim 1$ TeV. Eq. (30) can be met only for shifted and smooth FHI.

- FHI in nmSUGRA: Acceptable n_s's can be obtained by restricting the parameter $k_{4S} > 0$ involved in V_{HI} to rather low values (of order 10^{-3}). Enforcing the validity of Eq. (31), the reduction of n_s below around 0.95 is prevented. The status of Eq. (30) is as in mSUGRA.

- FHI in nnmSUGRA: By constraining the parameters $k_{4S} < 0$ and $k_{6S} > 0$ of V_{HI} to values of order 10^{-2} and 10^{-1} respectively, we can achieve n_s's compatible with data. Eq. (30) can be attained in all the models of FHI. We remark a sizable enhancement of $|\alpha_s|$ whereas r remains well below its WMAP7 upper bound in [11].

- FHI in hSUGRA: The two extra parameters (α and β) contained in the Kähler potential of the extra fields give us the chance to eliminate the mass-squared term from V_{HI} for natural k_{4S}'s and k_{6S}'s and generate suitable c_{4K}'s aiming at reducing n_s to an acceptable level. Eq. (31) can be satisfied only if $c_{4K} \geq 0$ which gives observationally less interesting n_s's. Eq. (30) can be met as in mSUGRA.

Given that from the imposed requirements Eqs. (26), (27) and (29), the convergence and the boundedness of V_{HI} are fulfilled by all the considered settings of FHI, we are left with a subset of requirements – hierarchically represented by Eqs. (28), (30) and (31) – which, in conjunction with the naturalness inequality $0.1 \leq |k_{4S}| \leq 10$, can be employed in order to rate the analyzed models. Note that the latter criterion does not apply in mSUGRA where $k_{4S} = 0$ by definition. Also, we do not include the naturalness of k_{6S} in our test, since it is more or less assured in both relevant models (nnmSUGRA and hSUGRA) and so, it does not influence decisively our comparisons. Our four-point test is displayed schematically in Table 5. We respectively use the symbol $\surd$, $\times$ or $\sim$ when the corresponding requirement is satisfied, is not satisfied or is partially and/or less naturally satisfied. From our final score, we can infer that no model can be regarded as totally satisfactory, since at least one shortcoming is encountered in all cases. However, smooth FHI in nnmSUGRA or hSUGRA can be qualified as the most promising model – no $\times$ is signed. On the other hand, the most compelling implementation of standard [shifted] FHI is within nnmSUGRA [nmSUGRA] since just one $\times$ is listed. In our last statement we take into account that the four criteria are imposed hierarchically – e.g., the attainment of Eq. (28) is considered as more important than the achievement of Eq. (31).

Throughout our investigation we concentrated on the predictions derived from the realizations of FHI, assuming that we had suitable initial conditions for FHI to take place. For this reason, we paid special attention to the monotonicity of V_{HI}, which is crucial for a relatively natural attainment of FHI. In general, it is not clear [22, 23] how the inflaton can reach the maximum of V_{HI} in the context of hilltop inflation. Probably an era of eternal inflation prior to FHI could be useful [28] in order the proper initial conditions to be set.

Let us finally note that a complete inflationary scenario should specify the transition to the radiation dominated era and also explain the origin of the observed baryon asymmetry. For FHI in mSUGRA or nmSUGRA this has been extensively studied – see, e.g.,

Table 5: Test performance of the studied models of FHI in SUGRA. The symbol $\surd$ [$\times$] denotes that the corresponding requirement is [is not] satisfied, whereas the symbol $\sim$ stands for a partial or less natural fulfilment of the requirement

REQUIREMENTS	TYPES OF FHI		
	STANDARD	SHIFTED	SMOOTH
	mSUGRA – see Eq. (34)		
Eq. (28)	$\sim$	$\sim$	$\times$
Eq. (30)	$\times$	$\surd$	$\surd$
Eq. (31)	$\surd$	$\surd$	$\surd$
	nmSUGRA – see Eq. (39)		
Eq. (28)	$\surd$	$\surd$	$\surd$
Eq. (30)	$\times$	$\surd$	$\surd$
Eq. (31)	$\sim$	$\sim$	$\surd$
$0.1 \leq \lvert k_{4S} \rvert \leq 10$	$\times$	$\times$	$\times$
	nnmSUGRA – see Eq. (46)		
Eq. (28)	$\surd$	$\surd$	$\surd$
Eq. (30)	$\surd$	$\surd$	$\surd$
Eq. (31)	$\sim$	$\times$	$\surd$
$0.1 \leq \lvert k_{4S} \rvert \leq 10$	$\times$	$\times$	$\sim$
	hSUGRA – see Eq. (56)		
Eq. (28)	$\surd$	$\sim$	$\surd$
Eq. (30)	$\times$	$\surd$	$\surd$
Eq. (31)	$\times$	$\times$	$\sim$
$0.1 \leq \lvert k_{4S} \rvert \leq 10$	$\surd$	$\surd$	$\surd$

Ref. [6, 7, 10]. Obviously our models preserve many of these successful features of this post-inflationary evolution which may further constrain their parameter space and help us to distinguish the most compelling version of FHI. Moreover, the proposed scenaria will be even more challenged by the measurements of the Planck satellite [38] which is expected to give results on n_s with an accuracy $\Delta n_s \simeq 0.01$ by the next spring.

Acknowledgments

The work of R.A. was supported by the Tomalla Foundation. We would like to thank G. Lazarides for helpful discussions.

References

[1] A.H. Guth, *Phys. Rev. D* 23, 347 (1981);
A.H. Guth and D.I. Kaiser, *Science* 307,884 (2005) [astroph0502328].

[2] D.H. Lyth and A. Riotto, *Phys. Rept.* 314, 1 (1999) [hep-ph/9807278];
A. Mazumdar and J. Rocher, *Phys. Rept.* 497, 85 (2011) [arxiv1001.0993].

[3] G. Lazarides, *Lect. Notes Phys.* 592, 351 (2002) [hep-ph/0111328];
G. Lazarides, *J. Phys. Conf. Ser.* 53, 528 (2006) [hep-ph/0607032].

[4] E.J. Copeland *et al.*, *Phys. Rev. D* 49, 6410 (1994) [astro-ph/9401011].

[5] G.R. Dvali, Q. Shafi and R.K. Schaefer, *Phys. Rev. Lett.* 73, 1886 (1994) [hep-ph/9406319].

[6] R. Jeannerot *et al.*, *J. High Energy Phys.* 10, 012 (2000) [hep-ph/0002151];
R. Jeannerot, S. Khalil and G. Lazarides, *J. High Energy Phys.* 07, 069 (2002) [hep-ph/0207244];
S. Khalil *et al.*, *Phys. Rev. D* 83, 063522 (2011) [arxiv1010.3657];
M. Civiletti *et al.*, *Phys. Rev. D* 84, 103505 (2011) [arxiv1104.4143].

[7] G. Lazarides and C. Panagiotakopoulos, *Phys. Rev. D* 52, 559 (1995) [hep-ph/9506325];
G. Lazarides *et al.*, *Phys. Rev. D* 54, 1369 (1996) [hep-ph/9606297];
R. Jeannerot, S. Khalil, and G. Lazarides, *Phys. Lett. B* 506, 344 (2001) [hep-ph/0103229];
M.U. Rehman and Q. Shafi, *Phys. Rev. D* 86, 027301 (2012) [arxiv1202.0011];
S. Khalil, Q. Shafi and A. Sil, arxiv1208.0731.

[8] G.R. Dvali, G. Lazarides and Q. Shafi, *Phys. Lett. B* 424, 259 (1998) [hepph9710314].

[9] G. Lazarides and Q. Shafi, *Phys. Rev. D* 58, 071702 (1998) [hepph9803397].

[10] G. Lazarides and Q. Shafi, *Phys. Lett. B* 258, 305 (1991);
K. Kumekawa, T. Moroi and T. Yanagida, *Prog. Theor. Phys.* 92, 437 (1994) [hepph9405337];
G. Lazarides, R.K. Schaefer and Q. Shafi, *Phys. Rev. D* 56, 1324 (1997) [hep-ph/9608256].

[11] E. Komatsu *et al.* [WMAP Collaboration], *Astrophys. J. Suppl.* 192, 18 (2011) [arxiv1001.4538];
http://lamb da.gsfc.nasa.gov/product/map/dr2/parameters.cfm.

[12] E.D. Stewart, *Phys. Rev. D* 51, 6847 (1995) [hepph9405389];
M. Bastero-Gil and S.F. King, *Nucl. Phys.* B549, 391 (1999) [hepph9806477];
G. German, G.G. Ross and S. Sarkar, *Phys. Lett. B* 469, 46 (1999) [hepph9908380].

[13] A.D. Linde and A. Riotto *Phys. Rev. D* 56, 1841 (1997) [hep-ph/9703209];
V.N. Şenoğuz and Q. Shafi, *Phys. Lett. B* 567, 79 (2003) [hep-ph/0305089].

[14] V.N. Şenoğuz and Q. Shafi, *Phys. Rev. D* 71, 043514 (2005) [hep-ph/0412102]; M.U. Rehman, Q. Shafi and J.R. Wickman, *Phys. Lett. B* 683, 191 (2010) [arxiv0908.3896];
K. Nakayama *et al.*, *J. Cosmol. Astropart. Phys.* 12, 010 (2010) [arxiv1007.5152].

[15] R.A. Battye *et al.*, *J. Cosmol. Astropart. Phys.* 09, 007 (2006) [astro-ph/0607339]; G. Lazarides, I.N.R. Peddie, and A. Vamvasakis, *ibid.* 78, 043518 (2008) [arxiv0804.3661];
R. Battye, B. Garbrecht and A. Moss, *Phys. Rev. D* 81, 123512 (2010) [arxiv1001.0769].

[16] G. Lazarides and C. Pallis, *Phys. Lett. B* 651, 216 (2007) [hepph0702260];
G. Lazarides and A. Vamvasakis, *ibid.* 76, 123514 (2007) [arxiv0709.3362];
G. Lazarides, *Int. J. Mod. Phys. A* 22, 5747 (2007) [arxiv0706.1436];
C. Pallis, *AIP Conf. Proc.* 1122, 368 (2009) [arxiv0812.0249].

[17] G. Lazarides and C. Pallis, *Phys. Rev. D* 82, 063535 (2010) [arxiv1007.1558];
C. Pallis, *PoS CORFU* 2011, 028 (2011) [arxiv1207.6351].

[18] K. Dimopoulos *et al.*, *J. Cosmol. Astropart. Phys.* 02, 018 (2012) [arxiv1111.1929];
G. Lazarides, *AIP Conf. Proc.* 1467, 166 (2012) [arxiv1205.4830].

[19] C. Panagiotakopoulos, *Phys. Lett. B* 402, 257 (1997) [hep-ph/9703443];

[20] H.P. Nilles, *Phys. Rept.* 110, 1 (1984);
S.P. Martin, " *Perspectives on supersymmetry II*", edited by G.L. Kane, p. 1-153 [hepph9709356].

[21] C. Panagiotakopoulos, *Phys. Lett. B* 459, 473 (1999) [hepph9904284];
C. Panagiotakopoulos, *Phys. Rev. D* 71, 063516 (2005) [hepph0411143].

[22] M. Bastero-Gil, S.F. King, and Q. Shafi, *Phys. Lett. B* 651, 345 (2007) [hep-ph/0604198];
B. Garbrecht *et al.*, *J. High Energy Phys.* 12, 038 (2006) [hep-ph/0605264].

[23] M.U. Rehman, V.N. Şenoğuz, and Q. Shafi, *Phys. Rev. D* 75, 043522 (2007) [hep-ph/0612023];
G. Lazarides and A. Vamvasakis, *Phys. Rev. D* 76, 083507 (2007) [arxiv0705.3786].

[24] C. Pallis, *"High Energy Physics Research Advances"*, edited by T.P. Harrison and R.N. Gonzales (Nova Science Publishers Inc., New York, 2008) [arxiv0710.3074].

[25] Q. Shafi and J.R. Wickman, *Phys. Lett. B* 696, 438 (2011) [arxiv1009.5340];
M.U. Rehman, Q. Shafi and J.R. Wickman, *Phys. Rev. D* 83, 067304 (2011) [arxiv1012.0309].

[26] R. Armillis, G. Lazarides and C. Pallis, *work in preparation.*

[27] C. Pallis, *J. Cosmol. Astropart. Phys.* 04, 024 (2009) [arxiv0902.0334].

[28] L. Boubekeur and D. Lyth, *J. Cosmol. Astropart. Phys.* 07, 010 (2005) [hepph0502047];
K. Kohri, C.M. Lin and D.H. Lyth, *J. Cosmol. Astropart. Phys.* 12, 004 (2007) [arxiv0707.3826];
C.M. Lin and K. Cheung, *J. Cosmol. Astropart. Phys.* 03, 012 (2009) [arxiv0812.2731].

[29] W. Buchmuller, L. Covi and D. Delepine, *Phys. Lett. B* 491, 183 (2000) [hepph0006168].

[30] S.C. Davis and M. Postma, *J. Cosmol. Astropart. Phys.* 04, 022 (2008) [arxiv0801.2116];
S. Antusch *et al.*, *J. Cosmol. Astropart. Phys.* 01, 040 (2009) [arxiv0808.2425].

[31] J. Rocher and M. Sakellariadou, *J. Cosmol. Astropart. Phys.* 03, 004 (2005) [hep-ph/0406120];
R. Jeannerot and M. Postma, *J. High Energy Phys.* 05, 071 (2005) [hep-ph/0503146].

[32] T.W.B. Kibble, *J. Phys. A* 9, 387 (1976).

[33] G. Ballesteros *et al.*, *J. Cosmol. Astropart. Phys.* 03, 001 (2006) [hep-ph/0601134].

[34] M.Yu. Khlopov and A.D. Linde, Phys. Lett. B 138, 265 (1984);
J. Ellis, J.E. Kim, and D.V. Nanopoulos, *ibid.* 145, 181 (1984).

[35] See e.g. L.E. Ibanez and D. Lust, *Nucl. Phys.* B382, 305 (1992) [hep-th/9202046];
D. Lust, S. Reffert and S. Stieberger, *Nucl. Phys.* B727, 264 (2005) [hep-th/0410074].

[36] K. Enqvist, L. Mether and S. Nurmi, *J. Cosmol. Astropart. Phys.* 014, 11 (2007) [arxiv0706.2355];
S. Nurmi, *J. Cosmol. Astropart. Phys.* 01, 016 (2008) [arxiv0710.1613].

[37] R. Allahverdi *et al.*, *Phys. Rev. Lett.* 97, 191304 (2006) [hepph0605035];
J.C. Bueno-Sanchez *et al.*, *J. Cosmol. Astropart. Phys.* 01, 015 (2007) [hepph0608299].

[38] Planck collaboration, http://www.rssd.esa.int/index.php?project=Planck.

In: Recent Advances in Cosmology
Editors: Anderson Travena and Brady Soren
ISBN: 978-1-62417-943-3

Chapter 7

RELATIVE VELOCITIES, GEOMETRY AND EXPANSION OF SPACE

Vicente J. Bolós[1,*]**, *Sam Havens***[2,†] ***and David Klein***[2,‡]
[1]Dpto. Matemáticas para la Economía y la Empresa,
Facultad de Economía, Universidad de Valencia,
Valencia, Spain
[2]Department of Mathematics and Interdisciplinary Research
Institute for the Sciences, California State University,
Northridge, California, US

Abstract

What does it mean to say that space expands? One approach to this question is the study of relative velocities. In this context, a non local test particle is "superluminal" if its relative velocity exceeds the local speed of light of the observer. The existence of superluminal relative velocities of receding test particles, in a particular cosmological model, suggests itself as a possible criterion for expansion of space in that model. In this point of view, superluminal velocities of distant receding galaxy clusters result from the expansion of space between the observer and the clusters. However, there is a fundamental ambiguity that must be resolved before this approach can be meaningful. The notion of relative velocity of a nonlocal object depends on the choice of coordinates, and this ambiguity suggests the need for coordinate independent definitions. In this chapter, we review four (inequivalent) geometrically defined and universal notions of relative velocity: Fermi, kinematic, astrometric, and spectroscopic relative velocities. We apply this formalism to test particles undergoing radial motion relative to comoving observers in expanding Robertson-Walker cosmologies, and include previously unpublished results on Fermi coordinates for a class of inflationary cosmologies. We compare relative velocities to each other, and show how pairs of them determine geometric properties of the spacetime, including the scale factor with sufficient data. Necessary and sufficient conditions are given for the existence of superluminal recessional Fermi speeds in general Robertson-Walker cosmologies. We conclude with a discussion of expansion of space.

*E-mail address: vicente.bolos@uv.es
†E-mail address: samhavens@gmail.com
‡E-mail address: david.klein@csun.edu

1. Introduction

General relativity restricts the speed of a test particle to be less than the speed of light relative to an observer at the exact spacetime point of the test particle, but for test particles and observers located at different space-time points, the theory provides no *a priori* definition of relative velocity. Different coordinate charts give rise to different relative velocities.

This is perhaps most convincingly illustrated by the Milne universe, characterized by scale factor $a(t) = t$, where t is cosmological time (for the metric see (1) below, with $k = -1$). According to Hubble's law, whose formulation uses standard curvature coordinates,

$$\dot{d}(t) \equiv v_H = Hd,$$

where $H \equiv \dot{a}(t)/a(t)$ is the Hubble parameter, d is the proper distance at fixed time t from the observer to a comoving test particle, and the overdot signifies differentiation with respect to t. For the Milne universe, $H = 1/t > 0$, and thus at any time t and at sufficiently large proper distance d, the relative speed, v_H, of the test particle necessarily exceeds the local speed of light for the observer. However, with a simple change of coordinates, $\tau = t\cosh\chi$ and $\rho = t\sinh\chi$, the metric (1) is transformed to the Minkowski metric, and the Milne universe becomes the forward light cone of Minkowski spacetime. In these coordinates there are no superluminal speeds.

The ambiguity illustrated by this example has analogs in all spacetimes, and this feature led to consideration of the need for a strict definition of "radial velocity" within the solar system at the General Assembly of the International Astronomical Union (IAU), held in 2000 (see [1, 2]).

Thereafter, a series of papers [3, 4, 5] appeared addressing the general question of relative velocities and culminated in the introduction of four geometrically defined (but inequivalent) notions of relative velocity: Fermi, kinematic, astrometric, and the spectroscopic relative velocities. All four relative velocities have physical justifications and have been used to study properties of spacetimes (see [6, 7, 8]).

Two distinct notions of simultaneity play roles in the four definitions of relative velocities: "spacelike simultaneity" (also described as "Fermi simultaneity", see [9]) and "lightlike simultaneity." The Fermi and kinematic relative velocities are defined in terms of spacelike simultaneity, according to which events are simultaneous if they lie on the same Fermi space slice determined by a fixed Fermi time coordinate. For a test particle undergoing radial motion, the Fermi relative velocity, v_{Fermi} is the rate of change of proper distance of the test particle away from the central observer along the Fermi space slice with respect to proper time of the observer. The kinematic relative velocity is found by first parallel transporting the 4-velocity u' of the test particle at the spacetime point q_{s}, along a radial spacelike geodesic (lying on a Fermi space slice) to a 4-velocity denoted by $\tau_{q_{\text{s}}p}u'$ in the tangent space of the observer at spacetime point p, whose 4-velocity is u. The kinematic relative velocity v_{kin} is then the unique vector orthogonal to u, in the tangent space of the observer, satisfying $\tau_{q_{\text{s}}p}u' = k(u + v_{\text{kin}})$ for some scalar k (which is easily shown to be uniquely determined).

The spectroscopic (or barycentric) and astrometric relative velocities can be found, in principle, from spectroscopic and astronomical observations. Mathematically, both rely on the notion of "lightlike simultaneity", according to which two events are simultaneous if they both lie past-pointing horismos (which is tangent to the backward light cone) at the spacetime point p of the central observer. The spectroscopic relative velocity v_{spec} is calculated analogously to v_{kin}, described in the preceding paragraph, except that the 4-velocity u' of the test particle is parallel transported to the tangent space of the observer along a null geodesic lying on the past-pointing horismos of the observer, instead of along the Fermi space slice. The astrometric relative velocity, v_{ast}, of a test particle whose motion is purely radial is calculated analogously to v_{Fermi}, as the rate of change of the affine distance, which corresponds to the *observed* proper distance (through light signals at the time of observation) with respect to the proper time of the observer, as may be done via parallax measurements. We describe this more precisely in the sequel, and complete definitions for arbitrary (not necessarily radial) motion may be found in [5].

In this chapter, we review and explain these four notions of relative velocities in the context of test particles receding radially from comoving observers in Robertson-Walker cosmologies. This particular scenario lends itself to a consideration of a possible meaning for the expansion of space. The existence of a superluminal relative velocity of receding test particles, in a particular cosmological model, is a possible criterion for expansion of space in that model*. In this framework, superluminal velocities of distant receding galaxy clusters, in the actual universe, are the result of the expansion of space between the observer and the clusters.

Let us make precise the concept "superluminal". Let $v(q)$ be a relative velocity of a test particle at an event q with respect to an observer at an event p, and let $c(q)$ be the corresponding relative velocity of a photon at the same event q (note that any well-posed concept of relative velocity can be extended to photons) with the same spatial direction as the particle and with respect to the same observer at p. Then $\|v(q)\| < \|c(q)\|$ always, but $\|v(q)\|$ can exceed the local speed of light at p, which we take as $c = 1$ throughout the chapter. In this case, we say that the particle is "superluminal", but that does not mean that it travels faster than light. In the particular case of the relative velocities introduced above, $\|c(q)\|$ is always 1 for the kinematic and spectroscopic velocities, and so, there do not exist superluminal spectroscopic or kinematic velocities. On the other hand, the Fermi and astrometric relative velocities of a photon can be less than 1, equal to 1, or greater than 1, allowing for the possibility superluminal velocities.

Much of the material we present here is based on references [7] and [10]. Exact Fermi coordinates were found in [7] for expanding Robertson-Walker spacetimes and Fermi charts were shown to be global for non inflationary scale factors†. These coordinates were then used to calculate the (finite) diameter of the Fermi space slice, as a function of the observer's proper time, and Fermi velocities of (receding) comoving test particles. Reference [10] extended the results of [7] by finding general formulas for all four relative velocities for test particles undergoing arbitrary radial motion in expanding Robertson-Walker spacetimes, finding relationships among these relative velocities, and showing that their ratios determine geometric properties of the spacetime. Those results were illustrated in the de Sitter universe, the radiation-dominated universe, the matter-dominated universe, and more generally, for cosmologies for which the scale factor, $a(t) = t^{\alpha}$ for $0 < \alpha < 1$.

*One must, of course, take into account an ambiguity. Different space slices are associated with different coordinate systems.

†A scale factor $a(t)$ is non inflationary if $\ddot{a}(t) \leq 0$ for all t.

Previously unpublished results [11] are also included in Section 8 of this chapter, which provides a proof that Fermi coordinates extend to the cosmological event horizon in inflationary cosmologies with scale factors of the form $a(t) = t^\alpha$ for $\alpha > 1$. This allows for interesting comparisons between inflationary and non inflationary expanding cosmologies, which we discuss in the final section.

2. The Robertson-Walker Metric

The Robertson-Walker metric in curvature-normalized coordinates (or Robertson-Walker coordinates) is given by the line element

$$\mathrm{d}s^2 = -\mathrm{d}t^2 + a^2(t)\left(\mathrm{d}\chi^2 + S_k^2(\chi)\,\mathrm{d}\Omega^2\right), \tag{1}$$

where $\mathrm{d}\Omega = \mathrm{d}\theta + \sin^2\theta\mathrm{d}\varphi$, $a(t)$ is a positive and increasing scale factor, with $t > 0$, and $S_k(\chi) = \sin(\chi)$ if $k = 1$, χ if $k = 0$, or $\sinh(\chi)$ if $k = -1$. There is a coordinate singularity in (1) at $\chi = 0$, but this will not affect the calculations that follow. Since our purpose is to study radial motion with respect to a central observer, it suffices to consider the 2-dimensional Robertson-Walker metric given by

$$\mathrm{d}s^2 = -\mathrm{d}t^2 + a^2(t)\mathrm{d}\chi^2, \tag{2}$$

for which there is no singularity at $\chi = 0$. We assume throughout that $a(t)$ is a smooth, increasing function of $t > 0$.

3. Notation

We denote a central observer located at $\chi = 0$ by the world line, $\beta(\tau) = (\tau, 0)$, and a test particle by $\beta'(\tau') = (t(\tau'), \chi(\tau'))$. Each spacetime path is parameterized by its own proper time, and we will always assume that $\chi > 0$ for the latter. We have to remark that in the case $k = 1$, the coordinate χ is upper bounded by π; thus, for simplicity, we assume throughout that $k = -1$ or 0, but our methods also work for the case $k = 1$ by restricting χ to $]0, \pi[$.

Our aim is to study the relative velocities of β' with respect to, and observed by, β. Denote the 4-velocity of β by U and identify $U = \frac{\partial}{\partial t} = (1, 0)$. Similarly, denote the 4-velocity of β' by $U' = \dot{t}\frac{\partial}{\partial t} + \dot{\chi}\frac{\partial}{\partial \chi} = (\dot{t}, \dot{\chi})$, where the overdot indicates differentiation with respect to τ', the proper time of β'. From $g(U', U') = -1$, we find,

$$\dot{t} = \sqrt{a^2(t)\dot{\chi}^2 + 1}. \tag{3}$$

Vector fields will be represented by upper case letters, and vectors (in the tangent space at a single spacetime point) by lower case letters. Following this notation, the 4-velocity of β at a fixed event $p = (\tau, 0)$ will be denoted by $u = (1, 0)$. The Fermi, kinematic, spectroscopic, and astrometric relative velocity vector fields (to be defined below) are denoted respectively as V_{kin}, V_{Fermi}, V_{spec} and V_{ast}. We shall see that they are vector fields defined on the spacetime path, β, i.e., the central observer. By definition, all of these relative velocities are spacelike and orthogonal to U, so that they are each proportional to the unit vector field $\mathcal{S} := \frac{1}{a(t)} \frac{\partial}{\partial \chi}$‡.

It is possible to compare the four relative velocities for a given test particle. Direct comparisons may be made of the Fermi and kinematic relative velocities, because of the common dependence of these two notions of relative velocity on spacelike simultaneity. Similarly, direct comparisons of the astrometric and spectroscopic relative velocities are also possible. However, a comparison of all four relative velocities made at a particular instant by the central observer β is possible only with data from two different spacetime events (q_s and q_ℓ in Figure 1) of the test particle. Such a comparison, to which we refer as an *instant comparison*, therefore lacks physical significance, unless the evolution of the test particle β' can be deduced from its 4-velocity at one spacetime point, e.g. for comoving or, more generally, geodesic test particles.

It will also be possible to compare all four relative velocities at a fixed spacetime event q_ℓ of the test particle through observations from two different times of the central observer, identified as τ and τ^* in Figure 1. In the sequel, we refer to such a comparison of the relative velocities as a *retarded comparison*.

In all that follows, we use the following notation for vectors at a given spacetime point $p = (\tau, 0)$ and a spacetime point $p^* = (\tau^*, 0)$ in the past of p (see Figure 1): $v_{\text{kin}} = V_{\text{kin}\ p}$, $v_{\text{Fermi}} = V_{\text{Fermi}\ p}$, $v^*_{\text{kin}} = V_{\text{kin}\ p^*}$, $v^*_{\text{Fermi}} = V_{\text{Fermi}\ p^*}$, $v_{\text{spec}} = V_{\text{spec}\ p}$ and $v_{\text{ast}} = V_{\text{ast}\ p}$. So, in an instant comparison we compare v_{kin}, v_{Fermi}, v_{spec}, and v_{ast}, while in a retarded comparison we compare v^*_{kin}, v^*_{Fermi}, v_{spec}, and v_{ast}.

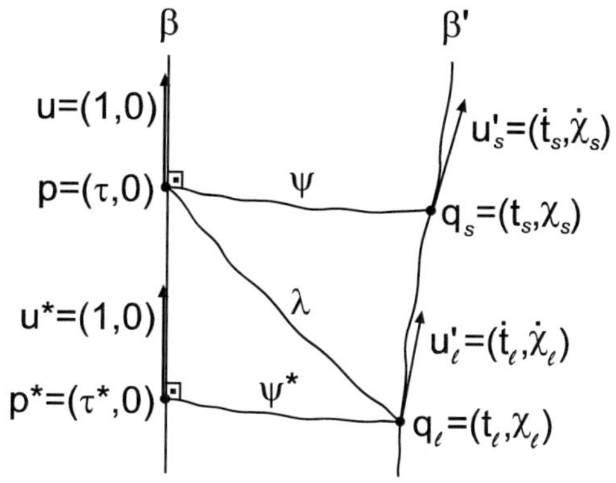

Figure 1. Scheme of the elements involved in the study of the relative velocities of a test particle β' with respect to the central observer β. The curves ψ and ψ^* are spacelike geodesics orthogonal to the 4-velocity of β, and λ is a lightlike geodesic.

‡This should not to be confused with the relative position vector field S used in [5]; in fact, $\mathcal{S}$ is the normalized version of S.

4. Spacelike Simultaneity and Fermi Coordinates

Let $p = (\tau, 0)$ be an event of the central observer β with 4-velocity u. An event q is *spacelike* (or *Fermi*) *simultaneous* with p if $g(\exp_p^{-1} q, u) = 0$[§]. The *Fermi space slice* $\mathcal{M}_\tau$ consists of all the events that are spacelike simultaneous with p[¶].

More explicitly, let ρ denote proper length along a spacelike geodesic orthogonal to β. Then, the vector field,

$$X := \frac{\partial}{\partial \rho} = \frac{dt}{d\rho}\frac{\partial}{\partial t} + \frac{d\chi}{d\rho}\frac{\partial}{\partial \chi} = -\sqrt{\left(\frac{a(\tau)}{a(t)}\right)^2 - 1}\frac{\partial}{\partial t} + \frac{a(\tau)}{a^2(t)}\frac{\partial}{\partial \chi}, \tag{4}$$

is geodesic, spacelike, unit, and X_p is orthogonal to the 4-velocity $u = (1, 0)$ at p; i.e., X_p is tangent to $\mathcal{M}_\tau$. Let $q_{\mathrm{s}} = (t_{\mathrm{s}}, \chi_{\mathrm{s}})$ be the unique event of $\beta' \cap \mathcal{M}_\tau$. Then there exists an integral curve of X from p to q_{s} (the geodesic ψ in Figure 1), and so, using (4) we can find a relationship between τ, t_{s} and χ_{s}:

$$\int_\tau^{t_{\mathrm{s}}} \frac{a(\tau)}{a^2(t)} \frac{-1}{\sqrt{\left(\frac{a(\tau)}{a(t)}\right)^2 - 1}} \mathrm{d}t = \int_0^{\chi_{\mathrm{s}}} \mathrm{d}\chi \Longrightarrow \int_{t_{\mathrm{s}}}^{\tau} \frac{a(\tau)}{a(t)} \frac{1}{\sqrt{a^2(\tau) - a^2(t)}} \mathrm{d}t = \chi_{\mathrm{s}}. \tag{5}$$

From (5) we see that $t_{\mathrm{s}} < \tau$, and more generally, it follows from (4) that the time coordinate t is a decreasing function of proper length ρ along any spacelike geodesic orthogonal to the central observer's worldline. Moreover, since we only consider positive coordinate times, we have to impose $t_{\mathrm{s}} > 0$ and then it is necessary that $0 < \chi_{\mathrm{s}} < \chi_{\mathrm{smax}}$, where

$$\chi_{\mathrm{smax}} := \int_0^\tau \frac{a(\tau)}{a(t)} \frac{1}{\sqrt{a^2(\tau) - a^2(t)}} \mathrm{d}t. \tag{6}$$

The metric (1) may be expressed in the Fermi coordinates of the central observer β (or Fermi observer in this context). By design of Fermi coordinates, $\tau = t$ on the path $\beta(t)$ of the Fermi observer (where $\rho = 0$), but the two time coordinates differ away from that path. For expanding Robertson-Walker spacetimes, it was shown in [7] that Fermi coordinates are global when the spacetime is non inflationary, and we show in Section 8 that Fermi coordinates extend to the cosmological event horizon when the scale factor takes the form $a(t) = t^\alpha$ for $\alpha > 1$.

In any case, for two spacetime dimensions, it may be shown that the metric (2) expressed in Fermi coordinates (τ, ρ) takes the form,

$$\mathrm{d}s^2 = g_{\tau\tau}(\tau, \rho)\mathrm{d}\tau^2 + \mathrm{d}\rho^2. \tag{7}$$

General formulas for $g_{\tau\tau}$ are provided in [7, 10], but we include in Section 5 a self-contained description of $g_{\tau\tau}$ in terms of the Fermi and kinematic relative velocities of test particles, relative to the central observer.

---

§The exponential map, $\exp_p v$, denotes the evaluation at affine parameter 1 of the geodesic starting at the point p, with initial derivative v.

¶See [7]; $\mathcal{M}_\tau$ is also called the *Landau submanifold* and denoted by $L_{p,u}$ in [3].

Setting $ds^2 = 0$ in (7), shows that the velocity of a distant radial photon with spacetime coordinates (τ, ρ), relative to the Fermi observer β, is given by $|d\rho/d\tau| = \sqrt{-g_{\tau\tau}(\tau, \rho)}$. Thus, the metric (7) may be understood as a natural generalization of the Minkowski metric when the speed of a photon depends on its spacetime coordinates, and may expressed as $ds^2 = -c(\tau, \rho)^2 d\tau^2 + d\rho^2$ where $c(\tau, \rho)$ is the Fermi speed of a photon at the spacetime point (τ, ρ) relative to the Fermi observer located at $(\tau, 0)$.

5. Fermi and Kinematic Relative Velocities

Based on the previous section, we may write the 4-velocity of the radially moving test particle, β', using Fermi coordinates as, $u'_{\mathrm{s}} = \dot{\tau}\frac{\partial}{\partial \tau}\Big|_{q_{\mathrm{s}}} + \dot{\rho}\frac{\partial}{\partial \rho}\Big|_{q_{\mathrm{s}}} = (\dot{\tau}, \dot{\rho})$, where the overdot signifies differentiation with respect to proper time of β'.

Let $p = (\tau, 0)$ be an event of β from which we measure relative velocities. The Fermi relative velocity of u'_{s} with respect to $u = (1, 0)$ is given by,

$$v_{\mathrm{Fermi}} = \frac{d\rho}{d\tau}\mathcal{S}_p = \frac{\dot{\rho}}{\dot{\tau}}\mathcal{S}_p, \tag{8}$$

where $\mathcal{S}_p = \partial/\partial\rho|_p$. For computing v_{Fermi}, observe first from (4), that,

$$\rho = \rho(t_{\mathrm{s}}, \chi_{\mathrm{s}}) = \int_{t_{\mathrm{s}}}^{\tau(t_{\mathrm{s}}, \chi_{\mathrm{s}})} \frac{a(t)}{\sqrt{a^2(\tau(t_{\mathrm{s}}, \chi_{\mathrm{s}})) - a^2(t)}}\, dt, \tag{9}$$

where the function $\tau(t_{\mathrm{s}}, \chi_{\mathrm{s}})$ is defined implicitly by (5). Then, applying (9) in (8) we have,

$$v_{\mathrm{Fermi}} = \frac{\dot{\rho}}{\dot{\tau}}\mathcal{S}_p = \frac{\frac{\partial \rho}{\partial t_{\mathrm{s}}}\dot{t}_{\mathrm{s}} + \frac{\partial \rho}{\partial \chi_{\mathrm{s}}}\dot{\chi}_{\mathrm{s}}}{\frac{\partial \tau}{\partial t_{\mathrm{s}}}\dot{t}_{\mathrm{s}} + \frac{\partial \tau}{\partial \chi_{\mathrm{s}}}\dot{\chi}_{\mathrm{s}}}\mathcal{S}_p.$$

In order to define kinematic relative velocity, we need some additional notation. Let $\tau_{q_{\mathrm{s}}p}$ represent the parallel transport from q_{s} to p by the unique geodesic joining q_{s} and p, named ψ in Figure 1. Following the description provided in the introduction, the kinematic relative velocity v_{kin} of u' with respect to the central observer's four-velocity u is given by (see [5]),

$$v_{\mathrm{kin}} = \frac{1}{-g\left(\tau_{q_{\mathrm{s}}p}u'_{\mathrm{s}}, u\right)}\tau_{q_{\mathrm{s}}p}u' - u. \tag{10}$$

It is a general property of the kinematic relative velocity that $\|v_{\mathrm{kin}}\| < 1$. Using the relations $g\left(\tau_{q_{\mathrm{s}}p}u'_{\mathrm{s}}, \frac{\partial}{\partial\rho}|_p\right) = g\left(u'_{\mathrm{s}}, \frac{\partial}{\partial\rho}|_{q_{\mathrm{s}}}\right)$ and $g\left(\tau_{q_{\mathrm{s}}p}u'_{\mathrm{s}}, \tau_{q_{\mathrm{s}}p}u'_{\mathrm{s}}\right) = -1$, we can find $\tau_{q_{\mathrm{s}}p}u'_{\mathrm{s}}$, and then obtain the kinematic relative velocity of u'_{s} with respect to u as,

$$v_{\mathrm{kin}} = \frac{1}{\sqrt{-g_{\tau\tau}(\tau, \rho)}}\frac{d\rho}{d\tau}\mathcal{S}_p. \tag{11}$$

Comparing (8) and (11), we see that the kinematic and Fermi relative velocities of an arbitrary test particle at the spacetime point (τ, ρ) determine $g_{\tau\tau}(\tau, \rho)$. We state this in the form of a proposition:

Proposition 5.1. *For a Robertson-Walker spacetime with scale factor $a(t)$ that is a smooth, increasing, unbounded function of t, the kinematic and Fermi speeds of any test particle undergoing radial motion with respect to a comoving observer determine the Fermi metric tensor element $g_{\tau\tau}$ at the spacetime point of the particle, via,*

$$g_{\tau\tau}(\tau,\rho) = -\frac{\|v_{\mathrm{Fermi}}\|^2}{\|v_{\mathrm{kin}}\|^2}.$$

The following corollary now follows from the observations made in the final paragraph of the preceding section.

Corollary 5.1. *With the same assumptions as above, the Fermi relative velocity of a radially moving test particle at position (τ,ρ) within a Fermi coordinate chart satisfies*

$$\|v_{\mathrm{Fermi}}\| < \sqrt{-g_{\tau\tau}(\tau,\rho)},$$

and can therefore exceed the central observer's local speed of light ($c = 1$) if and only if $-g_{\tau\tau}(\tau,\rho) > 1$.

Returning to the curvature-normalized coordinates, and exploiting the symmetry of this spacetime through the killing field $\partial/\partial\chi$, the kinematic relative velocity may also be expressed explicitly in terms of $\dot{\chi}_{\mathrm{s}}$. From $g\left(\tau_{q_{\mathrm{s}}p}u'_{\mathrm{s}}, X_p\right) = g\left(u'_{\mathrm{s}}, X_{q_{\mathrm{s}}}\right)$ and $g(\tau_{q_{\mathrm{s}}p}u'_{\mathrm{s}}, \tau_{q_{\mathrm{s}}p}u'_{\mathrm{s}}) = -1$ we can obtain $\tau_{q_{\mathrm{s}}p}u'_{\mathrm{s}}$, and then find that,

$$v_{\mathrm{kin}} = \frac{\dot{t}_{\mathrm{s}}\sqrt{\frac{a^2(\tau)}{a^2(t_{\mathrm{s}})} - 1} + a(\tau)\dot{\chi}_{\mathrm{s}}}{\sqrt{\left(\dot{t}_{\mathrm{s}}\sqrt{\frac{a^2(\tau)}{a^2(t_{\mathrm{s}})} - 1} + a(\tau)\dot{\chi}_{\mathrm{s}}\right)^2 + 1}}\mathcal{S}_p,$$

where $\dot{t}_{\mathrm{s}} = \sqrt{a^2(t_{\mathrm{s}})\dot{\chi}_{\mathrm{s}}^2 + 1}$ by (3).

We conclude this section with examples to illustrate the proposition.

Example 5.1.

a) For Milne universe, $-g_{\tau\tau}(\tau,\rho) \equiv 1$ and Fermi coordinates are just Minkowski coordinates, so the Fermi chart is global, and all Fermi relative speeds are subluminal.

b) For the de Sitter universe, $-g_{\tau\tau}(\tau,\rho) = \cos^2(H_0\rho)$, with $H_0\rho < \pi/2$ (see [7, 12, 13]) where H_0 is the Hubble constant. The Fermi chart is valid up to the cosmological horizon of this spacetime. Thus, all Fermi relative velocities are less than the local speed of light.

c) For the radiation-dominated universe, i.e., for the case that $a(t) = \sqrt{t}$ in (1), the Fermi chart is global and

$$-g_{\tau\tau}(\tau,\rho) = \frac{1}{\sigma}\left(1 + \sqrt{\sigma - 1}\,\sec^{-1}\sqrt{\sigma}\right)^2,$$

where $\sigma \geq 1$ is a parameter that depends on ρ and τ. It may be shown that for any $\tau > 0$, the least upper bound of ρ is $\frac{\pi}{2}\tau$ and that $\sqrt{-g_{\tau\tau}(\tau,\rho)} \to \frac{\pi}{2}$ asymptotically as $\rho \to \frac{\pi}{2}\tau$ (see [7]). Thus, Fermi relative speeds can exceed the local speed of light in this spacetime, but are bounded above by $\frac{\pi}{2}$.

6. Lightlike Simultaneity and Optical Coordinates

As in the preceding sections, let $p = (\tau, 0)$ be an event of the central observer β. An event is *lightlike simultaneous* with p if it lies on the past-pointing horismos E_p^- (which is tangent to the backward light cone at the spacetime point p). The vector field,

$$Y := \frac{\partial}{\partial \delta} = \frac{dt}{d\delta}\frac{\partial}{\partial t} + \frac{d\chi}{d\delta}\frac{\partial}{\partial \chi} = -\frac{a(\tau)}{a(t)}\frac{\partial}{\partial t} + \frac{a(\tau)}{a^2(t)}\frac{\partial}{\partial \chi}, \tag{12}$$

is geodesic, lightlike, and the integral curve λ such that $\lambda(0) = p$ is a past-pointing null geodesic, parameterized in such a way that that δ is the affine distance from p to $\lambda(\delta)$ (see [5, Proposition 6]). Let $q_\ell = (t_\ell, \chi_\ell)$ be the unique event of $\beta' \cap E_p^-$. Then λ is the unique geodesic from p to q_ℓ, and so, using (12) we can find a relationship between τ, t_ℓ and χ_ℓ:

$$-\int_\tau^{t_\ell} \frac{1}{a(t)}\,\mathrm{d}t = \int_0^{\chi_\ell} \mathrm{d}\chi \quad \Longrightarrow \quad \int_{t_\ell}^{\tau} \frac{1}{a(t)}\,\mathrm{d}t = \chi_\ell. \tag{13}$$

From (13) we see that $t_\ell < \tau$, and more generally, it follows from (12) that t is a decreasing function of affine distance δ. Since we consider only positive coordinate times, $t_\ell > 0$, and then it is necessary that $0 < \chi_\ell < \chi_{\ell\max}(\tau)$, where

$$\chi_{\ell\max}(\tau) := \int_0^{\tau} \frac{1}{a(t)}\,\mathrm{d}t \tag{14}$$

is the *particle horizon* for the observer β at p.

In the framework of lightlike simultaneity, it will be convenient to use optical (also called observational) coordinates with respect to the observer β. Referring to Figure 1, we set the optical coordinates of the point $q_\ell = (t_\ell, \chi_\ell)$ to be (τ, δ). From (13), $\tau(t_\ell, \chi_\ell)$ is determined implicitly, and differentiation gives,

$$\frac{\partial \tau}{\partial t_\ell} = \frac{a(\tau)}{a(t_\ell)}, \qquad \frac{\partial \tau}{\partial \chi_\ell} = a(\tau). \tag{15}$$

It follows from (12) that,

$$\delta = \delta(t_\ell, \chi_\ell) = \int_{t_\ell}^{\tau(t_\ell, \chi_\ell)} \frac{a(t)}{a\left(\tau(t_\ell, \chi_\ell)\right)}\,\mathrm{d}t. \tag{16}$$

Differentiating (16) and using (15), gives

$$\frac{\partial \delta}{\partial t_\ell} = \frac{a(\tau)}{a(t_\ell)} - \frac{a(t_\ell)}{a(\tau)} - \delta\frac{\dot{a}(\tau)}{a(t_\ell)}, \qquad \frac{\partial \delta}{\partial \chi_\ell} = a(\tau) - \delta\dot{a}(\tau), \tag{17}$$

where $\dot{a}(t)$ is the derivative of $a(t)$. Now using (15) and (17), we may express the Robertson-Walker metric in optical coordinates (with respect to β) in the form

$$\mathrm{d}s^2 = \tilde{g}_{\tau\tau}\mathrm{d}\tau^2 + 2\mathrm{d}\tau\mathrm{d}\delta \equiv -2\left(1 - \frac{\dot{a}(\tau)}{a(\tau)}\delta - \frac{1}{2}\frac{a^2(t_\ell)}{a^2(\tau)}\right)\mathrm{d}\tau^2 + 2\mathrm{d}\tau\mathrm{d}\delta, \tag{18}$$

where $t_\ell(\tau, \delta)$ is given implicitly by (16).

7. Astrometric and Spectroscopic Relative Velocities

Based on the previous section, we may write the 4-velocity of the radially moving test particle with worldline β' in optical coordinates as, $u'_\ell = \dot{\tau}\frac{\partial}{\partial \tau}|_{q_\ell} + \dot{\delta}\frac{\partial}{\partial \delta}|_{q_\ell} = (\dot{\tau}, \dot{\delta})$, where the overdot represents differentiation with respect to proper time of β'.

Let $p = (\tau, 0)$ be the event of β from which we measure the velocities. The astrometric relative velocity of u'_ℓ with respect to $u = (1, 0)$ is given by

$$v_{\text{ast}} = \frac{d\delta}{d\tau}\mathcal{S}_p = \frac{\dot{\delta}}{\dot{\tau}}\mathcal{S}_p. \tag{19}$$

From (18), (19) and the requirement that $g\left(u'_\ell, u'_\ell\right) = -1$, we obtain,

$$v_{\text{ast}} = \frac{\dot{\delta}}{\dot{\tau}}\mathcal{S}_p = \frac{1}{2}\left(-\tilde{g}_{\tau\tau} - \frac{1}{\dot{\tau}^2}\right)\mathcal{S}_p. \tag{20}$$

There is no upper bound for $\|v_{\text{ast}}\|$ in the case of a radially approaching test particle (i.e., for the case $d\delta/d\tau < 0$) because $\dot{\tau}$ can be chosen to be arbitrarily close to zero in (20). However, a sharp upper bound for the case of a radially receding test particle is given by $\|v_{\text{ast}}\| < -\tilde{g}_{\tau\tau}/2$, where the right side of this inequality is the relative speed, $d\delta/d\tau$, of a distant radially receding photon. Since it follows from (18) that $-\tilde{g}_{\tau\tau} < 2$ when $\dot{a}(\tau) \geq 0$, we have the following general result.

Proposition 7.1. *In any expanding Robertson-Walker spacetime, the astrometric relative velocity of a radially receding test particle is always less than the central observer's local speed of light ($c = 1$).*

The spectroscopic relative velocity, discussed briefly in the introduction, is defined in a way analogous to the definition of the kinematic relative velocity given in (10). However, for the case of radial motion, the following equivalent formula, more convenient for our purposes, was deduced in [5]:

$$v_{\text{spec}} = \frac{\left(\frac{\nu'}{\nu}\right)^2 - 1}{\left(\frac{\nu'}{\nu}\right)^2 + 1}\mathcal{S}_p, \tag{21}$$

where ν, ν' are the frequencies observed by u, u'_ℓ, respectively, of a photon emitted from the spacetime point q_ℓ. It is clear from (21) that $\|v_{\text{spec}}\|$ must be less than 1. The frequency ratio is given by,

$$\frac{\nu'}{\nu} = \frac{g(\mathrm{P}_{q_\ell}, u'_\ell)}{g(\mathrm{P}_p, u)}, \tag{22}$$

where $\mathrm{P} := -\partial/\partial\delta$ is the 4-momentum tangent vector field of the emitted photon. Using (18) in (22), we find,

$$\frac{\nu'}{\nu} = \dot{\tau}, \tag{23}$$

and thus, applying (23) in (21), we obtain,

$$v_{\text{spec}} = \frac{\dot{\tau}^2 - 1}{\dot{\tau}^2 + 1}\mathcal{S}_p. \tag{24}$$

In order to find expressions for the astrometric and spectroscopic relative velocities in terms of curvature-normalized coordinates, we make use of (15) to obtain

$$\dot{\tau} = \frac{\partial \tau}{\partial t_\ell}\dot{t}_\ell + \frac{\partial \tau}{\partial \chi_\ell}\dot{\chi}_\ell = a(\tau)\left(\sqrt{\dot{\chi}_\ell^2 + a^{-2}(t_\ell)} + \dot{\chi}_\ell\right), \tag{25}$$

where we have used the identification, $\dot{t}_\ell = \sqrt{a^2(t_\ell)\dot{\chi}_\ell^2 + 1}$, which follows from (3). Combining this last expression with (24) yields,

$$v_{\text{spec}} = \frac{a^2(\tau)\left(\sqrt{\dot{\chi}_\ell^2 + a^{-2}(t_\ell)} + \dot{\chi}_\ell\right)^2 - 1}{a^2(\tau)\left(\sqrt{\dot{\chi}_\ell^2 + a^{-2}(t_\ell)} + \dot{\chi}_\ell\right)^2 + 1}\mathcal{S}_p.$$

Similarly, combining (25) with (20) gives,

$$v_{\text{ast}} = \left(1 - \frac{\dot{a}(\tau)}{a(\tau)}\delta - \frac{a^2(t_\ell)}{2a^2(\tau)}\left[1 + \left(\sqrt{a^2(t_\ell)\dot{\chi}_\ell^2 + 1} + a(t_\ell)\dot{\chi}_\ell\right)^{-2}\right]\right)\mathcal{S}_p.$$

Moreover, combining (24) with (20) yields a relationship between the spectral and astrometric relative velocities:

$$v_{\text{spec}} = \frac{1 + \tilde{g}_{TT}(\tau, \delta) \pm \|v_{\text{ast}}\|}{1 - \tilde{g}_{TT}(\tau, \delta) \mp \|v_{\text{ast}}\|}\mathcal{S}_p, \tag{26}$$

where in the case that $d\delta/d\tau > 0$ (i.e., in the case of a receding test particle), the positive sign in the numerator and negative sign in the denominator are chosen, and the opposite choices of signs are taken when $d\delta/d\tau < 0$ (i.e., in the case of an approaching test particle).

Now using (26) we may formulate the following proposition.

Proposition 7.2. *For a Robertson-Walker spacetime with scale factor $a(t)$ that is a smooth, increasing function of t, a measurement of the astrometric and spectroscopic speeds of a receding test particle relative to the comoving observer determine the metric tensor element $\tilde{g}_{TT}$ in optical coordinates at the spacetime point of the particle via,*

$$\tilde{g}_{TT}(\tau, \delta) = \frac{1 - \|v_{\text{ast}}\|}{1 + \|v_{\text{spec}}\|}\|v_{\text{spec}}\| - \frac{1 + \|v_{\text{ast}}\|}{1 + \|v_{\text{spec}}\|}.$$

8. Fermi Coordinate Charts for Power Law Cosmologies

It was shown in [7] that the Fermi coordinate chart for a comoving observer covers the entire Robertson-Walker spacetime, provided the scale factor $a(t)$ is smooth, increasing, unbounded, and for all $t > 0$, $\ddot{a}(t) \leq 0$, i.e., for non inflationary cosmologies (with or without a big bang).

Cosmologies with scale factors of the form $a(t) = t^\alpha$, with $0 < \alpha \leq 1$ fall within this category and have global Fermi coordinates for comoving observers. Included are the radiation dominated universe ($\alpha = 1/2$) and the matter dominated universe ($\alpha = 2/3$).

Here we consider scale factors of the form $a(t) = t^\alpha$, with $\alpha > 1$. Power law cosmologies with $\alpha > 1$ have been used to model dark energy, and astronomical measurements have been made to support their consideration (see [14]). These cosmologies are inflationary ($\ddot{a} > 0$), and have the additional property that they contain spacetime points from which the central observer can never receive light signals, i.e., these cosmologies include *cosmological event horizons*. The χ-coordinate at time t_s of the event horizon is given by,

$$\chi_{\mathrm{horiz}}(t_\mathrm{s}) := \int_{t_\mathrm{s}}^{\infty} \frac{\mathrm{d}t}{a(t)},$$

or more explicitly,

$$\chi_{\mathrm{horiz}}(t_\mathrm{s}) = \int_{t_\mathrm{s}}^{\infty} \frac{\mathrm{d}t}{t^\alpha} = \frac{t_\mathrm{s}^{1-\alpha}}{\alpha - 1} < +\infty.$$

We note that in contrast to the case $0 < \alpha < 1$, there is no particle horizon when $\alpha > 1$, i.e., the integral in (14) is infinite, so that the astrometric and spectroscopic relative velocities are well-defined for arbitrarily large χ-coordinates. Let,

$$\mathcal{V} := \{(t, \chi) : t > 0 \text{ and } 0 < \chi < \chi_{\mathrm{horiz}}(t)\}.$$

Intuitively, $\mathcal{V}$ is the set of all eventually observable events. For $0 < \alpha < 1$, $\mathcal{V} =]0, +\infty[\times]0, +\infty[$ represents all spacetime points distant from the central observer.

We will show that the set of spacetime points with coordinates in $\mathcal{V}$ is a maximal chart for Fermi coordinates for a central observer in an inflationary power law cosmology. For that purpose, we extract the integral from (5) and define,

$$\chi_\mathrm{s}(\tau) := \int_{t_\mathrm{s}}^{\tau} \frac{a(\tau)}{a(t)} \frac{1}{\sqrt{a^2(\tau) - a^2(t)}} \mathrm{d}t. \tag{27}$$

In geometric terms, $\chi_\mathrm{s}(\tau)$ is the value of the χ-coordinate of the point with t-coordinate t_s on the spacelike geodesic orthogonal to β with initial point $(\tau, 0)$ on the central observer's worldline. With the change of variables, $\tilde{\sigma} = (a(\tau)/a(t))^2$ (with τ held fixed), expression (27) becomes,

$$\chi_\mathrm{s}(\tau) = \frac{1}{2} \int_{1}^{\sigma(\tau)} \dot{b}\left(\frac{a(\tau)}{\sqrt{\tilde{\sigma}}}\right) \frac{1}{\sqrt{\tilde{\sigma}}\sqrt{\tilde{\sigma} - 1}} \mathrm{d}\tilde{\sigma}, \tag{28}$$

where $b(t)$ is the inverse function of $a(t)$, (so that $b(a(t)) = t$), and,

$$\sigma(\tau) := \left(\frac{a(\tau)}{a(t_\mathrm{s})}\right)^2. \tag{29}$$

It follows from (4) that $t_\mathrm{s} \leq \tau$ and that t_s decreases with increasing proper distance along spacelike geodesics orthogonal to the central observer worldline β. Then $\sigma(\tau) \geq 1$ and for a given initial point $(\tau, 0)$, σ may be used as a (non affine) parameter for this geodesic.

It was shown in [7] that for a class of scale factors that includes the inflationary power laws considered here, the spacelike geodesic ψ depicted in Figure 1, may be parameterized as, $\psi(\sigma) = (t(\tau,\sigma), \chi(\tau,\sigma))$, where $\sigma \geq 1$, and,

$$t(\tau,\sigma) = b\left(\frac{a(\tau)}{\sqrt{\sigma}}\right), \tag{30}$$

$$\chi(\tau,\sigma) = \frac{1}{2}\int_1^\sigma \dot{b}\left(\frac{a(\tau)}{\sqrt{\tilde{\sigma}}}\right)\frac{1}{\sqrt{\tilde{\sigma}}\sqrt{\tilde{\sigma}-1}}\,\mathrm{d}\tilde{\sigma}. \tag{31}$$

Our strategy to prove that the set of spacetime points with coordinates in $\mathcal{V}$ is a chart for Fermi coordinates involves first showing that (τ,σ) are coordinates on that chart. To carry this through, we require a sequence of lemmas.

Lemma 8.1. *For $a(t) = t^\alpha$ with $\alpha > 1$, and any $\tau > t_\mathrm{s} > 0$,*

$$\chi_\mathrm{s}(\tau) < \chi_\mathrm{horiz}(t_\mathrm{s}).$$

Proof. Substituting $a(t) = t^\alpha$ into (28) gives,

$$\chi_\mathrm{s}(\tau) = \frac{1}{2\alpha\tau^{\alpha-1}}\int_1^{\sigma(\tau)}\frac{1}{\tilde{\sigma}^{1/2\alpha}\sqrt{\tilde{\sigma}-1}}\,\mathrm{d}\tilde{\sigma}. \tag{32}$$

The substitution $x^{2\alpha} = \tilde{\sigma}$ then gives,

$$\chi_\mathrm{s}(\tau) = \frac{1}{\tau^{\alpha-1}}\int_1^{\tau/t_\mathrm{s}}\frac{1}{x}\frac{x^{2\alpha-1}}{\sqrt{x^{2\alpha}-1}}\,\mathrm{d}x. \tag{33}$$

Integration by parts in (33) yields,

$$\begin{aligned}
\chi_\mathrm{s}(\tau) =& \frac{1}{\alpha\tau^{\alpha-1}}\left[\frac{\sqrt{(\tau/t_\mathrm{s})^{2\alpha}-1}}{\tau/t_\mathrm{s}} + \int_1^{\tau/t_\mathrm{s}}\frac{\sqrt{x^{2\alpha}-1}}{x^2}\,\mathrm{d}x\right] \\
<& \frac{1}{\alpha\tau^{\alpha-1}}\left[\left(\frac{\tau}{t_\mathrm{s}}\right)^{\alpha-1} + \int_1^{\tau/t_\mathrm{s}} x^{\alpha-2}\,\mathrm{d}x\right] \\
<& \frac{1}{\alpha\tau^{\alpha-1}}\left[\left(\frac{\tau}{t_\mathrm{s}}\right)^{\alpha-1} + \frac{1}{\alpha-1}\left(\frac{\tau}{t_\mathrm{s}}\right)^{\alpha-1}\right] \\
=& \frac{t_\mathrm{s}^{1-\alpha}}{\alpha-1} = \chi_\mathrm{horiz}(t_\mathrm{s}).
\end{aligned}$$

∎

Lemma 8.2. *For $a(t) = t^\alpha$ with $\alpha > 1$, and any $t_\mathrm{s} > 0$,*

$$\lim_{\tau\to+\infty}\chi_\mathrm{s}(\tau) = \chi_\mathrm{horiz}(t_\mathrm{s}).$$

Proof. From L'Hpital's and Leibniz's rules applied to (32), we have,

$$\lim_{\tau \to +\infty} \chi_{\mathrm{s}}(\tau) = \lim_{\tau \to +\infty} \left(\frac{t_{\mathrm{s}}^{1-\alpha}}{\alpha - 1} \right) \left(\frac{\tau^{\alpha}}{\sqrt{\tau^{2\alpha} - t_{\mathrm{s}}^{2\alpha}}} \right) = \frac{t_{\mathrm{s}}^{1-\alpha}}{\alpha - 1} = \chi_{\mathrm{horiz}}(t_{\mathrm{s}}).$$

■

Lemma 8.3. *For $a(t) = t^{\alpha}$ with $\alpha > 1$, and any $\tau > t_{\mathrm{s}} > 0$,*

$$\frac{d\chi_{\mathrm{s}}}{d\tau}(\tau) > 0.$$

Proof. For convenience, rewrite (32) as $\chi_{\mathrm{s}}(\tau) = \dfrac{f(\tau)}{g(\tau)}$ where,

$$f(\tau) := \int_{1}^{\sigma(\tau)} \frac{1}{\tilde{\sigma}^{1/2\alpha}\sqrt{\tilde{\sigma} - 1}}\, \mathrm{d}\tilde{\sigma}, \quad \text{and} \quad g(\tau) := 2\alpha\tau^{\alpha-1}.$$

By the quotient rule, $\dfrac{d\chi_{\mathrm{s}}}{d\tau} > 0$ if and only if

$$\chi_{\mathrm{s}}(\tau) = \frac{f(\tau)}{g(\tau)} < \frac{f'(\tau)}{g'(\tau)}, \tag{34}$$

By Leibniz's rule, the quotient on the right in (34) is,

$$\frac{f'(\tau)}{g'(\tau)} = \left(\frac{t_{\mathrm{s}}^{1-\alpha}}{\alpha - 1} \right) \left(\frac{\tau^{\alpha}}{\sqrt{\tau^{2\alpha} - t_{\mathrm{s}}^{2\alpha}}} \right). \tag{35}$$

Since the last term on the right in (35) is always greater than 1, we have,

$$\frac{f'(\tau)}{g'(\tau)} > \frac{t_{\mathrm{s}}^{1-\alpha}}{\alpha - 1} = \chi_{\mathrm{horiz}}(t_{\mathrm{s}}),$$

Therefore $\dfrac{d\chi_{\mathrm{s}}}{d\tau} > 0$ for any τ such that $\chi_{\mathrm{s}}(\tau) < \chi_{\mathrm{horiz}}(t_{\mathrm{s}})$. The result now follows from Lemma 8.1..

■

Lemma 8.4. *For $a(t) = t^{\alpha}$, the map $F :]0, \infty[\times]1, \infty[\to \mathcal{V}$ given by*

$$F(\tau, \sigma) := (t(\tau, \sigma), \chi(\tau, \sigma)),$$

where the functions t and χ are defined by (30) *and* (31), *respectively, is a diffeomorphism.*

Proof. Let $(t_{\mathrm{s}}, \chi_{\mathrm{s}}) \in \mathcal{V}$ be arbitrary but fixed. To prove that F is a bijection, we must show that there exists a unique pair $(\tau_0, \sigma_0) \in]0, +\infty[\times]1, +\infty[$ such that $F(\tau_0, \sigma_0) = (t_{\mathrm{s}}, \chi_{\mathrm{s}})$. From (30), it follows that σ_0 is uniquely determined by τ_0 and,

$$\sigma_0 = \left(\frac{a(\tau_0)}{a(t_s)} \right)^2 .$$

So it remains only to find τ_0. To that end, let $\sigma(\tau)$ be given by (29). It then follows from (28) and (31) that

$$\chi(\tau, \sigma(\tau)) = \chi_s(\tau).$$

Since by assumption, $\chi_s < \chi_{\text{horiz}}(t_s)$, it follows from Lemmas 8.2. and 8.3. that there is a unique $\tau_0 > t_s$ such that $\chi(\tau_0, \sigma(\tau_0)) = \chi((\tau_0, \sigma_0) = \chi_s$. Thus, F is a bijection.

The Jacobian determinant $J(\tau, \sigma)$ for the transformation F was calculated in [7] for a general class of scale factors including the power law scale factors considered here, and is given by,

$$J(\tau, \sigma) = \frac{\dot{a}(\tau)}{2\sigma} \dot{b}\left(\frac{a(\tau)}{\sqrt{\sigma}} \right) \left(\frac{\dot{b}\left(\frac{a(\tau)}{\sqrt{\sigma}}\right)}{\sqrt{\sigma - 1}} + \frac{a(\tau)}{2\sqrt{\sigma}} \int_1^{\sigma} \frac{\ddot{b}\left(\frac{a(\tau)}{\sqrt{\tilde{\sigma}}}\right)}{\tilde{\sigma}\sqrt{\tilde{\sigma} - 1}} \, d\tilde{\sigma} \right). \tag{36}$$

Thus, from (36),

$$J(\tau, \sigma(\tau)) =$$

$$\frac{a(\tau)\dot{a}(\tau)}{2\sigma(\tau)^{3/2}} \dot{b}\left(\frac{a(\tau)}{\sqrt{\sigma(\tau)}} \right) \left[\frac{\dot{b}\left(\frac{a(\tau)}{\sqrt{\sigma(\tau)}}\right)\sqrt{\sigma(\tau)}}{a(\tau)\sqrt{\sigma(\tau) - 1}} + \frac{1}{2} \int_1^{\sigma(\tau)} \frac{\ddot{b}\left(\frac{a(\tau)}{\sqrt{\tilde{\sigma}}}\right)}{\tilde{\sigma}\sqrt{\tilde{\sigma} - 1}} \, d\tilde{\sigma} \right]. \tag{37}$$

The first term in the square brackets in (37) may be rewritten:

$$\frac{\dot{b}\left(\frac{a(\tau)}{\sqrt{\sigma(\tau)}}\right)\sqrt{\sigma(\tau)}}{a(\tau)\sqrt{\sigma(\tau) - 1}} = \frac{a(\tau)\dot{b}(a(t_s))}{a(t_s)^2\sqrt{\sigma(\tau)}\sqrt{\sigma(\tau) - 1}}. \tag{38}$$

Now, applying Leibniz's rule to (28) and (29) yields,

$$\frac{d\chi(\tau)}{d\tau} = \dot{a}(\tau) \left[\frac{a(\tau)\dot{b}(a(t_s))}{a(t_s)^2\sqrt{\sigma(\tau)}\sqrt{\sigma(\tau) - 1}} + \frac{1}{2} \int_1^{\sigma} \frac{\ddot{b}\left(\frac{a(\tau)}{\sqrt{\tilde{\sigma}}}\right)}{\tilde{\sigma}\sqrt{\tilde{\sigma} - 1}} \, d\tilde{\sigma} \right]. \tag{39}$$

Combining (37), (38) and (39) gives,

$$J(\tau, \sigma(\tau)) = \frac{a(\tau)\dot{b}(a(t_s))}{2\sigma(\tau)^{3/2}} \frac{d\chi(\tau)}{d\tau}. \tag{40}$$

Thus, applying Lemma 8.3. in (40), $J(\tau, \sigma(\tau)) > 0$ for all τ. Now given any $\tau > 0$ and $\sigma > 1$ there exists a positive $t_s < \tau$ such that $\sigma(\tau) = (\tau/t_s)^{2\alpha} = \sigma$. Therefore $J(\tau, \sigma) > 0$ for all (τ, σ), and by the inverse function theorem, F is a diffeomorphism.

■

To finish the construction of Fermi coordinates, we return to (9) in the form,

$$\rho = \int_{t_s}^{\tau} \frac{a(t)}{\sqrt{a^2(\tau) - a^2(t)}}\, \mathrm{d}t, \tag{41}$$

which gives the proper distance along the spacelike geodesic (ψ in Figure 1) orthogonal to β, from the initial point $(\tau, 0) \in \beta$ to the unique point whose t-coordinate is t_s. The radius $\rho_{\mathcal{M}_\tau}$ of the Fermi slice, $\mathcal{M}_\tau$, at proper time τ of the central observer is obtained by replacing t_s by zero in (41). The result for $a(t) = t^\alpha$ is,

$$\rho_{\mathcal{M}_\tau} = \rho_{\mathcal{M}_\tau}(\alpha) = \frac{\sqrt{\pi}\,\Gamma\left(\frac{1+\alpha}{2\alpha}\right)}{\Gamma\left(\frac{1}{2\alpha}\right)}\tau, \tag{42}$$

which holds for all $\alpha > 0$ (see [7]). The change of variables, $\tilde{\sigma} = (a(\tau)/a(t))^2$, applied to the integral in (41) results in the following formula for the Fermi coordinate ρ, which also appears in [7]:

$$\rho = \rho_\tau(\sigma) = \frac{a(\tau)}{2} \int_1^{\sigma} \dot{b}\left(\frac{a(\tau)}{\sqrt{\tilde{\sigma}}}\right) \frac{1}{\tilde{\sigma}^{3/2}\sqrt{\tilde{\sigma} - 1}}\, \mathrm{d}\tilde{\sigma}. \tag{43}$$

It is clear from (43) that for a fixed value of τ, $\rho_\tau(\sigma)$ is an increasing function of σ, and therefore has an inverse function, $\sigma_\tau(\rho)$. Define,

$$\mathcal{U}_{\text{Fermi}} := \{(\tau, \rho) : \tau > 0 \text{ and } 0 < \rho < \rho_{\mathcal{M}_\tau}\},$$

and let $G(\tau, \sigma) := (\tau, \rho(\sigma))$. Then G is a diffeomorphism with inverse $G^{-1}(\tau, \rho) = (\tau, \sigma_\tau(\rho))$. Define $H : \mathcal{V} \to \mathcal{U}_{\text{Fermi}}$ by

$$H(t, \chi) := G \circ F^{-1}(t, \chi).$$

Then H is a diffeomorphism. Taking into consideration that χ and ρ are both positive, radial coordinates, we can now state the main result of this section:

Theorem 8.1. *For a comoving observer in a Robertson-Walker cosmology whose scale factor is $a(t) = t^\alpha$ with $\alpha > 1$, the maximal domain of Fermi coordinates is the set of all spacetime points whose curvature coordinates take values in $\mathcal{V}$, i.e., Fermi coordinates extend to the cosmological event horizon. The range of Fermi coordinates is $\mathcal{U}_{\text{Fermi}}$.*

9. Comparisons of Relative Velocities in Cosmologies with Power Law Scale Factors

Consider a test particle comoving with the Hubble flow, $\beta'(\tau') = (\tau', \chi)$, where $\chi > 0$ is constant. Referring to Figure 1, we have that $q_s = (t_s, \chi)$ and $q_\ell = (t_\ell, \chi)$; moreover $u'_s = \frac{\partial}{\partial t}\big|_{q_s} = (1, 0)$ and $u'_\ell = \frac{\partial}{\partial t}\big|_{q_\ell} = (1, 0)$.

In (5), t_s is implicitly defined as a function of (τ, χ), and similarly in (13), t_ℓ is implicitly defined as a function of (τ, χ). From now on, it will be convenient to regard not only t_s and t_ℓ as functions of (τ, χ), but also the four relative velocities. However, it is important to recognize that in this context χ is a parameter that labels a comoving test particle (with fixed coordinate χ), and τ is the time of observation by the central observer β. The relative velocities are vectors in the tangent space of the point $p = (\tau, 0)$ for test particles with coordinates (t_s, χ) in the case of the Fermi and kinematic relative velocities, and with coordinates (t_ℓ, χ) in the case of the astrometric and spectroscopic relative velocities. Since all the velocities are proportional to $\mathcal{S}_p$ and in the same direction, we will find expressions only for the moduli of the relative velocities.

In [10], comoving test particles in cosmologies with a variety of scale factors were studied. These include power law scale factors of the form

$$a(t) = t^\alpha$$

with $0 < \alpha \leq 1$. Here, we relax that restriction and allow $\alpha > 1$.

In preparation for the study of the relative velocities, it is convenient to define a parameter

$$v = v(\tau, \chi) := \dot{a}(\tau)\chi = \alpha\chi\tau^{\alpha-1}. \tag{44}$$

This parameter will be useful in the description of the relative velocities because their moduli depend on (τ, χ) by means of v. In (44), the overdot represents differentiation with respect to τ, and v is the *Hubble speed* of a comoving test particle with curvature-normalized coordinates (τ, χ).

9.1. Spacelike Simultaneity

For notational convenience in this subsection, let,

$$C_\alpha := \frac{\sqrt{\pi}\,\Gamma\left(\frac{1+\alpha}{2\alpha}\right)}{\Gamma\left(\frac{1}{2\alpha}\right)}. \tag{45}$$

It follows from (42) and (45) that for any $\alpha > 0$,

$$C_\alpha = \frac{\rho_{\mathcal{M}_\tau}(\alpha)}{\tau}, \tag{46}$$

where $\rho_{\mathcal{M}_\tau}(\alpha)$ is the proper radius of the spaceslice $\mathcal{M}_\tau$ for scale factor $a(t) = t^\alpha$.

From (6), it follows that $\chi_{\text{smax}}(\tau) = +\infty$ for $\alpha > 1$, and so for this case, χ and v have no upper bounds. In contrast, $\chi_{\text{smax}}(\tau) = \frac{\tau^{1-\alpha}}{1-\alpha}C_\alpha$ is finite and v is bounded by $v_{\text{smax}} := \frac{\alpha}{1-\alpha}C_\alpha$ for $0 < \alpha < 1$ (see [10]).

By (5) we have,

$$\left(\frac{t_s}{\tau}\right)^{1-\alpha} {}_2F_1\left(\frac{1}{2}, \frac{1-\alpha}{2\alpha}; \frac{1+\alpha}{2\alpha}; \left(\frac{t_s}{\tau}\right)^{2\alpha}\right) = C_\alpha + \frac{\alpha-1}{\alpha}v, \tag{47}$$

where ${}_2F_1(\cdot, \cdot; \cdot; \cdot)$ is the Gauss hypergeometric function.

Define the function $F_\alpha(z) := z^{\frac{1-\alpha}{\alpha}}\, {}_2F_1\left(\frac{1}{2}, \frac{1-\alpha}{2\alpha}; \frac{1+\alpha}{2\alpha}; z^2\right)$ where $0 < z < 1$ and $\alpha > 0$, $\alpha \neq 1$[‖]. It is bijective and by (47),

$$t_s(\tau, \chi) = G_\alpha(v)\tau, \tag{48}$$

with $G_\alpha(v) := \left(F_\alpha^{-1}\left(C_\alpha + \frac{\alpha-1}{\alpha}v\right)\right)^{1/\alpha}$, where the superscript $^{-1}$ denotes the inverse function. Using techniques analogous to those used in [10], from (48) we find:

[‖]The case $\alpha = 1$ corresponds with the Milne universe and it is studied in [10].

Proposition 9.1. *The kinematic and Fermi speeds of a comoving test particle relative to a comoving central observer in a Robertson-Walker cosmology with scale factor* $a(t) = t^{\alpha}$ *and* $\alpha > 0$, $\alpha \neq 1$ *are given by*

$$\|v_{\mathrm{kin}}\| = \sqrt{1 - G_{\alpha}^{2\alpha}(v)}, \tag{49}$$

and

$$\|v_{\mathrm{Fermi}}\| = G_{\alpha}^{1-\alpha}(v)\sqrt{1 - G_{\alpha}^{2\alpha}(v)} - \frac{\alpha - 1}{\alpha}\left(1 - G_{\alpha}^{2\alpha}(v)\right) v, \tag{50}$$

where v *is the parameter given by* (44). *Moreover, from* (49), (50), *and taking into account* (46), *we have,*

$$\lim_{v \to v_{\mathrm{smax}}} \|v_{\mathrm{kin}}\| = 1$$

and

$$\lim_{v \to v_{\mathrm{smax}}} \|v_{\mathrm{Fermi}}\| = C_{\alpha} = \frac{\rho_{\mathcal{M}_{\tau}}}{\tau}, \tag{51}$$

where $v_{\mathrm{smax}} = \frac{\alpha}{1-\alpha} C_{\alpha}$ *if* $0 < \alpha < 1$, *and* $v_{\mathrm{smax}} = +\infty$ *if* $\alpha > 1$.

It follows from (51) that the limiting Fermi speeds exceed 1 only for $0 < \alpha < 1$. More generally, it follows from (50) that $\|v_{\mathrm{Fermi}}\| < 1$ for any comoving particle when $\alpha > 1$, i.e., there are no superluminal Fermi velocities for inflationary power law scale factors. Examples are presented in Figure 2.

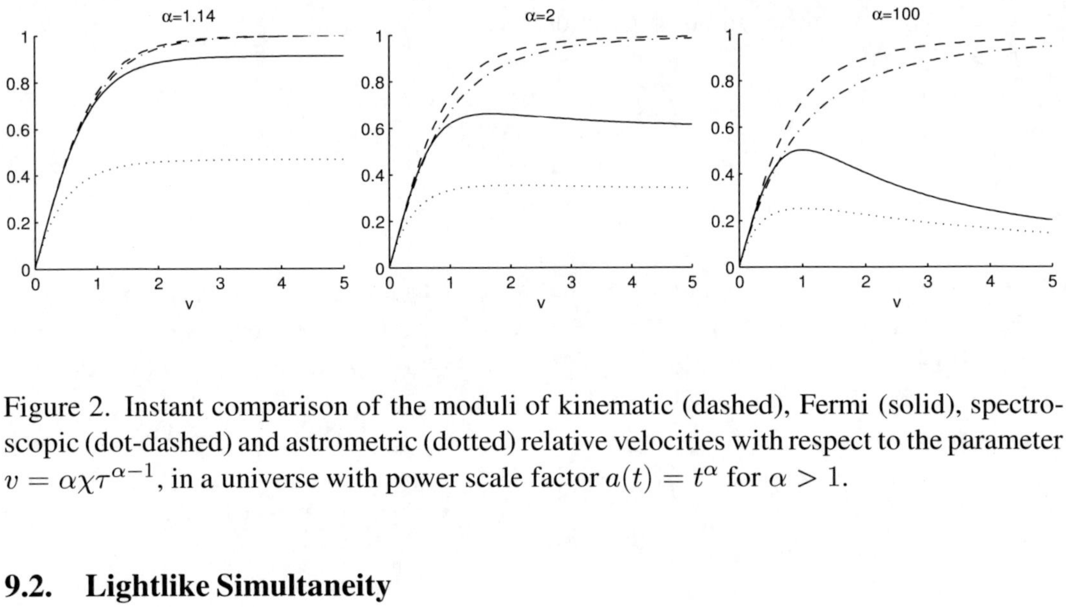

Figure 2. Instant comparison of the moduli of kinematic (dashed), Fermi (solid), spectroscopic (dot-dashed) and astrometric (dotted) relative velocities with respect to the parameter $v = \alpha\chi\tau^{\alpha-1}$, in a universe with power scale factor $a(t) = t^{\alpha}$ for $\alpha > 1$.

9.2. Lightlike Simultaneity

From (14), it follows that $\chi_{\ell\mathrm{max}}(\tau) = +\infty$ for $\alpha > 1$, and so for this case, χ and v have no upper bounds in the framework of lightlike simultaneity. In contrast, $\chi_{\ell\mathrm{max}}(\tau) = \frac{\tau^{1-\alpha}}{1-\alpha}$ is finite and v is bounded by $v_{\ell\mathrm{max}} := \frac{\alpha}{1-\alpha}$ for $0 < \alpha < 1$ (see [10]).

By (13)

$$t_{\ell}(\tau, \chi) = \left(1 - \frac{1-\alpha}{\alpha} v\right)^{\frac{1}{1-\alpha}} \tau. \tag{52}$$

Using techniques analogous to those used in [10], from (52) we find:

Proposition 9.2. *The spectroscopic and astrometric speeds of a comoving test particle relative to a comoving central observer in a Robertson-Walker cosmology with scale factor $a(t) = t^\alpha$ and $\alpha > 0$, $\alpha \neq 1$ are given by*

$$\|v_{\text{spec}}\| = \frac{1 - \left(1 + \frac{\alpha-1}{\alpha}v\right)^{\frac{2\alpha}{1-\alpha}}}{1 + \left(1 + \frac{\alpha-1}{\alpha}v\right)^{\frac{2\alpha}{1-\alpha}}}, \tag{53}$$

and

$$\|v_{\text{ast}}\| = \frac{1}{1+\alpha}\left(1 - \left(1 + \frac{\alpha-1}{\alpha}v\right)^{\frac{1+\alpha}{1-\alpha}}\right) + \frac{\alpha-1}{\alpha}v\left(1 + \frac{\alpha-1}{\alpha}v\right)^{\frac{2\alpha}{1-\alpha}}, \tag{54}$$

where v is the parameter given by (44). *Moreover, from* (53) *and* (54) *we have,*

$$\lim_{v \to v_{\ell\max}} \|v_{\text{spec}}\| = 1$$

and

$$\lim_{v \to v_{\ell\max}} \|v_{\text{ast}}\| = \frac{1}{1+\alpha}, \tag{55}$$

where $v_{\ell\max} = \frac{\alpha}{1-\alpha}$ if $0 < \alpha < 1$, and $v_{\ell\max} = +\infty$ if $\alpha > 1$.

It follows from (55) that the limiting astrometric speeds do not exceed 1. Spectroscopic and astrometric velocities are also represented in Figure 2.

Remark 9.1. For "retarded comparisons" (described in Section 3), the kinematic and Fermi relative velocities of u'_ℓ must be calculated relative to $u^* = \frac{\partial}{\partial t}\big|_{p^*} = (1, 0)$, i.e., the 4-velocity of β at $p^* = (\tau^*, 0)$. By (5), we have that $\tau^* = \tau^*(t_\ell(\tau, \chi_\ell), \chi_\ell)$ is defined implicitly from

$$\int_{t_\ell(\tau,\chi_\ell)}^{\tau^*} \frac{a(\tau^*)}{a(t)} \frac{1}{\sqrt{a^2(\tau^*) - a^2(t)}}\, \mathrm{d}t = \chi_\ell, \tag{56}$$

where $t_\ell(\tau, \chi_\ell)$ is given implicitly by (13).

Concluding Remarks

We have found general expressions for the Fermi, kinematic, astrometric, and spectroscopic velocities of test particles experiencing radial motion relative to an observer comoving with the Hubble flow (called the central observer) in any expanding Robertson-Walker cosmology. Specific numerical calculations and formulas were given for cosmologies for which the scale factor, $a(t) = t^\alpha$, including both inflationary ($\alpha > 1$) and non inflationary universes ($0 < \alpha \leq 1$). These include the radiation-dominated and matter-dominated universes, and models for dark energy (see [14]).

It follows from Propositions 5.1. and 7.2. that in principle, knowledge of either pair of the relative velocities for moving test particles at each spacetime point uniquely determines the geometry of the two dimensional spacetime via (7) and (18), and therefore the scale factor $a(t)$. Since the affine distance (i.e., the optical coordinate δ) can be measured by parallax, and the frequency ratio can be found by spectroscopic measurements, the astrometric and spectroscopic relative velocities can, in principle, be determined solely by physical measurements, and so, they could confirm or contradict assumptions about the value of $a(t)$ for the actual universe.

Of the four relative velocities, only the Fermi relative velocity of a radially receding test particle can exceed the local speed of light of the observer (i.e., be superluminal), and this is possible at a spacetime point (τ, ρ), in Fermi coordinates, if and only if $-g_{\tau\tau}(\tau, \rho) > 1$.

Under general conditions, the Hubble velocity of comoving test particles also become superluminal at large values of the radial parameter, χ, and this is taken as a criterion for the expansion of space in cosmological models, and for the actual universe. By way of comparison, the Fermi relative velocity has both advantages and disadvantages to the Hubble velocity. For comoving particles, both velocities measure the rate of change of proper distance away from the observer with respect to the proper time of the observer. But for the Fermi velocity, the proper distance is measured along spacelike geodesics, while for the Hubble velocity the proper distance is measured along non geodesic paths. In this respect the Fermi velocity is more natural and more closely tied to the observer's natural frame of reference, i.e., to Fermi coordinates in which locally the metric is Minkowskian to first order in the coordinates. In addition, the notion of Fermi relative velocity, along with the other three relative velocities discussed in this chapter, are geometric and may be calculated in any spacetime, while the Hubble velocity is specific to Robertson-Walker cosmologies. The example of the Milne universe, discussed in the introduction, illustrates the limitations of the use of superluminal Hubble velocities as indicators of expansion of space.

Cosmological models with a scale factor of the form $a(t) = t^{\alpha}$ for $\alpha > 0$, provide test cases for the use of the Fermi relative velocities of comoving test particles for understanding of expansion of space in general, and the effect of inflation and event horizons, in particular. Fermi coordinates are global in the non inflationary case, i.e., for $0 < \alpha \leq 1$ (see [7]), and maximal Fermi charts were shown in Section 8 to extend up to (but not include) the cosmological event horizon, for the inflationary case with $\alpha > 1$. Perhaps surprisingly, superluminal relative Fermi velocities of comoving particles exist only for the non inflationary cases, $0 < \alpha < 1$. Although not discussed in this chapter, the situation for the de Sitter universe is analogous. There, the Fermi chart is valid only up to the cosmological horizon (see [7, 12, 13]), and Fermi relative velocities of comoving test particles are necessarily subluminal. One might expect that accelerating universes (in the sense that $\ddot{a} > 0$) would allow for greater relative velocities, rather than impose lower speed limits.

Some insight into this phenomenon comes from the geometry of the simultaneous space slices, $\{\mathcal{M}_\tau\}$, and in particular, the dependence of the proper radius, $\rho_{\mathcal{M}_\tau}$, on α. It was shown in [7] that for $0 < \alpha \leq 1$, superluminal relative Fermi velocities exist because "there is enough space" in the sense that the proper radius $\rho_{\mathcal{M}_\tau}$ of $\mathcal{M}_\tau$ satisfies the condition $\rho_{\mathcal{M}_\tau} > \tau$. For $0 < \alpha \leq 1$, Fermi speeds increase with proper distance from the central observer, reaching their limiting value $\rho_{\mathcal{M}_\tau}/\tau$ asymptotically.

It follows from (42) that $\rho_{\mathcal{M}_\tau}(\alpha)$ decreases monotonically to zero as $\alpha \to +\infty$**. This means that the spacelike geodesics orthogonal to the central observer's worldline reach the

**The Hubble radius and proper distance to the event horizon in curvature coordinates also decrease monotonically with α.

big bang, at $t = 0$, in a proper distance that decreases with α. Since $\rho_{\mathcal{M}_\tau}/\tau < 1$ for $\alpha > 1$, one might expect, on this basis, the disappearance of superluminal Fermi relative velocities in the inflationary case. We note, however, that in the inflationary cases, maximal Fermi speeds are achieved at proper distances less than $\rho_{\mathcal{M}_\tau}$.

Does space expand in the Robertson-Walker cosmologies with power law scale factors? An affirmative answer may be given in the following sense. Let α be fixed; for any comoving geodesic observer, the Fermi space slices of τ-simultaneous events, $\{\mathcal{M}_\tau\}$, that foliate spacetime up to the event horizon (for $\alpha > 1$), or the entire space-time (for $0 < \alpha \leq 1$), have finite proper radii, $\rho_{\mathcal{M}_\tau}$, that increase with the observer's proper time.

References

[1] M. Soffel, *et al.* The IAU 2000 resolutions for astrometry, celestial mechanics and metrology in the relativistic framework: explanatory supplement. *Astron. J.* 126 (2003), 2687–2706.

[2] L. Lindegren, D. Dravins. The fundamental definition of 'radial velocity'. *Astron. Astrophys.* 401 (2003), 1185–1202.

[3] V. J. Bolós, V. Liern, J. Olivert. Relativistic simultaneity and causality. *Internat. J. Theoret. Phys.* 41 (2002) 1007–1018.

[4] V. J. Bolós. Lightlike simultaneity, comoving observers and distances in general relativity. *J. Geom. Phys.* 56 (2006), 813–829.

[5] V. J. Bolós. Intrinsic definitions of "relative velocity" in general relativity. *Commun. Math. Phys.* 273 (2007), 217–236.

[6] D. Klein, P. Collas. Recessional velocities and Hubble's law in Schwarzschild-de Sitter space. *Phys. Rev.* D 81 (2010), 063518.

[7] D. Klein, E. Randles. Fermi coordinates, simultaneity, and expanding space in Robertson-Walker cosmologies. *Ann. Henri Poincaré* 12 (2011), 303–328.

[8] V. J. Bolós. A note on the computation of geometrically defined relative velocities. *Gen. Relativ. Gravit.* 44 (2012), 391–400.

[9] E. Fermi. Sopra i fenomeni che avvengono in vicinanza di una linea oraria. *Atti R. Accad. Naz. Lincei, Rendiconti, Cl. sci. fis. mat & nat.* 31 (1922), 21–23, 51–52, 101–103.

[10] V. J. Bolós, D. Klein. Relative velocities for radial motion in expanding Robertson-Walker spacetimes. *Gen. Relativ. Gravit.* 44 (2012), 1361–1391.

[11] S. Havens, Fermi coordinates and relative motion in inflationary power law cosmologies, *Masters Thesis*, Department of Mathematics, California State University, Northridge, In preparation.

[12] C. Chicone, B. Mashhoon. Explicit Fermi coordinates and tidal dynamics in de Sitter and Goedel spacetimes. *Phys. Rev.* D 74 (2006), 064019.

[13] D. Klein, P. Collas. Exact Fermi coordinates for a class of spacetimes. *J. Math. Phys.* 51 (2010), 022501.

[14] Zong-Hong Zhu, M. Hu, J. S. Alcaniz and Y.-X. Liu, Testing power-law cosmology with galaxy clusters. *Astron. & Astophys.* 483 (2008), 15-18.

INDEX

D

E

F

G

H

I

K

L

M

N

O

P

Q

R

S

T

U

V

W

Y